WESTCHESTER PUBLIC LIBRARY
3 1310 00148 7905

W9-BSS-309

THE ILLUSTRATED GUIDE TO

THE BIBLE

Oxford

THE ILLUSTRATED GUIDE TO
THE BIBLE

J. R. PORTER

New York
Oxford University Press
1995

The Illustrated Guide to the Bible

Published in the United States of America by Oxford University Press, Inc.
198 Madison Avenue, New York, N.Y. 10016

Conceived, created, and designed by
Duncan Baird Publishers, London, England

Library of Congress
Cataloging-in-Publication Data

Porter. J. R. (Joshua Roy), 1921–
The illustrated guide to the Bible / by J. R. Porter.
p. cm.
Includes bibliographical references and index
Summary: Examines the Biblical narratives in their historical, social, archaeological, and mythological contexts and provides a section with a book-by-book summary of the Bible.
ISBN 0-19-521159-6

Printing (last digit) 9 8 7 6 5 4 3 2 1

1. Bible--History of Biblical events.
2. Bible--Illustrations. 3. Jews--Civilization--To 70 AD.
4. Palestine--Civilization. [1. Bible--History of Biblical events. 2. Bible--Introductions.] I. Title.
BS635.2.P67 1995
220.6'1--dc20 95-7696
CIP
AC

Page 1:
The Road to Calvary, *one of the frescoes painted ca. 1305 by Giotto (1266–1337) for the Scrovegni chapel in Padua, Italy.*

Page 2:
A late 12th-century manuscript from England depicting the life of King David. Top: *David fights and kills Goliath (1 Sam. 17.49–51).* Center: *Saul tries to kill David (1 Sam. 18.10–11); Samuel anoints David (1 Sam. 16.13).* Bottom: *David's son Absalom is killed (2 Sam. 18.9–15); David weeps for Absalom (2 Sam. 18.33).*

Page 3:
Noah Releases the Dove *(Gen. 7.8); a Byzantine mosaic in St Mark's basilica, Venice, Italy.*

Editor: Peter Bently
Consultant editor: Michael D. Coogan
Assistant editors: Lucy Rix, Daphne Bien Tebbe
Designer: Paul Reid
Assistant designer: Gail Jones
Picture research: Julia Brown, Jan Croot
Commissioned artwork: Ron Hayward
Map illustration and decorative borders: Lorraine Harrison

Book-by-Book Summary compiled by Mike Darton

Typeset in Sabon 10/11pt
Color reproduction by Colourscan, Singapore
Printed in China by Imago Publishing Limited

NOTE
The abbreviations CE and BCE are used throughout this book:
CE Common Era (the equivalent of AD)
BCE Before the Common Era (the equivalent of BC)

CONTENTS

Part of the scroll of the book of Isaiah found at Qumran and dated to ca. 100 BCE. It is the longest and oldest of the so-called "Dead Sea Scrolls."

THE MAKING OF THE BIBLE

Categories of Literature in the Hebrew Bible

The word "Bible" comes from the medieval Latin *Biblia.* The singular word *Biblia* was originally a Greek plural, meaning "books," and this is a good description of the character of what we know as the Bible. It consists of a number of different writings or "books" dating from approximately 1000 BCE to 100 CE or a little later. The books are very diverse in nature, and most of them have undergone a considerable amount of editing, reinterpretation, adaptation, and expansion in the course of their transmission. This process can be elucidated only by the techniques of biblical criticism developed since the nineteenth century, the methods of which are used by a broad spectrum of scholars. Such study is highly complicated, its results often remain hypothetical, and individual scholars differ in their assessments of the evidence.

The books of the Hebrew Bible (also called the Hebrew Scriptures or Old Testament) can be classified into several broad categories, which do not necessarily correspond to those of earlier ages. The first seventeen books, from Genesis to Esther (in the order found in the Christian Bible), are typically viewed as narrative works. The first five, Genesis, Exodus, Leviticus, Numbers, and Deuteronomy, have always been recognized as a distinct section, the Torah or Pentateuch, forming a continuous narrative from the creation of the world (see pp.26–7) to Moses' death just before Israel's occupation of Canaan (see pp.66–7). The next seven books, Joshua through 2 Kings, recount the Israelite nation's development from the conquest of Canaan (see pp.68–71) through the Babylonian capture of Jerusalem (see pp.110–11), and are generally recognized as a coherent history, by one or more authors of the Deuteronomic school (see p.106). A similar work is 1 and 2 Chronicles, which appears to make use of the Deuteronomic history, while the books of Ezra and Nehemiah (see pp.116–17) continue Jewish history into the period after the return from the Exile. Esther, a kind of historical novel, tells the story of Jews being liberated from their oppressors in Persia (see pp.130–31).

The Pentateuch also represents a second major category of works in the Hebrew Bible: Law. The Ten Commandments (Decalogue) and the Book of the Covenant, originally separate law codes, appear in Exodus (20.1–17 and 20.22–23.19, respectively; see pp.60–63). Deuteronomy, too, contains the Decalogue (Deut. 6.6–21), as well as another extensive law code (Deut. 12–26; see p.107). Priestly regulations known as the Holiness Law make up Leviticus 17–26, and similar material plays a large part in the later chapters of Exodus and Numbers.

Prophetical books form a third important category in the Hebrew Bible. Known as the "Latter Prophets" in Jewish tradition (the "Former Prophets" being the books of Joshua, Judges, Samuel, and Kings), these are the books of Isaiah, Jeremiah, Ezekiel, plus the "Twelve Prophets" (Hosea through Malachi); the Christian tradition includes Daniel as well. They cover a period from the eighth century BCE to the early postexilic age, but other traditions attest to the deeds and words of earlier prophets. Because the dates and circumstances of individual prophets can usually be established with considerable certainty, the prophetic books provide valuable information about the political and social conditions of particular periods.

In most cases, the book bearing the name of a prophet purports to contain that prophet's own oracles. However, in the process of preserving and collecting the prophet's sayings, these original words were often combined with other material, such as biographical narrative. Also, in the course of literary transmission, the original oracles were often reinterpreted to adapt them to new circumstances. Other ele-

The title page to St. John's gospel, from the Lindisfarne Gospels, *produced at the monastery on Lindisfarne Island off northeast England, ca. 700* CE *(see p.14).*

Moses and Pharaoh's Daughter *(Exod. 2.10), from the* Golden Haggadah, *a Hebrew manuscript of 1320. The story of the Israelite Exodus from Egypt was central to the development of the Hebrew Bible.*

ments may also have been added, including later material from other prophets. This complex compositional process is especially apparent in the three major prophetic books, Isaiah, Jeremiah, and Ezekiel (see especially the discussion of Isaiah on p.105, p.112, and p.117).

Proverbs, Job, and Ecclesiastes, together with the apocryphal works Sirach (Ecclesiasticus) and the Wisdom of Solomon, represent another category of the Hebrew Bible: Wisdom literature (see pp.126–7). The beginnings of this genre are seen in the bulk of the material making up the book of Proverbs: brief, pithy, proverbial sayings which enshrine popular understanding of life and society. But other sections of Proverbs show the development of Wisdom as a more profound theoretical concept. Later Wisdom works, such as Job (see pp.128–9), Ecclesiastes, Sirach (Ecclesiasticus), and the Wisdom of Solomon, wrestle with some of the fundamental problems of the human condition, particularly the existence of evil and suffering.

There are many other ways of categorizing the books of the Old Testament: Daniel, for example, is also an apocalypse, a type of literature that is related to classical prophecy but goes far beyond it in its cosmic and eschatological speculations (see pp.136–7). One of the most fruitful results of modern biblical study has been the identification of more of the wide range of linguistic and literary resources that biblical authors drew upon in composing their works. These include myths, sagas, legends, etiologies (origin stories), popular etymologies, and legal works. This study has greatly illuminated the character of the most important surviving collection of Hebrew poetry, the Psalms (see p.121), which can now be classified into a number of different types, according to the occasions for which they were produced.

Categories of Literature in the New Testament

While the composition and collection of the material in the Hebrew Bible spans a period of over a thousand years, the writings of the New Testament were probably produced in the space of less than a century. They can be classified into two main categories: historical narratives with strong biographical features, such as the gospels and Acts; and letters. Revelation (see pp.254–6) is certainly a Christian apocalypse, possessing many of the features that had become traditional in the Jewish genre of literature known as "apocalyptic" (see above), but nevertheless it is in the form of a letter.

Many of the considerations that arise in the study of the Hebrew Scriptures are also applicable to the New Testament. Each of the gospels represents its author's own understanding and interpretation of the basic tradition about Jesus, and each has its own special material. The Synoptic Gospels (Matthew, Mark, and Luke; see pp.148–9), at least, are largely constructed from small units, such as healing stories, conflict stories, and parables. Many of the writings of the apostle Paul are genuine letters, but several of those which have been attributed to him and to other apostles are really different kinds of compositions. Often they reflect features of contemporary Hellenistic literature, such as the treatise, encyclical, diatribe, or what is known as *parenesis* (see pp.238–9).

The Evolution of the Pentateuch

The first five books of the Hebrew Bible, the Torah or Pentateuch, were traditionally regarded as the product of a single author, Moses, who is the main human character in much of the story. Parts of the material may, indeed, have been contributed by Moses, but it is now generally accepted that the Pentateuch is a complex, anonymous work, developed from traditions dating from before and, especially, after the time of Moses himself.

Since the late nineteenth century, the dominant theory of the origins of the Pentateuch has been Julius Wellhausen's "Documentary Hypothesis," although this has been increasingly questioned by scholars. The Documentary Hypothesis proposes that the books of the Pentateuch are in the main a combination of four separate literary sources, which were written at different times. The first source, the Yahwist (or "J," from *Jahwe*, the German spelling of Yahweh), may have originated in Jerusalem and is generally dated to the tenth or ninth century BCE. The Yahweist is so-called because this author uses the word "Yahweh," the personal name of God. The "E" document, which uses the Hebrew word *elohim* for God in Genesis, appeared a little later, in the ninth or eighth century BCE, and reflects the traditions of the northern Israelite kingdom of Israel. The book of Deuteronomy, or "D," a separate work which gives a distinctive picture of the figure of Moses, was written in the southern kingdom of Judah in the seventh century BCE, although it too preserves northern traditions (see p.106). There may also have been a priestly source, "P," which was added after the Babylonian Exile, in the sixth or fifth century BCE (see pp.26–7). The final stage in the evolution of the Pentateuch was the combination of all these sources into the existing whole, a task probably carried out by a priestly editor who worked in Jerusalem in the period after the return from Exile.

Form criticism, a type of analysis that looks at individual units of the text, suggests that a long period of transmission may lie behind the existing written material. It is possible, for example, that the myths, sagas, etiological tales, and sanctuary legends of Genesis were orally composed and transmitted, though some of these may well have been written down at an early stage. Scholars came to recognize that what had been thought of as the nation of "Israel" was in fact a confeder-

Israel in Egypt *(Exod. 1.11), from the* Brother Haggadah, *Barcelona, ca. 1350.* Haggadahs *are collections of Bible stories telling the events of the Exodus from Egypt. Readings from them are traditionally made during the Jewish Passover supper.*

THE PENTATEUCH IN RECENT SCHOLARSHIP

In recent times, biblical scholars have tended to concentrate on the purpose and background of the Pentateuch in its final form. They generally agree with the claim of the Documentary Hypothesis (see main text, left) that the last stage in the development of the Pentateuch took place in the postexilic period. But they have emphasized that the final editors were attempting to convey a message tailored to the needs, hopes, and fears of people of their own time. The Pentateuch may be viewed as aiming to encourage and warn Jewish exiles faced with the challenge of returning to the Promised Land, and its themes can be seen as providing a program for the restoration of Israel. These themes are: the abiding covenant with God; the promise of the land in perpetuity to Israel; and the trials of the people in the wilderness before the settlement of the Promised Land.

This view has led some scholars to challenge the Documentary Hypothesis. Some have even argued that there is nothing to show that the Pentateuch uses any existing bodies of material, such as "J" and "P." This probably goes too far. The Pentateuch is a literary and theological product of the postexilic age, but it may well have drawn on earlier material, even if these cannot be identified with certainty.

ation of originally separate groups, and they were able to isolate within the Bible the distinct traditions of particular tribes or groups (see also sidebar on p.9).

The Canon of the Hebrew Bible and Old Testament

Not all the literature that was produced in Judaism or the early Church by the end of the first century CE is included in the Bible. Many other Jewish religious writings exist: the Hebrew Bible mentions books that are now lost, such as the book of Jashar (Josh. 10.13; 1 Sam.1.18), apparently a collection of ancient Hebrew poetry. The New Testament speaks of letters that are also no longer extant.

The list of biblical books is known as the "canon," a Greek word meaning "rule," but how and why the books in the Bible acquired this unique authority remains a subject of much scholarly speculation.

Three distinct sections comprise the canon of the Hebrew Bible: the Law, or Torah, consisting of the first five books, the Pentateuch; the Prophets, including the "Former Prophets" (Joshua through 2 Kings) and the "Latter Prophets" (Isaiah, Jeremiah, Ezekiel, and the twelve short prophetic books known as the Twelve or Minor Prophets); and the Writings, a collection of the rest of the works in the canon (Ruth, Chronicles, Ezra, Nehemiah, Esther, Job, Psalms, Proverbs, Ecclesiastes, Song of Solomon, Lamentations, and Daniel).

The Law has always been regarded as having special authority in Judaism: the other books repeat, amplify, and explain the Law. It seems likely, then, that this was the first section to be recognized as canonical. The prophetical books were probably collected together about the time of the return from the Exile (see pp.114–17), when the Deuteronomic history (Deuteronomy through 2 Kings) was also produced. When the two were combined, they formed the next portion of the canon. The third section, the Writings, seems to have taken longer to become fixed, with various books being added to it over a long period. The threefold division of the Hebrew Bible seems to have been settled by the second century BCE, when it is referred to in the prologue of Sirach (Ecclesiasticus), dating from about 130 BCE.

The Septuagint, the Greek translation of the Hebrew Bible, includes apocryphal writings (see p.12) which are absent from the canon. This may mean that Hellenistic Jews recognized a wider canon than the Palestinian Jews. The Septuagint became the dominant Christian version of the Old Testament, and so these additional books were at first included in the Christian canon. The different order of books in the Septuagint, which has become standard in Christian Bibles, represents a specifically Christian theological arrangement: Israel's history is viewed as reaching its fulfillment in the coming of Christ. In the Christian arrangement, the Pentateuch is followed by most of the

The Vision of the Lamb (Rev. 5), *from the* Lambeth Apocalypse, *produced in England, ca. 1250. The Apocalypse of John (more usually called Revelation) was a relatively late addition to the New Testament canon (see p.12). The passage illustrated here is Revelation 5.6, which is quoted and interpreted in the text below the picture.*

books in the third section of the Hebrew canon, and the Prophets bring the work to a close. With this arrangement, God's promise in Malachi 4.5–6 to send the prophet Elijah provides a transition to the New Testament, which begins with the appearance in the gospels of the second Elijah: John the Baptist (see pp.152–3).

The Canon of the New Testament

The first Christians inherited the Jewish canon and knew no other. Their own writings were intended to give pastoral guidance and instruction to emerging congregations and were not envisaged as sharing the canonical authority of the Jewish Scriptures.

The idea of a New Testament canon, parallel to that of the Hebrew Bible, began to emerge in the second century CE. By this time, the four gospels, with their basically similar pattern, were increasingly recognized as authoritative, and Paul's letters had been collected. The concept of a canon – a fixed body of authoritative works – was first applied to Christian writings by Marcion, who died around 160 CE. He rejected the entire body of Jewish Scripture and accepted only Luke's gospel and an expurgated version of ten of Paul's letters. In reaction to Marcion, the mainstream Church emphasized the authoritative character of all four gospels and the thirteen letters attributed to Paul, thus implicitly excluding the many other Christian writings that had appeared by this time.

The main outlines of the New Testament canon were now fixed, but uncertainty remained about various other books, such as the letter to the Hebrews (see pp.250–51), Revelation (see pp.254–6), and some of

THE APOCRYPHA

The Greek word *apocrypha* refers to some fourteen or fifteen writings whose canonical status has long been disputed. The term means "things hidden away," but, in fact, the books in question were never considered secret.

Some apocryphal books were originally in Hebrew, but they were excluded from the Hebrew canon and, until recent discoveries, survived only in Greek. Because they were included in the Septuagint, the Bible of the early Church, they were accepted without question by Greek-speaking Christians for most of the first four centuries CE. A great change occurred in 382 CE, when the Pope commissioned the scholar and churchman Jerome (see picture, below right) to make a new translation of the Bible into Latin. Jerome's translation is known as the Vulgate, from the Latin *vulgata editio*, roughly meaning "edition for common circulation." Jerome was convinced that only the Hebrew canon could be regarded as authentic, and hence he rejected the books found only in Greek, labeling them "apocryphal." But his views were not accepted, and the Vulgate retained the apocryphal books in the Old Testament.

In the sixteenth century, during the Reformation, Protestants universally agreed with Jerome. They either removed the apocryphal books from the Bible altogether, or placed them in a separate section between the Hebrew Scriptures and the New Testament. In recent years, a new appreciation of the apocryphal writings has developed, as scholars have increasingly realized their importance for both biblical interpretation and the understanding of the development of Judaism and early Christianity.

St. Jerome in his Study *(1475/6), by Antonello da Messina (ca. 1430–1479). Jerome first used the term "apocrypha" for non-canonical writings. Because he was a papal adviser, Jerome was conventionally depicted by medieval artists as a cardinal, although this office did not exist in Jerome's own day. The lion that is shown in the background recalls the legend in which Jerome removed a thorn from a lion's paw.*

the non-Pauline letters. A decisive stage was reached when Athanasius of Alexandria listed all twenty-seven books of the existing New Testament in a letter of 367 CE. Athanasius was the first person to use the word "canon" to refer to the Bible. His canon was widely accepted in the eastern Church, and in succeeding years it was also authorized by a number of synods in the West. Thereafter the New Testament was generally accepted as the second part of the Christian biblical canon.

The Text of the Hebrew Bible and New Testament

Until the end of the first century CE, the main contents of the Hebrew Bible were transmitted in various forms or families: there was no single text. But after the fall of Jerusalem in 70 CE, as part of the restoration of Judaism under the Pharisees, one particular textual tradition emerged and became accepted as the official Hebrew Bible. As it now exists, it is known as the Masoretic text, from the learned scribes, the Masoretes, who compiled it from ca. 900 CE. They took their name from the Masorah, a body of notes that they appended to the text. They also added Hebrew vowel signs to a Hebrew text that had until then been written only with consonants. The work of the Masoretes gradually replaced other forms of the text, and so the earliest manuscripts of the complete Hebrew Bible date only from the tenth century CE. However, some of the biblical documents from Qumran, the so-called "Dead Sea Scrolls" (see p.145), some of which are basically the same as the Masoretic text, confirm its early antecedents.

Although the New Testament is a quarter of the length of the Hebrew Bible, its textual history is much more complicated, largely because there were so many more early witnesses to the text. About five and a half thousand manuscripts are known, ranging from the late second or early third century CE to the thirteenth century. Compared to the relatively small number of manuscripts of other writings from the ancient world, this number is remarkable. The oldest evidence is provided by a number of papyri from the second and third centuries CE, but many of them are fragmentary and contain only portions, sometimes quite brief, of the New Testament. The earliest manuscripts were in the form of a scroll, but around the beginning of the second century CE, the codex, or book form, began to be used in the Church, and parchment or vellum (calfskin) took the place of papyrus. This led to the production of a number of handsome Greek texts, containing the entire New Testament, and sometimes also the Septuagint. The most complete of these is the fourth-century CE Codex Sinaiticus, discovered in Saint Catherine's monastery in Sinai. Other important manuscripts of the same date or a little later include the Codex Alexandrinus, Codex Vaticanus, and Codex Bezae.

New Testament scholars are faced with a situation similar to the

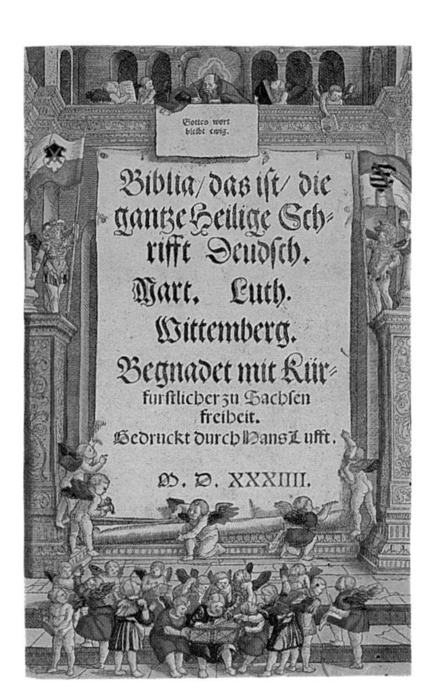

The title page of the 1534 edition of Luther's translation of the Bible. Martin Luther (1483–1546), was declared an outlaw in 1521 by the authorities of the Holy Roman Empire (a confederation of central European states) for his reformist religious views. Justifiably fearing for Luther's life, one of his aristocratic patrons organized his "kidnapping" and hid him for his own safety for several months at the Wartburg castle near Eisenach in Thuringia, Germany. It was there, in disguise and going by the name "Squire George," that Luther undertook his translation of the New Testament from the original Greek. The work was not Luther's alone; he succeeded in assembling a committee of advisers to check and, if necessary, improve his work. The Old Testament was completed after Luther left the castle.

one that would have existed in the case of the Hebrew Bible before the production of the standard Masoretic text. Confronted with the large number of variations among the New Testament manuscripts, they have been led to distinguish three main textual families. One is associated with Alexandria in Egypt, a second is described as "Western," because the chief witnesses for it are Latin sources, and a third, later, form is known as the "Byzantine." The majority of manuscripts belong to this last tradition, and they became the basis for the most generally accepted *textus receptus* ("received text"), being reproduced in the first printed editions of the Greek New Testament, and in subsequent editions until the late nineteenth century.

Modern editions of the New Testament seek to reconstruct the original writer's words by comparing variant readings and attempting to decide which are the most authentic. However, the significance of the manuscript variants should not be exaggerated: all the manuscripts witness to the same basic New Testament writings, most of which were probably written within a century of Jesus' death, by authors who belonged to the first generations of Christians. These writings are distinguished from the many apocryphal gospels (see p.149), acts of the Apostles, and letters, which were generally written much later than the New Testament canon and lack its authenticity (see p.149).

Translations of the Bible

One of the most striking phenomena in connection with the Bible is the tremendous extent of its influence through translations into many different languages. The earliest of these was the translation of the Hebrew into Greek, the Septuagint, an undertaking unparalleled in the ancient world (see p.132). Each of the three great early translations, the Septuagint, the Latin Vulgate, and the Syriac Peshitta, became the standard Bible of one of the three main areas of medieval Christendom, respectively the Greek Orthodox, the Roman Catholic, and the Syrian.

In Western Europe, the Latin Vulgate remained the dominant version of the Bible throughout the Middle Ages. But, from an early period, there were numerous renderings of Scripture into vernacular languages. Sometimes these took the form of glosses on the Latin text, as with the famous gospels produced at the monastery of Lindisfarne off the coast of northeast England (ca. 950 CE). There were also many separate vernacular versions of the Bible in circulation. However, these renderings were generally somewhat free paraphrases, rather than strict translations, and they were only of certain biblical books.

In Eastern Europe, the first translations of the Bible into the Slavonic languages were made by the Greek missionaries Cyril and Methodius in the 860s. The language of their translations, known as Old Church Slavonic, became the liturgical language of the Orthodox Church in

Slavonic-speaking lands. An edition of the Old Church Slavonic Bible published in St Petersburg, Russia, in 1751 (the Bible of Elizabeth) is the standard Bible of the Russian and other Slav Orthodox churches.

The sixteenth-century Reformation in Western Europe was a great motivating force behind many vernacular translations. The most important may well have been Martin Luther's translation of the Bible into German from the original Greek and Hebrew (from 1522). The influence of Luther's work was probably even more profound than that of the English Bible of 1611 (see sidebar, right), with which it has been compared. It was the first real literary work in modern German prose, and it created the vocabulary of literary German. It also served as the model for translations into other European languages, including English. None of these versions (in this context, "version" means "translation," as in "Authorized Version") achieved the same authority as the German and English standard Bibles. Nor did they have as great an influence on the cultures within which they appeared.

Protestant versions of the Scriptures led the way, but Catholics soon responded to a demand for Bibles in the vernacular. German Catholic Bibles appeared in Luther's own lifetime, and English Catholic exiles in Europe published an English New Testament in 1582 and an Old Testament in 1609–1610. French Catholic translations vied in popularity with Protestant ones: the seventeenth-century Catholic Bible of de Sacy was widely used by Protestants as well, and in some respects was the nearest French equivalent of the Luther or King James Bibles.

It has been estimated that the Bible has been published, either complete or in part, in more than two thousand languages, representing eighty percent of the world's population. Many translations have resulted from the Church's missionary work outside Europe, which, in modern times, began with Roman Catholic missions in the seventeenth century. Protestant missionaries were notably active in the era of European imperial expansion, and further work was undertaken by the Bible societies founded in the eighteenth and nineteenth centuries. The first translation of the Bible into an indigenous language of the New World was produced in 1663 by a Protestant missionary, John Eliot, in the now extinct Massachusetts Native American language.

In recent times Bible translation has been most vigorously pursued in Asia and Africa. In Asia, translators are often dealing with languages that already possess a considerable literature – for example, Japanese, Burmese, and the main Chinese and Indian languages. In Africa, on the other hand, they may face the task of creating a written language from a purely oral tradition. The same problem faces translations into the indigenous languages of South America. Translators also have to find terms to convey biblical concepts and images to peoples for whom they may be misleading, unfamiliar, or even unintelligible. The work of Bible translation continues to be a major task.

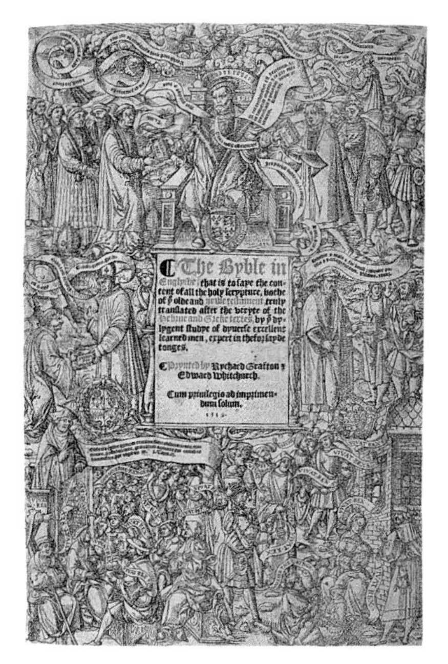
¶ The Byble in Englyshe, that is to saye the content of all the holy scrypture, bothe of ye olde and newe testament, truly translated after the veryte of the Hebrue and Greke textes, by ye dylygent studye of dyuerse excellent learned men, expert in the forsayde tonges.

¶ Prynted by Rychard Grafton & Edward Whitchurch.

Cum priuilegio ad imprimendum solum.

The title page of the Great Bible, *1539.*

THE BIBLE IN ENGLISH

The first complete English Bible was produced in 1382 from the Latin Vulgate by John Wycliffe, John Purvey, and Nicholas of Hereford. In 1526, William Tyndale produced the first English New Testament from Greek, and later translated Genesis–2 Chronicles from Hebrew.

Several complete English Bibles, all indebted to Tyndale, appeared in the reign of Henry VIII (1509–1547), notably the Great Bible (see above), which was placed in all English churches by royal order. The Catholic Mary I (1553–1558) banned English Bibles, but one was produced by exiled Protestants in Geneva. The "Geneva Bible" (published 1560), the first English Bible with verse numbers, was very popular, but its extreme Protestant annotations offended the bishops of Elizabeth I (1558–1603). In 1568 they revised the Great Bible, and this "Bishop's Bible" was the basis of the "Authorized" or "King James Version" (KJV), commissioned by King James I and published in 1611. The KJV took four decades to replace the Geneva Bible in popularity, but thereafter its cultural influence was immense. By 1870 biblical scholarship had indicated many places where the translation could be improved, and the KJV was revised. But it remained the basis of the Revised Version and of the American Standard Version and its successor, the New Revised Standard Version. Many recent translations have aimed to be independent of the KJV, notably the New and Revised English Bible, the Good News Bible, and the New International Version.

The Gezer Calendar (ca. 1000 BCE), an inscription on limestone in ancient Hebrew script recording the annual agricultural cycle (see opposite page).

HISTORY, ARCHAEOLOGY, AND THE BIBLE

The archaeological investigation of sites in the ancient Near East has made a notable contribution to biblical studies, often illuminating, or even confirming, the scriptural record. Excavations have uncovered a vast range of literary texts and inscriptions, many of which provide fuller information about events and personalities, Israelite and non-Israelite, mentioned in the Hebrew Bible. Ancient Near Eastern religious texts, such as myths and rituals, contribute to a more profound understanding of many aspects of Israel's religion and worship, and legal documents place much of Israelite life in its wider Near Eastern context. Discoveries, such as buildings, pottery, tools, and weapons, from both within and outside Palestine, shed light on the daily life of the Israelites. And, of course, archaeology has determined the location of many sites referred to in biblical literature.

Previous generations of scholars often linked archaeology very directly with the Bible: they spoke of "biblical archaeology" and saw it as a means of establishing the credibility of Scripture. It was felt that the archaeological evidence could demonstrate the historical veracity of, for example, the Ancestors, the Exodus, and the Israelite occupation of Canaan (see pp.68–71).

But these claims have not carried conviction. Scholars now recognize the complex relationship between archaeology and the biblical record, and there is no consensus about the course and historical setting of Israel's early existence. Many archaeologists concerned with the geographical area relating to the Hebrew Scriptures tend to see their work as an independent discipline, preferring to speak of "Syro-Palestinian archaeology." Some even seek to provide a picture of this region's cultural development *without* any direct reference to the biblical text. This may be an equally one-sided approach. The history of any civilization must include the individual events that shaped it, and in the case of Israel, the Hebrew Bible is the only witness to these events. Archaeological evidence will rarely correlate neatly with the biblical record, but, with careful interpretation, it can often become a valuable supplement.

Archaeology in Palestine

Archaeological investigation in Palestine, the primary setting of most of the biblical narratives, has been ever more intensively pursued during the later twentieth century. The excavation of ancient towns there, and the discovery of both written and non-written remains, has revealed a great deal about the circumstances of Israelite daily life. Details of ordinary living and customs found in the Hebrew Bible have been invested with a new reality. The tenth-century "Gezer Calendar," discovered in 1908 (see picture on opposite page), has shown how Israelite farmers pursued their agricultural operations; and, at other sites, the layout of houses and streets, and the mass of domestic artifacts, have revealed a good deal about how the Israelites actually lived. Sometimes, findings will directly illuminate the biblical text. For example, the discovery of a weight labeled *pîm* has helped scholars understand 1 Samuel 13.21, the only verse in the Bible that includes the term; as the names of different weights (see picture, right) were used to indicate value, it is clear that the verse is referring to a payment made to the Philistines. Other examples of such finds are the remarkable carved ivories of ca. 800 BCE found at the site of the royal palace of Samaria, capital of the northern kingdom of Israel (which existed 922–721 BCE). These illuminate such passages as 1 Kings 22.39, which refers to the "ivory house" of King Ahab.

Ancient Israelite weights of the 7th century BCE; the amounts are inscribed on them in Hebrew.

Discoveries that show the nature of popular Israelite religion are of particular significance. At Arad, excavations have uncovered the only known Israelite temple and temple cult objects from the era of David and Solomon. Most remarkably, in the eighth-century site of Kuntillet Ajrud, archaeologists have found a painting of a half-nude female figure with an inscription mentioning "Yahweh of Samaria and his Asherah." Although the exact significance of this discovery is a matter of dispute, it may reflect the continuing veneration in Israel of Asherah, the great Canaanite mother goddess, in spite of the official state religion, which recognized Yahweh as the one and only God.

The Bible and History

The books of the Bible are all historical documents, produced by real people writing for their contemporaries against the background of their own times. One of the main aims of scholarship is to discover the historical setting of biblical texts and to show how the writers were responding to the circumstances of a given time and place. One aid in this task is the presence of references in the Bible to historical events, for example in the prophetical books. These references can often be corroborated by, and correlated with, evidence from extra-biblical sources.

A carved ivory plaque discovered at Samaria in northern Palestine, ca. 800 BCE. It depicts the Egyptian god Horus.

Other approaches are concerned less with the historical truth behind the scriptural narratives than with the way in which the narrative framework determines the significance of the material as a whole. The chronological narrative in the Pentateuch and the historical books is the product of the collection and arrangement of a mass

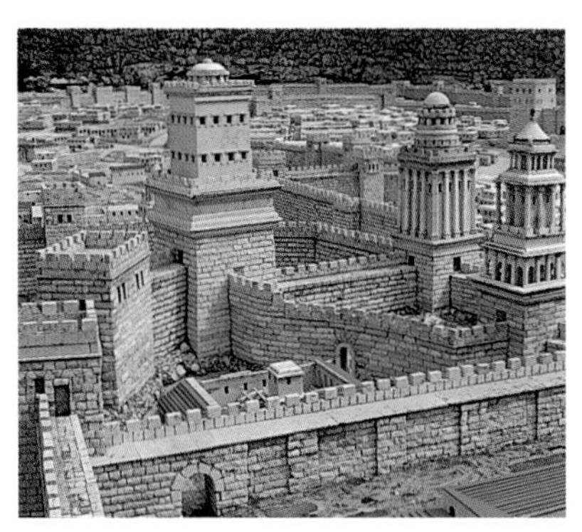

Top: *Part of the few visible remains of Jerusalem in Jesus' time. This tower was part of the royal palace complex built by Herod the Great in 23 BCE (see map on p.190). It may well be the tower of Phasael, the largest of the palace's three towers according to the 1st-century Jewish historian, Josephus. The masonry blocks in the lower two-thirds of the tower are dressed in typically Herodian style; the top dates from a much later period.*

Above: *A reconstruction of the palace, with the tower of Phasael prominent.*

of traditions and documents into a single corpus, and the choice of events recorded may well have been determined by the need to reflect and support the faith of the community. Many scholars today would consider that the Hebrew Bible's presentation of Israel's history is a reflection of a concern to understand the nation's destiny, and to preserve Israel's unity and identity in the light of the crisis brought about by the Exile.

Some scholars have concluded that it is virtually impossible to know what really happened in the course of Israel's history. But such thoroughgoing skepticism is hardly justified, even if there are historical inaccuracies in the biblical texts. Such inaccuracies are no greater or more significant than those found in other ancient writings.

With modern understanding, scholars can see that much that appears as history is actually myth, legend, or folklore, and evaluate it accordingly. For example, the picture of the Ancestors of Israel in Genesis as we have it (see pp.36–51) may well be a construction of the exilic age, but still it is possible to discuss the historical value of the traditions that lie behind it. What precisely happened at the Exodus from Egypt may be impossible to discover, but still it is possible to believe that some actual event triggered off the enormous significance that the Exodus story assumed for Israel's faith. From at least David's time onward, Israel's story begins to relate clearly to events in the wider Near East, and the biblical narrative, in spite of its exaggerations, errors, and unique interpretation of events, gains a solid basis in historical fact.

History and Divine Revelation

The Bible has been seen as fostering the view that the way in which God has made himself known to humanity (divine revelation) has not been through ideas, concepts, or theological propositions, but rather through concrete acts, such as the deliverance of the Israelites from Egypt. This view of history as the medium for divine activity has often been claimed to be unique to the Bible. However, the idea of historical events as divine revelation was part of the common thinking of the ancient Near East and Hellenistic world. Much material in the Hebrew Bible seems to have little to do with revelation through history – for example, the Wisdom literature (see pp.126–7), many of the Psalms (see p.121), and arguably parts of the New Testament. Fundamentally, it has to be asked whether an event was revelatory in itself or became so as a result of its interpretation. For example, what makes the Exodus so central to the Hebrew Bible is not the event in itself, but its significance as God's act. God himself communicates this to Moses before the crossing of the sea (see pp.58–9), and his explanation determines the meaning of the episode for the nation thereafter.

The Historical Jesus

Many of the historiographical issues arising from the Old Testament are also raised by the New Testament, especially by the gospels. Since the nineteenth century, it has become common to distinguish between the "Christ of faith," as presented in the New Testament, and the "historical Jesus," who can be discovered only by delving behind the biblical record. The gospels were produced to awaken faith in Jesus as the Messiah and Savior. Nevertheless, some have claimed that the gospel writers were not interested in giving an account of the actual life and teaching of Jesus but only in the traditions about him that were relevant to their own life and faith. In the past, the quest for the historical Jesus became somewhat discredited, because those who undertook the task produced very varied descriptions of their subject that generally reflected their own preconceptions. Recently, however, the quest has been revived, for two main reasons. To start with, our knowledge of first-century Judaism has increased greatly, leading to attempts to understand Jesus in his Jewish context. Also, it is now appreciated that the gospel writers and the early Church *were* interested in what Jesus did and taught in his earthly life. The gospels have a certain biographical character, not wholly unlike the work of Classical historians such as Herodotus and Tacitus, and their value for reconstructing at least the main outlines of the Jesus of history need not be doubted.

ARCHAEOLOGY AND THE NEW TESTAMENT

Scholars have tended to focus on the relevance of archaeological finds to the Hebrew Bible, leaving the New Testament comparatively neglected, but this is beginning to change. Once more, archaeology sheds valuable light on the social, economic, and political setting of the world of the gospels and the New Testament letters. Excavations in such towns as Nazareth, Capernaum, and Jericho have unearthed evidence of conditions there at the time Jesus that would have visited them. Archaeology, and the evidence of Josephus and the Mishnah, have also been able to provide a reasonably accurate picture of the Jerusalem of Jesus' day. Excavations have uncovered the pools of Bethzatha and Siloam, sites mentioned in association with cures by Jesus (John 5.1–9, 9.1–7). These places, mentioned only in the Fourth (John's) Gospel, indicate that the author had as detailed a knowledge of Jerusalem's topography as the authors of the three Synoptic Gospels. This has suggested to some scholars that the gospel was written earlier than the end of the first century CE, when it is otherwise generally thought to have been written.

The ruins of the Agora ("Marketplace") of Athens, where the apostle Paul preached (Acts 17.16). Like the Roman Forum, the Agora was a bustling civic and commercial center. The reconstructed stoa (colonnaded building) on the left of the picture gives an impression of the structures that stood here in Paul's time. The great cities of classical antiquity have long been investigated by archaeologists, revealing the places known to the apostle Paul and others in their missionary work. More recently, scholars have begun to use the archaeological evidence from these discoveries in their studies of the sociology of the early Christian communities. In particular, several scholars have made highly effective use of archaeological data from Corinth, both written and non-written, to better understand the actual situation to which Paul's Corinthian correspondence is addressed *(see p.243).*

MYTHOLOGY AND SCRIPTURE

It is generally recognized today that myths and mythological concepts figure prominently in the Bible. The Hebrew Bible (Old Testament) in particular can be set in the wider context of ancient Near Eastern religion, in which mythical thinking played an important part. The Bible contains what may be called myths in the strict sense: sacred stories set in an indeterminate "time of beginnings," in which the actors are divine or superhuman beings. These stories seek to account for the fundamental origins of the world, humanity, and social institutions. Such myths occupy a very small portion of the Bible – they are confined almost exclusively to the first eleven chapters of the book of Genesis (see pp.26–35) – but they are significant because they express how the Israelites understood the nature of the universe.

The mythic character of Genesis 1–11 is confirmed by comparing the material in these chapters with the mythologies of some of the cultures surrounding Israel. There are parallels between the account of the creation in Genesis 1 and the Babylonian creation epic (see p.27), and between Genesis and the theology of the ancient Egyptian cult center of Memphis. There are similarities between some themes in the narrative of the Garden of Eden, such as the molding of humans from soil (see p.27), the tree of life (see p.28), and the serpent (see p.29), and the Mesopotamian epics of Atrahasis and Gilgamesh. These two epics furnish close parallels with the biblical story of Noah and the Flood (see pp.32–3), and the tale of the building of the Tower of Babel (see p.34) also has its roots in Mesopotamian sources. Of course, in Genesis, the ancient sacred myths have been revised in the light of Israel's own distinctive theology.

The language and structure of myths can be used to recount and interpret historical situations. Particularly significant in the Hebrew Bible is the widespread Near Eastern myth of a great conflict between a deity, generally a storm god and a water, or sea monster. For example, the Babylonian creation epic tells how Marduk, the supreme god of Babylon, killed a female dragon, Tiamat, the primal saltwater ocean. There is no conflict between God and an enemy in the first creation story in Genesis (see pp.28–29), but the idea of a watery chaos appears in Genesis 1.2: "darkness covered the face of the deep, while a wind from God swept over the face of the waters." Tehom, the Hebrew word for "the deep" in this passage, may be related to the name Tiamat. A great wind is important in other creation accounts, and other passages in the Bible show that the ancient myth was well-

Noah releasing the animals from the ark after the Flood, a mosaic from St. Mark's Basilica in Venice, Italy (built 1071). The rainbow illustrates God's words to Noah: "I have set my bow in the clouds, and it shall be a sign of the covenant between me and the earth" (Gen. 9.13). This passage echoes the Babylonian creation myth, where the god Marduk hangs his bow in the sky after his defeat of Tiamat, a monster representing the waters of chaos (see also p.23 and p.32).

known in Israel. The deity's victory is often viewed as the prelude to his creation of the universe. There are numerous references to this myth in the Bible and it is found in the religion of the Canaanites, Israel's immediate neighbors, from whom it may well have been derived (see p.22).

A good example of mythologized history is the story of the Israelite Exodus from Egypt (Exod. 1–15; see pp.58–9). Whether the events narrated in these chapters have any basis in fact is virtually impossible to know. As it stands, the Bible interprets the episode in terms of the storm god's victory over the sea, especially in the culminating passage Exod. 15.1–18. This is a great poem, which may originally have been a liturgy commemorating the Israelites' deliverance from Egypt. In many cultures, the mythical events of primordial time are often reenacted through regular ritual celebrations. Some of Israel's festivals reflect this idea. For example, the feast of Tabernacles or Booths (see p.120) is considered by many scholars to contain elements representing the renewal of creation, Yahweh's victory over the forces of chaos, and his enthronement as king. This is the same pattern that is found in

A Canaanite bronze statuette, probably depicting the god Baal wielding a thunderbolt; ca.1200 BCE.

the case of Marduk in the Babylonian creation epic, or Baal in the Ugaritic texts. Myths are also a fruitful source for religious poetry, and for the inspired mystical utterances of prophecy (see p.94) and the genre known as "apocalyptic" (see pp.136–7). There are numerous echoes of mythological themes, particularly the deity's conflict with the chaos monster, in the Psalms, the prophetical books, and the apocalyptic writings (see also section on Canaanite Mythology, below).

Mythology and the New Testament

The extent to which the New Testament has a mythic character is much disputed. Revelation, as an apocalyptic work, certainly encompasses a wide range of mythical concepts, although it draws much of its inspiration from other apocalyptic books, especially Daniel and Ezekiel (see pp.254–6). The remainder of the New Testament also inherits many of the mythical themes and resonances of the religion of Israel. These were probably mediated through the apocalyptic writings (see, for example, box on p.191), although in the main less obviously than in the case of Revelation. Attention has often been drawn to the New Testament's mythological conception of the universe, derived from the Hebrew Scriptures, and to the cosmic dimension to the presentation of Jesus, who is seen as a being outside time and space – creator, primordial man, and victorious king – which is so central to the thought of Paul and the Fourth Gospel.

Clay figurines representing Ashtoreth or Astarte, the Canaanite fertility goddess; ca. 1200 BCE.

Canaanite Mythology

The discovery this century of a series of mythical poetic texts at the site of Ugarit (modern Ras Shamra), on the coast of Syria, provided first-hand evidence of the religion of the Canaanites. They gave details of the deities of the Canaanite pantheon, such as El, Baal, and Asherah – all names which occur in the Bible (see p.72). From the Ugaritic evidence, it appears that many of the characteristics of these deities were appropriated by the Israelite god Yahweh. The most important Canaanite myth deals with the fight for world kingship between Baal, the great storm god, and the chaos monster Yam, "the Sea," a being also called "River," "Dragon," "Serpent," and "Leviathan" in Ugaritic texts. Yam was the equivalent of the Mesopotamian chaos monster Tiamat (see p.20). Baal was victorious, and his triumph guaranteed his control of the rains, the regular round of the seasons, and the world order. Much the same is said about God's defeat of the sea monster in Psalm 74.16–17 (see p.23).

After his triumph, Baal proclaimed himself king and was rewarded by the supreme god, El, with a palace: this was the great temple of Baal at Ugarit. The Baal epic was probably meant to be recited at a great autumn festival celebrating the end of the agricultural season and the coming of the new year. It is clear that the outlines of this myth were familiar to the Israelites. A number of the Psalms that speak of Yahweh's kingship and his victory over the waters of chaos may also be liturgical texts designed to accompany an annual celebration in the Temple of Jerusalem.

A shell plaque depicting a god or hero fighting a seven-headed sea dragon; southern Mesopotamian, ca. 2450 BCE.

The Chaos Monster Myth in the Bible

The myth of God's defeat of the primordial chaos monster appears in various forms in the Bible. Psalm 74.13 describes God's adversary as the "sea," which is cleft in two, and as a many-headed sea dragon or "Leviathan," which God crushes. Verses 16–17 indicate that this took place before God's creation of the ordered world. The chaos monster is also called Leviathan and identified with the sea in Job 3.8. In apocalyptic writings, the ushering in of the new age at the end of time is often portrayed as the renewal of the Creation, and the defeat of Leviathan takes place before this renewal (for example, Isa. 27.1). Elsewhere, the sea monster has the name "Rahab" (Ps. 89.10; Isa. 51.9). This title is also applied to Egypt, to indicate that God has rendered that country powerless: Isaiah 30.7 describes Egypt as "Rahab who sits still." The identification of the chaos monster with the primordial waters accounts for the fact that, in apocalyptic writings, the forces of evil emerge as creatures from the sea (Dan. 7.4; 2 Esdras 11.1). The beast from the sea also occurs in the New Testament book of Revelation, in which imagery drawn from the ancient myth features prominently. In Revelation are the conflict with the dragon, the final disappearance of the hostile sea in God's new creation, and Rome as a great whore "enthroned over many waters."

THE HEBREW SCRIPTURES AND APOCRYPHA

Left: Scenes from the Exodus, *from the* Brother Haggadah, *a Hebrew manuscript produced by the Jewish community in Barcelona, Spain, ca. 1350.* Top: *the Israelites leave Egypt (Exod. 12.50–51); the prominent bearded figures on the left are Moses and Aaron who are both holding on to a staff, one of the symbols of their office* *(see pp.56–7)*. Bottom: *Pharaoh and his pursuing army (Exod. 14.8–9).*

I ORIGINS

The first eleven chapters of the book of Genesis present a series of origin stories which prepare the ground for the appearance of the Israelites, the chosen people of God. Genesis is not concerned with the idea of creation out of nothing, but rather with how the world, as perceived by human beings, came into being. The concept of absolute beginning only occurs later in the Bible (for example in 2 Maccabees 7.28). The origin stories in Genesis describe the creation of the cosmos and the first people, the destruction of the world in a great flood, the re-emergence of humanity, and the origin of the different peoples and languages of the world.

THE CREATION

The opening chapters of the Bible describe God's creation of the ordered world and of humankind. It is widely agreed today that there are in fact two creation stories in Genesis. The first account, which runs from Genesis 1.1 to 2.4a, is usually said to be the work of an anonymous priestly writer ("P") of the sixth century BCE. This author's account draws on ancient mythological traditions and aims primarily to glorify the majesty and greatness of God rather than to give a scientific explanation of how the universe began. The creation story has a seven-day structure and ends with the institution of the Sabbath, the seventh day, when God rests to contemplate all that he has made. There are repeated refrains after each of the first six days ("And there was evening and there was morning, the first day," and so on), which suggest that the story was originally a hymn or poem. It may have formed part of the liturgy of a great annual festival in the Temple of Jerusalem before the time of the Exile in Babylon

God in the Act of Creation, *from the French* Bible Historiale *of 1411. Western European artists sometimes depicted God as a cosmic architect, measuring out the universe with a pair of compasses. The image is derived from passages such as Job 38.4–7. He is surrounded by flames, an image which evokes his manifestation in a pillar of flame, as in Exodus 13.21.*

(see pp.112–13). The festival probably celebrated God's victory over the primal chaos, a theme which was derived from older Near Eastern mythology (see p.23). The Bible retains some features of the ancient myth, such as the primeval waters, but the idea of a literal, physical, battle between the deity and chaos has disappeared. God overcomes chaos and establishes the ordered world simply through his free and unrestrained word or command, as expressed in the typical formula: "And God said, 'Let the earth bring forth living creatures.' " The belief that a deity's words possessed an almost magical power was widespread throughout the ancient Near East. The creative divine word is celebrated in Mesopotamian hymns and also in Egyptian myth, where the ordered universe is said to have come into existence through the word of the god Ptah.

This clay tablet of ca. 700 BCE is inscribed with part of the much older Babylonian creation myth, and relates the fight between Marduk, the great god of Babylon, and Tiamat, the dragon of chaos. It is written in Mesopotamian cuneiform ("wedge-shaped") script.

God separates the elements of the universe to produce an orderly, habitable cosmos in which plants, animals, and humans (see sidebar, right) can live. He divides the light from the darkness (Gen. 1.4); the waters below the dome of the sky from those above the dome of the sky (Gen. 1.6–8); the dry land from the seas (Gen. 1.9); and the day from the night (Gen. 1.14). Creation by separation occurs in early Sumerian texts and in the Babylonian creation epic, where the supreme god Marduk defeats Tiamat, the dragon of chaos, and cuts her in two, making one half of her corpse into the sky and the other half into the earth.

The second, and older, account of creation runs from Genesis 2.4b to 2.25 and is generally attributed to an author who is called the Yahwist or Jahvist ("J"), because this writer adds *Yahweh*, the personal name of God (see p.54), to the more abstract term *Elohim* ("God"). This creation story is basically similar in outlook to the first one, but its form is very different. It reflects the interests and concerns of a peasant society: before creation, the world is an arid desert without either rain or the farmer to make it suitable for life (Gen. 2.5). Genesis 2.6 describes how, before the existence of rain, "a stream would rise from the earth, and water the whole face of the ground." In the Hebrew original this "stream" is described by a word of Mesopotamian origin that denotes the underground waters that play a prominent part in Sumerian cosmology.

God molds the first man (*adam* in Hebrew) "from the dust of the ground" and breathes into him "the breath of life." This divine breath gives the man a unique closeness to the deity that corresponds to the idea in the first creation account that humans are "in the image of God." The concepts of the breath of life and of creation in the divine image are found together in a document written in the twenty-second century BCE for the Egyptian pharaoh Meri-ka-re. The formation of humans from soil or clay is common in the creation mythology of the peoples surrounding the Israelites. In Egyptian myth, for example, the god Khnum is said to have made gods, humans, and animals on a potter's wheel and then, like God in the Bible, breathed life into them. The first creation story in Genesis ended with the religious institution of the Sabbath; the second ends with the social institution of marriage: "Therefore a man leaves his father and his mother and clings to his wife, and they become one flesh" (Gen. 2.24).

In the garden of Eden (see p.28), God presents the animals to the man to be named (Gen. 2.19), an act which on one level serves as a simple explanation of how animals got their names, while also implying that the man controls them. It corresponds to the passage in Genesis 1 in which God grants humankind dominion over all creatures.

THE ORIGIN OF HUMANKIND

The first creation story relates how humans were created on the sixth day as the crown of creation. God does not create people through his simple command (there is no formula: "Let there be humans"), but he prefaces the creation of humanity with the words "Let us make humankind" (Gen. 1.26). The plural here probably indicates that God is taking counsel with the divine beings who, elsewhere in the Bible, form his heavenly court, for example in Nehemiah 9.6 and Job 1.6. Humanity is created "in the image of God." This probably means that, unlike other creatures, humans can communicate with God and respond to his word.

The creative act of separation (see main text) is present in God's creation of humankind as male and female. This establishes the essential complementary nature of the sexes: one cannot exist without the other, and both are equal in the eyes of God. God blesses humanity and grants it dominion over the future of the world and other creatures (Gen. 1.28).

The creation of humanity is central to the second creation story (see main text). In this older narrative the first woman is created after the first man (Adam), from his body (Gen. 2.21–22), a situation which no doubt reflects the dominant position of males in a patriarchal society. But the implied inferior status of women becomes explicit only after the fall from grace in Eden (see p.28), when God tells the woman: "your desire shall be for your husband, and he shall rule over you" (Gen. 3.16). Adam then gives his partner a name, Eve, an act that echoes his naming of the beasts, which expressed his dominion over animalkind. However, this account stresses, even more strongly than the first, the unity of the sexes and their mutual need. God says: "It is not good that man should be alone; I will make him a helper as his partner" (Gen. 2.18). Only woman proves suitable for this role.

THE GARDEN OF EDEN

THE FALL OF HUMANKIND: THE TWO TREES

The first man and woman live in a state of primeval leisure and innocence in the garden of Eden until, tempted by the serpent (see box on opposite page), they eat the fruit of the "tree of the knowledge of good and evil" in defiance of God's wishes. It was one of two trees in the garden, the other being the "tree of life," that is immortality (Gen. 2.9).

These trees are thought to represent two originally separate traditions which the biblical author has brought together to express his view of the origin of the human condition. The first tree belongs to a distinctly Israelite tradition. The "knowledge of good and evil" probably means the capacity for rational and ethical judgment. This attribute belongs supremely to God, who forbids humans to acquire it. After the first human pair has defied his command, God condemns them and their descendants to a life of toil and pain. He drives them from the garden in order to deny them the fruit of the tree of life, which would bestow the other divine attribute, eternal life (Gen. 3.22–24). Thus humans are like God in one respect but, unlike him, they die. In this distinction lies the vast potential but also the limitation of human existence. A sacred tree that confers eternal life occurs in many mythologies and is common in Near Eastern iconography. Numerous myths of the region explain that humanity is mortal because one person failed to eat a magical food that would have made them live forever (one example is found in the Mesopotamian epic of Gilgamesh). But the story in Genesis is alone in stressing that humans continue to suffer death as a direct consequence of their disobedience to God's command.

According to the second chapter of Genesis, God planted a garden in a geographical region called "Eden" that is the source of a great river. The river flows into the garden and then divides into four branches, two of which, the Pishon and Gihon, are unidentified. The other two are the real rivers Tigris and Euphrates, which would therefore appear to locate Eden somewhere in the Mesopotamian region (the garden is described as "in the east," a phrase that is used to refer to Mesopotamia elsewhere in the Bible). However, although numerous more or less plausible suggestions have been put forward as to where the author intended Eden to be, the question is of secondary importance because the Bible's geography is essentially mythical and symbolic. The most significant point was that the fertility of the world known to the author depended on rivers flowing from the garden of God.

God created Eden as a habitat for newly created humanity (see p.27), but there are indications that the Israelites had a different, older, concept of Eden. In Ezekiel 28.13–19, it is referred to as the "garden of God" but also as a "holy mountain," full of precious stones. The description recalls the mountain between the source of two rivers that was said to be the abode of the god El, the head of the Canaanite pantheon (see p.72). This garden was the home of a supernatural being who was driven out for his presumption by a "guardian cherub." Such a myth probably lies behind the Genesis account, in which Adam and Eve are expelled from Eden and prevented from returning by cherubim (Gen. 3.24). Adam was probably in origin a supernatural or an angelic figure, as indeed he appears in some Jewish literature.

The garden of Eden occupied an important place in Jewish thought. For some of the prophets, it symbolized the renewal of Israel after the Babylonian Exile (see pp.112–15). The Greek translation of Genesis renders "Eden" as "paradise." Paradise, as the New Testament shows, came to be viewed as the final destiny of the righteous, who would return to the original state of perfection and harmony that was enjoyed by the first human pair in Eden.

A reconstruction of a mural from the palace of Mari on the Euphrates River in Syria (ca. 1750 BCE). Some of its features strikingly recall details of the Genesis description of the garden of Eden. In particular, the mural depicts two types of tree (compare Gen. 2.9), guarded by mythical winged animals or cherubim (compare Gen. 3.24). The lower half of the central panel shows figures holding jars from which flow four streams (compare Gen. 2.10). The upper scene may depict a king being invested by the Mesopotamian fertility goddess Ishtar: Eve has been associated with such divine figures (see box on opposite page).

The Serpent

Adam and Eve with the Serpent, *Spanish school, 12th century. Adam and Eve cover their nakedness after eating the fruit of the tree of knowledge (Gen. 3.7).*

The serpent in Genesis 3 is an ambiguous creature. In one respect, it is simply the snake, one of the animals made by God, and its physical characteristics and behavior are explained in Genesis 3.14–15. Here, in accordance with folktale, the serpent is represented as a cunning creature that causes trouble and is punished. It can also speak.

However, the serpent in Eden probably also represents a demythologized version of what was originally a much more significant being. Archaeologists have revealed that the serpent was widely venerated in Egypt and the Near East, not least in Canaan. There was also at one time a cult of the serpent in Israel, as demonstrated by the bronze serpent that is reported in the Bible to have once stood in the Temple at Jerusalem. (This serpent was destroyed by King Hezekiah: see 2 Kings 18.4.) The serpent was seen as the bringer of life, because it was able to shed and renew its skin annually: according to the Mesopotamian epic of Gilgamesh, it acquired this ability through eating the magical plant of immortality. The creature was particularly associated with fertility, and fertility goddesses are often shown accompanied by serpents. It has been suggested that "Eve," the name of the first woman in the biblical account, is related to a word for snake, and that she was originally a serpent goddess, the goddess of life and fertility.

In ancient tradition, the serpent is also a more sinister being that is associated with the underworld and its waters. In the Bible and elsewhere, the hostile monster that embodies primal chaos is frequently described as a serpent. When the author of Revelation identified "that ancient serpent" with the Devil or Satan (Rev. 12.9), he established this manifestation of the serpent as the enemy of God.

THE LAND OF DILMUN

It has been suggested that the biblical author meant to place Eden somewhere in the region of the northern end of the Persian Gulf. If so, it could be related to the land of Dilmun, which appears in the Sumerian myth of Enki and Ninhursag. In this myth, Dilmun, perhaps identifiable with present-day Bahrain, is a mythical place, a potential rather than a real world. Sickness and old age are unknown there, but it lacks the fresh water that is necessary to sustain plant and animal life. The god Enki therefore creates rivers with water from beneath the earth, and Dilmun becomes a rich and fruitful place. According to later Babylonian myth, a similar ideal land was the home of the few humans who became immortal.

CAIN AND ABEL

THE FAMILY OF ADAM AND EVE
The descendants of Adam and Eve, as presented in the two genealogies of Genesis chapters 4 and 5.

ADAM AND EVE

Cain	Seth
Enoch	Enosh
Irad	Kenan
Mehujael	Mahalalel
Methushael	Jared
Lamech	Enoch
	Methuselah
	Lamech
	Noah

The story of Adam's and Eve's descendants down to the time of Noah is recounted in Genesis 4–5. The account represents one or more traditions that were originally separate from the preceding three chapters of Genesis, as is revealed by a number of logical discrepancies. For example, the farmer Cain murders the shepherd Abel out of jealousy (Gen. 4.8) and God condemns Cain to be "a fugitive and wanderer on the earth." But God puts a mark on him "so that no one who came upon him would kill him." However, there is no indication of how, if Cain is the only surviving child of the first man and woman, there can be anyone else who might kill him. Similarly, Genesis 4.17 states that "Cain knew his wife," but does not explain where she came from. More generally, the narrative presupposes the existence of an organized society, civilization, and culture. There is no solution to such inconsistencies, because the biblical writer did not consider them to be of primary importance.

The narrative is otherwise carefully integrated with what precedes and follows it. It marks the culmination of the Eden story (see pp.28–9) and presents Cain, the first murderer, as the representative of primeval humanity. As in the story of Adam and Eve, the tale of Cain and Abel follows the pattern of wrongdoing, penalty, banishment, and mitigation of God's punishment. It begins a series of narratives in the first eleven chapters of Genesis which chart the increasing violence and presumption of humankind after the Fall and expulsion from Eden. In each case, humanity incurs severe divine judgment, and its very existence is threatened. But God always shows mercy and takes steps to ensure that, in spite of everything, the world and the human race will continue. This pattern determines the general shape and meaning of the story of Cain.

The account is constructed from various motifs, many of which are rooted in myth and folktale. One is the motif of God favoring a younger son over his elder brother, a situation which often leads to jealousy and violence. Genesis 4.4–5 recounts how each brother presents an offering to God, who prefers Abel's gift of animal flesh to Cain's gift of crops. Cain is angry at the rebuff and murders his brother out of jealousy. No reason is given for God's choice, although later writers suggested that God knew of Cain's inherent wickedness. As in the case of Isaac and Ishmael, Jacob and Esau, and Joseph and his brothers, the biblical author sees no need to account for the divine selection. However, the passage may allude to the Hebrew sacrificial system, in which animal sacrifice was more prominent and more highly valued than cereal offerings (see pp.122–3).

Other ancient ideas are present in the narrative. After the murder, Cain is no longer able to cultivate the land because his victim's blood is said to have made it barren, a common concept in the Bible and other ancient traditions. Another theme is the age-old tension between the pastoral and the agricultural ways of life, which is also seen in the Sumerian tale of Dumuzi and Enkimdu. Dumuzi the shepherd and Enkimdu the farmer are rivals for the hand of the goddess Inanna. As in the story of Cain and Abel, the deity prefers the shepherd to the farmer, although in this case the two rivals are reconciled.

The story does not present Cain in a wholly negative light. God sets a mark on Cain to protect him, and he is banished to "the land of Nod," a word which simply means "wandering" and indicates the life of a nomad. This part of the story serves to explain the existence of the Kenites, a nomadic people traditionally said to be descended from Cain. The Kenites lived on the fringes of settled civilization, but had close connections with the Israelites.

A clay mask from the city of Ur depicting an ancient Mesopotamian demon. Before Abel's murder, God warns Cain that "sin is lurking at the door" (Gen. 4.7). This may refer to a well-known Babylonian demon, the rabisu *or "croucher."*

After Eden: the decline of civilization

Between the stories of Cain and Noah there are two genealogies, one of Cain, and one of Adam. These are usually said to be by different authors, but the coincidence and similarity of some names (see sidebar on opposite page) suggest that both genealogies have a common source. The account of Cain and his descendants shows him as a founding father rather than the brooding murderer presented in the preceding narrative. He establishes the first city, called Enoch for his son, and his descendants are pictured as the originators of the human skills that characterize a settled existence: cattle-rearing, music, and metalworking (Gen. 4.17–22). The genealogy ends with the family of Lamech, a figure of violence who takes two wives and boastfully declares how he has killed a man (Gen. 4.23). He is probably meant to show how far humankind has degenerated morally from its primeval state, in spite of all the positive aspects of human development. A similar view of the development of humanity is found in the *Works and Days* by the Greek poet Hesiod, ca. 700 BCE.

Musicians playing horn and drum on a stone relief from Carchemish in northern Mesopotamia, ca. 800 BCE. According to the Bible, the first musician was Jubal, a son of Lamech and descendant of Cain.

Adam's genealogy (Gen. 5) through his third son Seth, serves a similar purpose to that of Cain, although nothing is said explicitly about the spread of evil among humankind. But the fact that the genealogy leads up to the picture of universal evil in the days of Noah suggests that it, too, intends to portray humanity's progressive moral degeneration. This genealogy has often been compared with the Sumerian "King List," a catalog of rulers before the great Flood. However, it is much closer to the work of Berossus, a Babylonian priest of the third century BCE, who lists ten rulers before the Flood and also gives an account of the deluge. Like Adam and his descendants, including Noah, each of these rulers has a fantastically long lifespan.

Tilling and sowing, a mural from Thebes, Egypt. Cain, "a tiller" (Gen. 4.2), was the ancestor of those who "have livestock" (Gen. 4.20).

THE FLOOD

The progressive corruption of the world after Eden culminates in God's decision to destroy "all flesh in which is the breath of life" by a great deluge, the story of which is told in Genesis 6–9. However, in accordance with the general pattern of divine judgment in Genesis, God mitigates his punishment. One man, Noah, "found favor in the sight of the Lord," because he "was a righteous man, blameless in his generation" (Gen. 6.8–9). God decides to spare Noah and his family to make a new start for the human race. He instructs Noah to build an ark (ship) to contain his family and at least one male and one female of every species of animal, so that the world can be repopulated after the Flood. Most authorities accept that the flood story as it stands is the work of two authors, "J" and "P" (see pp.26–7). This is revealed by some discrepancies in the story: for example, in Genesis 6.19 God tells Noah to take "two of every kind" of creature into the ark, but in 7.2 this becomes seven pairs of all clean animals, one pair of all unclean animals, and seven pairs of each species of birds. But both sources tell fundamentally the same tale, and they have been skillfully woven together to form a generally consistent whole.

The story functions as a kind of second creation, and the language it employs often recalls the original creation account. The flood waters come from both above the dome of the sky and below the earth, undoing the creative act of separation described in Genesis 1.6–7, and allowing the primitive state of chaos to return. When the deluge subsides, the appearance of the dry land is emphasized (Gen. 8.13–14), which recalls Genesis 1.9–10. The outcome of the narrative is the establishment of the world as we know it, when God promises that the earth will never again be destroyed. In one sense, the original creation is restored: the alternation of the seasons is assured (Gen. 8.22) and humanity will be fruitful and multiply (Gen. 9.1). But the post-Flood world does not mean a return to Eden. God now accepts that "the inclination of the human heart is evil from youth" (Gen. 8.21) and that the human behavior which caused the Flood can be controlled, but not abolished.

The dominion of humans over the beasts now means that animals will fear and dread humans, who can kill them for food (Gen. 9.2). However, humans may only eat meat from which the blood has been drained, and human life is stringently protected. These regulations form the core of what are called the Noachide ("of Noah") commands, which for the Jewish rabbis constituted a basic law for all humanity. The perpetuity of this new world order is guaranteed by a covenant, a unilateral promise by God that "never again shall all flesh be cut off by the waters of a flood, and never again shall there be a flood to destroy the earth" (Gen. 9.11). The term covenant, which makes its first appearance in Genesis 6.18, plays a central role in Israelite religion. Here, it brings humans and all other creatures into a direct relationship with God. The sign of the covenant is the "bow" which God sets in the clouds to remind him of his inviolable promise. This is usually taken to mean a rainbow, but it may also represent the weapon of a warrior god similar to Marduk in the Babylonian creation epic (see p.27). Marduk hangs his bow in the sky as a sign of his victory over Tiamat, the goddess of chaos. Perhaps at one time the end of the Flood was seen as the taming of the waters of chaos: Genesis 8.1 describes how a divine wind makes the floodwaters recede ("And God made a wind blow over the earth, and the waters subsided"), which recalls the Creation story (Gen. 1.2: "a wind from God swept over the waters"). In this case, the hanging up of the bow would signal the restoration of peace and harmony.

THE SONS OF GOD AND THE DAUGHTERS OF HUMANS

Genesis 6.1–8 forms the introduction to the story of the great Flood. It is in two sections, the first of which had no original connection with the Flood narrative. However, the author uses both sections to explain why God sent the deluge.

The first section (Gen. 6.1–4) gives a brief summary of an ancient myth that relates how divine beings mated with human women and produced offspring. In Hebrew thought, the human and the divine ought not to be confused, so these children violate the whole structure of the world. As punishment, God frustrates any attempt on humanity's part to attain divinity and immortality by declaring "my spirit shall not abide in mortals forever," and restricting their lifespan to one hundred twenty years. The children are described as "mighty men of old," who appear in a later chapter as the embodiment of absolute power in the ancient world. But they are also described as "Nephilim," which probably means "fallen ones." This is one source of the myth of the fallen angels that plays a prominent part in such books as Enoch and Jubilees, works that stand outside the Bible and Apocrypha. Jewish thought came to see these angels, rather than the sin of Adam, as the source of sin in the world.

The second section (Gen. 6.5–8) presents a picture of universal wickedness and evil that is the climax of humanity's increasing degeneracy (see main text).

Noah's Ark, *an illustration from the Nuremberg Bible of 1483. The shape of the ark has been the subject of much debate. In Genesis 6.15, God gives the dimensions of the ark as 300 by 50 by 30 cubits (one cubit is about 18 inches or 45 cm). This was often taken to mean that the vessel was rectangular, as here and on the coin below, rather than with a pointed prow and stern. The word for "ark" in Hebrew,* teba, *means "box" or "chest," and is otherwise used only of the basket in which the baby Moses was laid. In the epic of Gilgamesh (see box, below), Utnapishtim builds a box-like boat in the form of a perfect cube, although it differs from Noah's ark in other respects.*

Deluge myths of the ancient Near East

Stories of a great deluge occur widely in mythology throughout the world. The biblical story of Noah is a version of a flood myth that originated in Mesopotamia and was familiar throughout the ancient Near East from the time of the Sumerians (ca. 2500–2000 BCE). All the known versions of this myth have the same basic outline: a deity or deities send a deluge to destroy the world, but one righteous man is forewarned of the disaster and builds a ship in which he and his family survive the flood. After the flood has subsided, the world continues to exist and is repopulated.

Before God sends the Flood, the Bible tells us that human beings "began to multiply on the face of the ground" (Gen. 6.1). A similar expression occurs in the flood story in the Mesopotamian epic of Atrahasis, where the reason for the deluge is the gods' desire to curb human overpopulation. There are striking resemblances between the Genesis story and the account of a great flood related by Utnapishtim, who is the equivalent of Noah, in the Babylonian epic of Gilgamesh. For example, Utnapishtim survives the flood by building a ship. The vessel finally comes to rest on Mount Nisir, which is in the same region as "the mountains of Ararat," where Noah's ark comes to rest (Gen. 8.4). Utnaphishtim opens a window of his ship and sends out a dove, a swallow, and a raven to reconnoiter the ground, much as Noah opens the window of the ark and sends out a raven and a dove.

On emerging from the ship, Utnapishtim offers a sacrifice which propitiates the gods when they smell its sweet savor, and they agree that in future humanity should be punished if it behaves wickedly, but not destroyed. Noah likewise offers a sacrifice, and after smelling its "pleasing odor," Yahweh promises: "I will never again curse the ground because of humankind ... nor will I ever again destroy every living creature as I have done" (Gen. 8.21). Finally, both Utnapishtim and Noah are blessed and rewarded. Utnapishtim is granted immortality, while Noah lives for another three hundred fifty years and is granted a fecund progeny that will repopulate the whole world.

A Greek coin (artist's reconstruction below) of the 3rd century CE showing Noah and his wife, both outside and inside the ark.

THE TOWER OF BABEL

The story of the Tower of Babel (Gen. 11.1–9) begins with the ideal primeval situation in which everyone in the world speaks the same language. Humanity is presented as a small, homogenous group which travels from "the east" to the plain of Shinar, a region in what is now southern Iraq. There, people prepare to build a city and a tower "with its top in the heavens" (Gen. 11.4) to give them both identity and security. Yahweh, alarmed that "nothing that they propose to do will now be impossible for them" (Gen. 11.6), aborts the operation by causing the people to speak in different languages and scattering them over the earth. Genesis says that the city is named Babel from the Hebrew *balal* "to confuse," because Yahweh "confused the language of all the earth" (Gen. 11.9).

The narrative portrays the development of civilization, the change from nomadism to city life, and seeks to explain the existence of different languages and nations. It also attempts to account for the existence of a particular city, Babylon, and a particular tower: the description of the Tower of Babel in Genesis is probably based on the great ziggurat of Babylon (see opposite page and also p.37). The aim in building such a tower is to allow men to climb up to heaven and challenge the authority of the gods, an act of impiety and presumption that is frustrated by divine intervention.

The Building of the Tower of Babel, *by Abel Grimmer (1570–ca. 1619).*

Babylon

The materials used to build the Tower of Babel were the same as those employed for the construction of the great ziggurat of Babylon and similar ziggurats, according to ancient building inscriptions. The golden age of Babylon arrived when the Assyrian empire fell in 612 BCE. It became the dominant imperial power in the Near East, and the restoration of the city by Nebuchadrezzar II (605–562 BCE) transformed it into one of the seven wonders of the world in Greek eyes. Babylon fell to the Persian king Cyrus in 539 BCE and while it remained an important city under the later Achaemenid, Seleucid, and Parthian empires, it never again played an independent political role.

Babylon becomes in the Bible much more than a historical reality. After Nebuchadrezzar destroyed Jerusalem and exiled many of its inhabitants (see pp.110–13) it was viewed as the embodiment of all evil, a kingdom of wickedness set against God and his chosen people, a powerful and complex symbol of pride, oppression, wealth, luxury, sexual license, and idolatry. As such, Babylon is doomed to final destruction.

In Isaiah 14, the king of Babylon is described, in terms of an ancient myth, as Lucifer, the morning star, who aspired to scale heaven and challenge God himself. For his arrogance, Lucifer is cast into the underworld. The underlying theme of this story – the penalty paid for presumption – is similar to that of the Tower of Babel, as well as to that of the "fallen ones" (see p.32). The book of Isaiah mocks the idols which played so prominent a part in Babylonian religion and sees the city's chief deities, Marduk and Nebo, being carried away into captivity (Isaiah 46.1–2). The prophet depicts Babylon as a female prisoner of war, who was once a "lover of luxury" and a practitioner of sorceries and spells, but who is now stripped of all her splendor and forced to sit in the dust (Isaiah 47). For the prophet, the imminent capture of Babylon by Cyrus is evidence that the Persian king is the chosen agent of God's vengeance (Isaiah 45.1–7).

The name "Babylon" came to be applied to any empire that was seen as the enemy of God and the persecutor of his servants. The author of the book of Daniel tells stories of the humiliation of Nebuchadrezzar and the death of Babylon's last ruler, Belshazzar (see pp.134–5). Their fates serve as models for the destiny awaiting the Greek ruler Antiochus IV Epiphanes, the persecutor of the Jews in his own day (see pp.138–9). Subsequent Jewish writers identified Babylon primarily with the Roman empire. This identification is applied to the situation of the early Church in the book of Revelation, written in the shadow of the emperor Nero's persecution of Christians (see pp.254–6). There, Babylon is Rome, "the mother of whores and every obscenity" (Rev. 17.3–6), that will very soon suffer complete and dramatic destruction (Rev. 18).

Ironically, the deportation to Babylon of the leading citizens of the kingdom of Judah in the reign of Nebuchadrezzar signaled a more positive evaluation of Babylon in Jewish life. The presence of the exiled Jews in Babylon made the city a great Jewish intellectual center. The massive Babylonian Talmud acts, even to this day, as an authoritative source for Jewish religion.

An inscribed basalt block describing the restoration of Babylon ca. 670 BCE by the Assyrian king Esarhaddon (681–669 BCE).

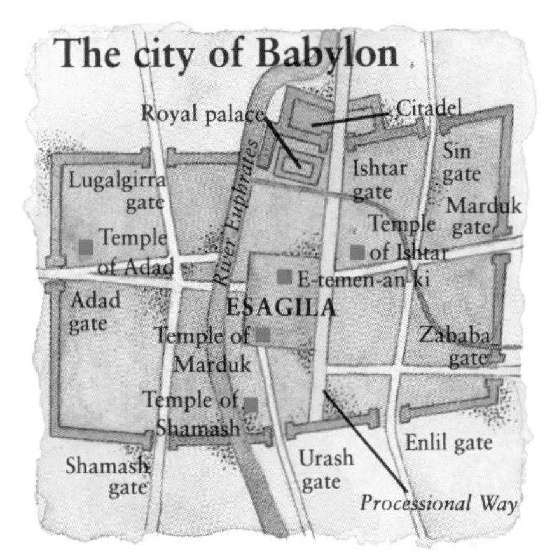

Babylon in the time of Nebuchadrezzar II. At its heart was Esagila, a huge sanctuary complex which contained the temple of Marduk, and E-temen-an-ki, the great ziggurat of Babylon, probably the prototype of the Tower of Babel.

II THE ANCESTORS

The end of the first stage of the biblical story is marked by the episode of the Tower of Babel. The succeeding chapters of Genesis move from the universal perspective of cosmic origins to the origins of a single people, the Israelites. Israel's beginnings are recounted through the lives of the founding ancestors of the people. The patriarchs Abraham, Jacob, and Joseph are each the focus of a major story cycle, and there is also a considerably more concise account of Isaac. It seems likely that these four individuals originally represented different tribal groups, with their own specific traditions. But they have been united into a single family to present a family history traced through four successive generations.

THE CALL OF ABRAHAM

The episode of the Tower of Babel (see p.34) leaves humanity divided into competing nations, and God's choice now falls on one particular people. In Genesis 12.1–2 he speaks to Abraham, the son of Terah and a descendant of Noah's son Shem: "Go from your country and your kindred and your father's house to the land that I will show you. I will make of you a great nation." The subsequent obedience of Abraham (who is called Abram in the early part of the story) reverses humanity's previous disobedience and establishes him as the ancestor of Israel.

In the present century, many experts have become confident enough in their knowledge of the ancient Near East to place the ancestors of Israel in the second millennium BCE. Recently, however, this conclusion has been challenged. Some features of the biblical stories seem to indicate a period later than the second millennium. A number of alleged parallels between the archaeological evidence and the Bible narrative have turned out, on more careful examination, to be much less close than has often been claimed. Also, many customs and beliefs that supposedly characterized the second millennium BCE continued into later centuries. This reexamination has led some authorities to claim that the traditions of the ancestors were largely produced by the Israelite community of a much later date, perhaps even during the period of the Exile (587/6–539 BCE). But this does not mean that the ancestors never existed. Later generations may have retained authentic memories of these figures, who play a significant part in Israel's history and religious development.

The story of Abraham begins in the ancient Sumerian town of Ur in southern Mesopotamia, which was an important city until its destruction in 1740 BCE. The decision of Abraham's father Terah to leave Ur with his family (Gen. 11.31) has been linked with a supposed migration from the region of the nomadic Amorite people as a result of the upheavals of the time. But Ur revived under the Neo-Babylonian or Chaldean empire (ca. 625–539 BCE), and the Genesis description "Ur of the Chaldeans" would appear to indicate this later period. Terah set out for Canaan but settled

in the city of Haran, in northern Mesopotamia. Haran was linked with Ur by a long-established trade route, and both cities were important centers of the cult of the moon god Sin. Like Ur, it was prominent both in the second millennium BCE and under the Neo-Babylonians. In Haran, Abraham received God's call and apparently set out from there on his journey. In fact, the real home of Abraham's family seems not to have been Ur, but the region of Haran. His brother is called Haran, and he has relatives called Terah, Nahor, and Serug, which are also the names of towns near Haran. Later in the story, Abraham sends his servant to find a wife for his son Isaac, from among what Abraham calls "my kindred" who live in "my country." This turns out to be not southern Mesopotamia but Nahor, where the servant finds Rebekah, Abraham's great-niece (Gen. 24). Later still, Isaac's and Rebekah's son Jacob takes refuge in Haran with his uncle Laban (Gen. 28), and marries into his family.

Numerous features of the story of the ancestors (such as their wanderings with livestock and living in tents) hint at the life of pastoral nomads. It has been suggested that the biblical authors were city dwellers who regarded nomads with some disdain, and sought to stress that the founders of Israel originally came from areas of settled urban civilization.

The much-restored remains of the great ziggurat or sacred tower of ancient Ur, in southern Iraq. Such structures probably inspired the Tower of Babel (see pp.34–5).

God's promises to the ancestors

The main concern of the biblical writers who related the story of the ancestors was not historical accuracy but the theological significance of the founders of the nation. The most common motif running through the narrative of Genesis 12–50 is God's pledge for the future. He makes two promises to the ancestors: that they will possess the land of Canaan and that they will become a great nation. These are brought together in God's announcement to Abraham before he sets out on his journey from Haran. The promise of the land appears to be the more important: the entire narrative of the ancestors begins with the divine command to Abraham to go to the land of Canaan, and ends with the divine assurance that his descendants will return to the land "that he swore to Abraham, to Isaac, and to Jacob" (Gen. 50.24).

The ancestors, in fact, did not take possession of Canaan, which only happened later under Joshua (see pp.68–9). This delay in the fulfillment of the promise meant that the family line had to be continued, so that there would be someone to whom the Promised Land could be given in due course. A prominent theme in the whole saga of Abraham and his successors is the need to produce a direct heir (see p.42).

A caravan of migrating Asiatics, from an Egyptian wall painting of ca. 1890 BCE. The wanderings of Abraham have sometimes been linked to the migration of nomadic Near Eastern peoples.

ABRAHAM'S TRAVELS

THE JEOPARDY OF THE MATRIARCH
When Abraham and Sarah arrive in Egypt, an episode occurs that reflects a folktale motif described as "the jeopardy of the female ancestor" (Gen. 12.11–20). Abraham fears that Sarah's beauty will cause the Egyptians to kill him so that someone can take her as a wife. He asks her to say that she is his sister. Sarah does so and becomes Pharaoh's wife. Plagues afflict the royal house, as a result of which Pharaoh discovers the deception. He is angry that Abraham has caused him to marry another man's wife unwittingly, but shows clemency. He orders Abraham and Sarah to leave Egypt with their possessions. The plot of this episode is repeated twice, once again in connection with Sarah (Gen. 20), and then with Rebekah (Gen. 26). Both of these later incidents involve a certain King Abimelech, probably a different ruler in each case.

The biblical account of Abraham's career may appear to be made up of a string of disconnected episodes, but there are several distinct themes running through them. The patriarch's story in many ways encapsulates the subsequent history of Israel. After God's call to Abraham, he leaves Haran together with his wife Sarah (who is called Sarai at this stage of the story; see p.42) and their nephew Lot, and the group journeys into Canaan. Abraham erects altars and invokes Yahweh at Shechem and Bethel. Later, after leaving Egypt, he puts up another altar at Mamre near Hebron. These places were ancient Canaanite places of worship, and there are two references (Gen. 12.6 and 13.18) to oaks (or terebinths), the sacred trees often found at Canaanite shrines. At Shechem and Mamre, Abraham erects altars in response to a theophany (a manifestation of the deity), in which God promises the land in perpetuity to the patriarch and his offspring. Establishing a new cult is one method of laying claim to territory, so Abraham's altar-building paves the way for Israel's eventual occupation of Canaan. It also sanctions the newcomers to maintain the holy places of the former inhabitants for their own worship.

Abraham, Sarah, and Lot move on to Egypt as the result of a famine. Abraham becomes a wealthy merchant, mainly on account of Pharaoh's patronage after Sarah enters his household (see sidebar, left). However, Abraham, Sarah, and Lot are eventually compelled to leave Egypt and return to Canaan. These travels recall the tradition of the migration of the Israelite tribes into Egypt at the time of Joseph, and the subsequent Exodus under Moses.

Once back in Canaan, Abraham and Lot decide to go their separate

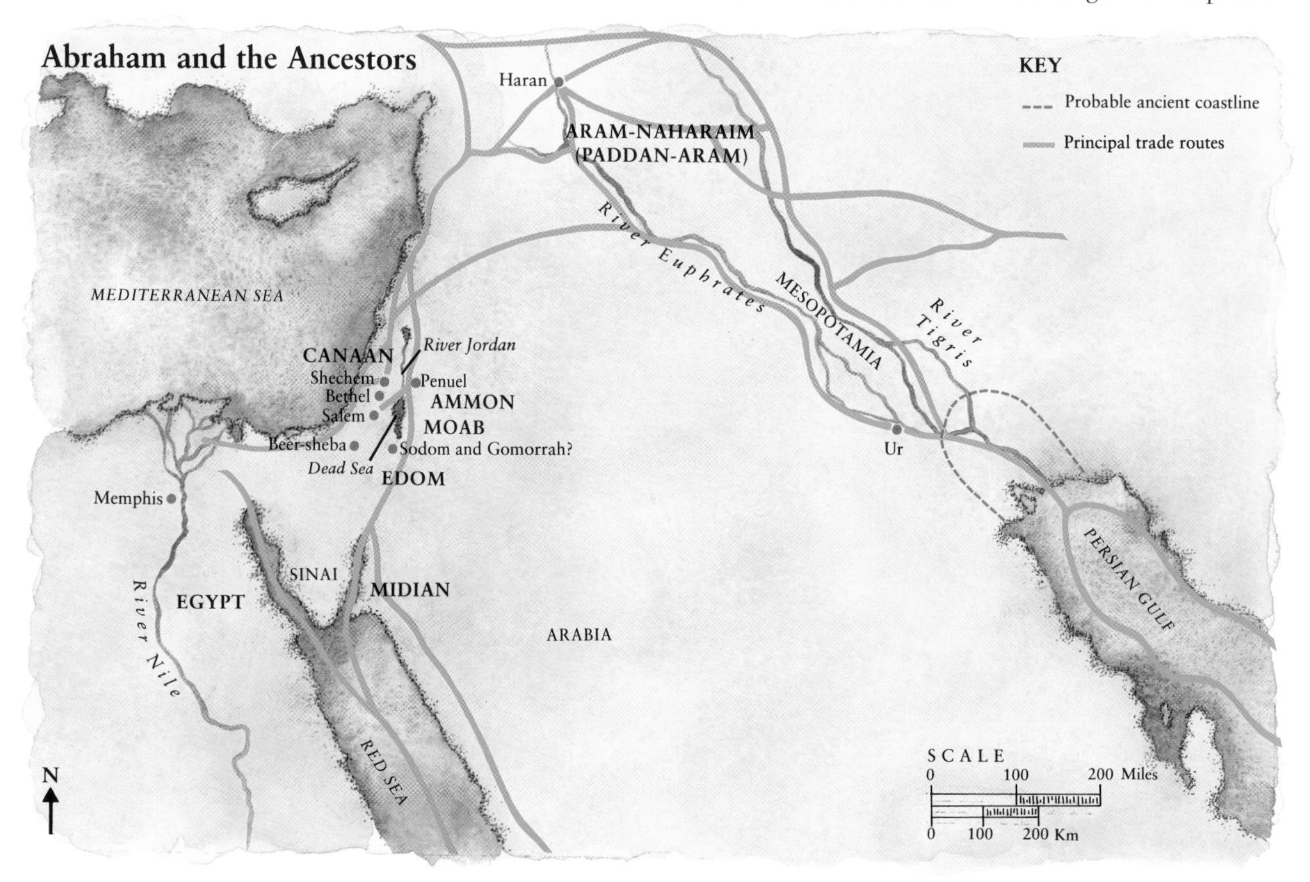

Abraham and Melchizedek

The narrative of Genesis 14 presents Abraham as a military commander, a picture that is very different from the other traditions about him. It tells of a coalition of four kings who campaign against five rulers of the area where Lot, Abraham's nephew, has settled. Two of the five rulers, the kings of Sodom and Gomorrah, are routed, and Lot is taken prisoner during the sack of Sodom, where he lives. Abraham collects three hundred eighteen retainers, defeats the four kings, and frees his nephew. Some of the names of the four kings are attested in non-biblical sources of the early second millennium BCE. But such a great military coalition is improbable, and the Bible account cannot be taken as historically reliable.

After his victory, Abraham is greeted and blessed by Melchizedek, the king of Salem (later Jerusalem) and priest of "God Most High," that is, El Elyon, the high god of pre-Israelite Jerusalem. Abraham identifies El Elyon with Yahweh and gives Melchizedek one-tenth of the booty he has won. The whole episode helps to explain the situation after the conquest of Jerusalem by David. El Elyon was assimilated to Yahweh, the Israelite monarch assumed the position of the former Canaanite priest-king, and the Israelites began to pay tithes for the upkeep of the ancient sanctuary. In the New Testament, the figure of Melchizedek is compared to Jesus in the letter to the Hebrews (see p.251).

The circumcision of two boys, from an ancient Egyptian relief of ca. 2200 BCE. The hieroglyphics above the man performing the operation on the left translate as: "Hold him, do not let him faint!"

ways, and Lot heads for the southern shores of the Dead Sea to settle in the plain of the Jordan River. It appears that Abraham originally intended to share Canaan with Lot, but Lot chooses part of the Jordan valley that lies outside the territory promised by God. Subsequently, in Genesis 19, Lot is said to be the ancestor of the Transjordanian Moabite and Ammonite peoples (although this Lot may originally have been unconnected with Abraham's nephew; see p.41). The parting of the patriarch and his nephew represents the separation of the direct descendants of Abraham from their ethnic relatives: Israel alone is the rightful heir of the divine pledge. God's promises to the ancestors (see p.37) are guaranteed by a covenant with Abraham, the first male ancestor. This covenant is found in two versions, one said to be by the biblical author "J," the other by the priestly author "P" (see p.26). The first, vividly described in Genesis 15, takes the form of God's unconditional commitment to give the patriarch all the land "from the river of Egypt to the great river, the river Euphrates," the ideal extent of Israelite rule. The covenant is effected in an ancient ritual in which an oath is confirmed by someone passing between the divided pieces of one or more sacrificial animals. In Genesis 15.17, Yahweh does this in the shape of a smoking brazier and a flaming torch. A similar rite is mentioned in a document of ca. 1775 BCE from the Mesopotamian city of Mari, which describes a treaty-making ceremony between two peoples.

"P"'s version of the covenant, in Genesis 17, involves an obligation not only on God's part but also on Abraham's. Abraham and his descendants are commanded to circumcise every male in the community, including slaves, on the eighth day after birth. Circumcision was an ancient custom practiced by many of Israel's neighbors. Originally, it was probably performed at the onset of puberty or before marriage, but the priestly writer reinterprets it as the mark of the covenant, the sign of belonging to God's chosen people. For this reason it should be performed at the earliest possible moment of a person's life (see also box on p.55).

A 19th-century photograph of the sacred oaks at Mamre, Hebron, reputed to be where Abraham erected an altar (Gen. 13.18) and where three divine beings appeared to him (Gen. 18.1; see also p.43).

SODOM AND GOMORRAH

One of the "pillars of salt," the craggy salt-rock formations that characterize the landscape around the Dead Sea. The sea is shown in the background.

God's destruction of "the cities of the Plain," which centers on the fate of Sodom and Gomorrah, is recounted in Genesis 18–19. The story was originally unconnected with the Abraham saga, and appears there because it is also concerned with Abraham's nephew, Lot. Its central theme recalls the story of the great Flood (see pp.32–3): the disaster is a divine punishment for the wickedness of the people, all of whom are destroyed except for one man. The downfall of Sodom and Gomorrah is mentioned more frequently in the Hebrew Scriptures and the New Testament than any other event in the book of Genesis. However, the language of the references suggests that they are not derived from the narrative in Genesis but reflect a popular tradition of a great calamity that occurred in primeval times. The story of what happened to Lot and his daughters, after the destruction of the cities (see sidebar on opposite page) also contains hints of a more cosmic myth, in particular the idea of the renewal of humanity following a universal disaster. This interpretation of the tale is found as early as the time of the Jewish commentators Philo (ca. 15 BCE–50 CE) and Josephus (37 CE–ca. 100 CE).

Just as the Flood story probably reflects the real inundations of the Tigris and Euphrates rivers, the story of Sodom and Gomorrah may have its source in some natural disaster in the region around the southern Dead Sea (see map on p.38). To some extent it serves to explain the bizarre appearance of the landscape in the area. The agents of destruction are described as "sulfur and fire" (Gen. 19.24), and there are large local deposits of sulfur and asphalt, which according to one theory might have been ignited, with great explosive effect, by an earthquake. However, there is nothing in the narrative to support another view, that the cities

Part of the southern shore of the Dead Sea. The sea is so salty in this region that white crusts of salt form on its surface, as can be seen on the right of this picture.

Abraham before the Lord

The discussion between God and Abraham about the fate of Sodom (Gen. 18.22–33) is a unique element in the book of Genesis. It reproduces the pattern of the various other dialogues between the patriarch and the deity, but it also represents an attempt at a theodicy – a vindication of God's goodness in a world where he permits evil to exist – by a theologian of the period after the Exile. In this era, God's justice was sometimes questioned, for example in the book of Job. Society was viewed in terms of the righteous and the wicked, as seen in Proverbs and the later Psalms, and there was an awareness of God's responsibility for the world outside Israel, as found in the book of Jonah.

The biblical author discusses these issues in the context of the destruction of Sodom, which indicates the deep impression that the event made on the Israelite consciousness. Abraham questions whether it is just that the fate that God plans for the city should also befall any righteous inhabitants. He addresses the problem in general terms and does not mention his relatives, Lot and his family, who live there. God says he will spare Sodom if there are "fifty righteous in the place." Abraham indulges in a piece of marketplace bargaining with the deity, who finally agrees that, if there are just ten righteous in the city, he will spare it. But the destruction of Sodom goes ahead, and so the reader is to assume that God could not find even this minimum number of righteous in the city (Lot and his family were warned by angels and escaped before the calamity, Gen. 19.15–16). Sodom is truly one mass of wickedness and fully merits its doom. The justice of the Lord is vindicated.

were overwhelmed by the waters of the Dead Sea. The fate of Lot's wife, who looks back at the destruction and is turned into a pillar of salt (Gen. 19.26), explains the origin of one of the natural pillars of salt that are on the top of Mount Sodom. It is also an instance of the widespread taboo, found in many cultures, on looking back. It occurs, for example, in the ancient Greek story of Orpheus and Eurydice.

The form of the narrative reproduces a widespread folktale of how a supernatural being goes incognito among mortals and receives hospitality from people who recognize a special quality in the stranger. There are two versions of the tale in Genesis. In one, the host is rewarded, often with the promise of a son and heir: this version underlies Abraham's reception of three divine beings immediately before the disaster (Gen. 18). In the other, attention is focused on the host's neighbors, who are severely punished for their lack of hospitality. This form is preserved in the story of Lot (Gen. 19), to which there is a close parallel in a narrative in the book of Judges (Judg. 19).

In the biblical tale, two of the three divine beings enter Sodom and are offered hospitality by Lot, who has settled in the city. But, while they are at his house, the men of Sodom – every male is implicated – gather in front of his house and demand that he hand over the strangers "so that we may know them" (Gen. 19.1–5). It is this threat of mass homosexual rape that triggers God's punishment. Elsewhere in the Bible, the "sin of Sodom" is pictured in more general terms as injustice, moral laxity, or disregard of the needy. Even the Genesis writer appears to be less concerned with the particular sexual proclivity of the men of Sodom than with the fact that it incites them to violate what is considered to be the sacred duty of hospitality.

The Angels Visiting Lot *(Gen. 19.1), from an illustrated Bible produced in 1279 for King Philip II of France.*

LOT AND HIS DAUGHTERS

After the destruction of Sodom and Gomorrah, Lot and his two daughters take refuge in a cave in the hills (Gen. 19.30). The daughters are concerned that they will not find husbands, so each in turn gets her father drunk and has intercourse with him. They both become pregnant and each bears a son.

Although the episode appears as a direct sequel to the destruction of the cities of the Plain, it is really an independent folktale, and the Lot here originally had no connection with Abraham's nephew. He may well have been identified with a local spirit of the cave where he was supposed to have taken refuge. The story is primarily an explanation of the origins of the peoples of Moab and Ammon, lands to the east of Israel. The elder daughter bears a son called Moab, who is the ancestor of the Moabites, and the younger has a son called Ben-ammi, the ancestor of the Ammonites. The sons' names are explained by popular etymology as meaning respectively "From father" and "Son of my kin."

The story may be an Israelite gibe at the supposed origin of two of their neighbors. But it seems more likely that the account is actually of Moabite and Ammonite origin, and that the action of Lot's daughters was originally seen as positive, a way to ensure that their descendants had no foreign blood. Whatever its source, it serves to demonstrate that these peoples were not direct descendants of Abraham, and had no claim to the land west of the Jordan (see p.39).

SARAH AND ISAAC

HAGAR AND ISHMAEL
As Abraham's firstborn child, Ishmael could claim to be his father's heir and thus frustrate God's purpose of continuing the line through Sarah. This danger appears to be averted when the pregnant Hagar is expelled from the household into the desert (Gen. 16). However, God commands Hagar to return. After the birth of Isaac, the threat that Ishmael may stake a claim to his father's inheritance recurs. Ishmael and his mother are finally driven out by divine permission. God allows this because he intends that a nation should spring from Ishmael, who now appears as a typical nomad, inhabiting the desert of Paran, on the border of Egypt. Hagar, herself an Egyptian, secures Ishmael a wife from Egypt. Hagar, in fact, is at the center of the story and is sympathetically depicted.

The divine promise to Ishmael is fulfilled when he becomes the ancestor of the twelve Arab tribes of the Syro-Arabian desert (Gen. 25.12–18), who are thus seen as having a close ethnic bond with the Israelites. Ishmael plays a significant part in Muslim tradition, not only as the ancestor of the North Arabian tribes, but also as a messenger and prophet. He and Abraham are said to have built the Holy House at Mecca. There are parallels between the stories of Ishmael and Esau (see pp.44–5).

The divine messengers who visited Abraham and Sarah (Gen. 18), part of a mosaic in the church of San Vitale, Ravenna, Italy (ca. 540 CE). Christians have found much symbolism in the episode; for example, in the Orthodox churches the messengers represent the Holy Trinity.

A central theme in the saga of Abraham is the repeated threat to God's promise to Abraham that he will be the founder of "a great nation" (see p.37). The Bible states frequently that his wife Sarah is infertile (for example, Gen. 11.30). Abraham decides to remain childless and make one of his household slaves his heir (Gen. 15.3), which may reflect an ancient Mesopotamian custom. Sarah, perhaps following another ancient custom, gives Abraham her Egyptian slave girl, Hagar, so that he may have an heir. In the early part of the story, the patriarch is called Abram (which may mean "exalted ancestor"), but at this in the narrative point God reassures him of the abundance of his future progeny and gives him a new name, Abraham, which Genesis interprets as "ancestor of a multitude" (Gen. 17.5). However, the precise meaning of Abraham and Abram is uncertain, and they are probably both variants of the same name.

However, it turns out that the chosen people are not to be descended from Ishmael, the son of Hagar (see sidebar, left). God promises a son to Sarah, although the matriarch is said to be ninety at this time and her husband ninety-nine. Until this point she has been called Sarai: now she receives the name Sarah as a token of God's renewal of his promise. Sarai and Sarah are variants of the same word, which is generally taken to mean "queen" or "princess."

The birth of Sarah's son is announced to Abraham in Genesis 17 by a divine oracle, which provides the model for several very similar birth oracles, in both the Hebrew Scriptures and in the New Testament, relating to barren women. God says that Sarah "shall give rise to nations; kings of peoples shall come from her" (Gen. 17.16). The literary form of the oracle is of Canaanite origin, and the account of Isaac's birth may have echoed the birth of a royal child, following well-attested "sacred marriage" rites of the ancient Near East. In these rites, a deity, who was represented by the king, had intercourse with a woman, generally the queen or a priestess. Love songs chanted on the occasion of the sacred marriage often praised the bride's beauty. Sarah's name has royal significance, and her great beauty is mentioned several times, for example in Genesis 12.11. It was for this reason that Pharaoh desired her, although she was about sixty-five at the time. Jewish tradition came to put great emphasis on Sarah's beauty, especially in a paraphrase of Genesis in Aramaic found among the Dead Sea Scrolls at Qumran (see p.153).

The idea of the sacred marriage has further echoes in Genesis 18. Three divine messengers or angels, who are identical with Yahweh himself, visit Abraham in human form (Gen. 18). One of them announces: "I shall return to you in due season and your wife shall have a son." Sarah overhears and laughs, but Genesis 21.1 relates that she indeed conceived after Yahweh "dealt with Sarah as he had said," a phrase that may have a sexual connotation. When the son is born, he is named Isaac, which means "he laughs." In Canaanite writings and elsewhere, laughter is said to accompany the birth of a child to a deity or royal figure.

The Sacrifice of Isaac

The threat to the continuation of Abraham's line did not disappear with the birth of Isaac. It surfaces again in the famous story in Genesis 22 usually called "the Sacrifice of Isaac," although in fact he is not sacrificed. The story, a masterpiece, tells how God asks Abraham to offer up his beloved son Isaac. The patriarch obeys, but at the last moment God halts the sacrifice and gives Abraham a ram to slaughter instead.

In rabbinic texts and in Jewish liturgy, the episode has the title *Aqedah*, "binding," and is interpreted as a perfect self-sacrifice by Isaac that atones for the sins of his descendants. In Genesis, however, no reason is given for God's demand that Abraham should sacrifice his son. The emphasis lies on the terrible challenge that the death of his son poses to the continuation of the line of Israel. It also presents Abraham as a paragon of righteous obedience, for which he is rewarded with a divine blessing.

In two respects the story serves as an explanatory myth. According to Hebrew law, the firstborn offspring of humans and animals had to be offered in sacrifice to God but, in the case of human beings, an animal victim could be substituted. The episode explains the origin of this practice and also the origin of the place-name Yahweh-yireh (traditionally rendered Jehovah Jireh). It is said to mean "Yahweh will provide," referring to the ram that is given by God. This follows a common pattern in the Bible, in which a particular place is authenticated as sacred through what is commonly known as a theophany, a manifestation of the deity.

The Sacrifice of Isaac, *a Turkish painting of 1583. An angel bearing a sacrificial ram calls out to Abraham at the very moment he is preparing to kill his son (Gen. 22.10–13).*

Detail from a 6th-century Byzantine painting of Rebekah at the well. It is here that she encounters Abraham's servant on his way to find a bride for Isaac (Gen. 24.10–15).

ISAAC AND REBEKAH

Isaac appears regularly alongside his father Abraham and his son Jacob as one of the founding fathers of Israel. However, there is comparatively little about him in Genesis, and events involving Isaac mainly double episodes in the Abraham saga. For example, the theme of "jeopardy of the female ancestor" (see sidebar on p.38) recurs in the case of Isaac's wife Rebekah (Gen. 26.6–11). Like Abraham in Genesis 21, Isaac in Genesis 26 has a dispute over water rights with the Philistines, but subsequently makes a covenant with the Philistine king Abimelech. Each episode is said to account for the name of the city of Beer-sheba.

There is one unique narrative dealing with Isaac, although he plays only a small part in it: the tale, in Genesis 24, of how he obtains a wife, Rebekah. The story is the longest in Genesis, and its main features reappear in the accounts of how Jacob and Moses found brides. Its narrative structure, which parallels the Canaanite epic of King Keret, involves a journey to a distant land, marriage negotiations, departure for home, and finally a wedding. Abraham appears not as a simple nomad but as a wealthy ruler who possesses camels, flocks, herds, gold, and silver. He has a large household of servants, one of whom is sent, laden with rich gifts, to find Isaac a bride from among Abraham's relatives. They live in the area of Haran, the town from which the patriarch had originally set out. In this story, the region is called Aram-naharaim, which indicates a period after the expansion of the Arameans in the early first millennium BCE.

Under God's direction, the servant's mission is brought to a successful conclusion and Rebekah is betrothed to Isaac. There are reminiscences of royal wedding ceremonies. Rebekah is decked out in jewelry and rich clothes, and urged to leave her family for her new husband's land (Gen. 24.53–58). She receives a blessing (Gen. 24.60) and is led away to meet her husband, accompanied by her maids. These events resemble the royal marriage song in Psalm 45.10–17. There is a romantic glow throughout, not least in the final scene where the couple meet, with the concluding statement that Isaac loved Rebekah.

Like Isaac's mother Sarah, Rebekah is said to be barren, but her infertility is cured after Isaac prays to God (Gen. 25.21), and she conceives twins. The story reflects the frequent motif in myth and folklore of the hostility between twins, even in the womb. The Genesis narrative is primarily concerned with accounting for two related peoples, the Edomites and Israelites. Rebekah's and Isaac's first child, Esau, emerges "red" and "hairy," the Hebrew words for which play respectively on the names Edom and Seir, another name for Edom. Esau's brother is born gripping Esau's heel and is called Jacob, which is said to mean "he takes by the heel." It can also mean "he supplants," which foreshadows the following episode, in which Esau sells Jacob his birthright for a stew (Gen. 27.27–34). The story contrasts the lifestyles of the nomadic hunter and the pastoral tent-dweller, and shows Esau's unworthiness to assume the legal status of the firstborn, which he dismisses with the words "of what use is a birthright to me?" The stew for which he sells his birthright is red, emphasizing again the origin of the name Edom.

Later, Jacob deceives Isaac, who is blind and near death, in order to receive his blessing in place of Esau (Gen. 27). The idea of Jacob as a trickster, and the rivalry between hunter and herder, color the narrative, but otherwise the story seems to represent a ritual ceremony of patriarchal blessing. There is a similar moment in Genesis 48, when Jacob blesses the sons of Joseph; on that occasion, too, the younger brother is preferred to the elder. Jacob brings Rebekah two goat kids for her to pre-

Masseboths, *sacred stone pillars, at the ancient Canaanite shrine of Gezer (see box on opposite page).*

pare for her husband. She tells Jacob to take the food to Isaac and receive his blessing (Gen. 27.9–10). Jacob tricks Isaac by wearing kidskins on his hands and neck to imitate Esau's hairiness: Isaac feels the skins and, convinced that Esau is before him, gives his blessing (Gen. 27.23). The kidskins are probably meant to be a slightly comic touch, but the episode may preserve memories of a sacrificial ceremony by which a younger son became the heir in place of his elder brother. The story also reveals the power of a verbal blessing for the ancient Israelites: even when Isaac knows that he has been deceived, he cannot cancel his blessing and can only compensate Esau with a rather second-rate promise of survival.

Esau plots to kill Jacob in revenge, but Rebekah learns of the threat and urges Jacob to leave home. Jacob flees and begins a perilous journey, a common feature in folktales of heroes. As the one designated to carry on the line of Israel. Jacob journeys to his distant kin to find a wife. Isaac tells him to wed one of the daughters of Laban, Rebekah's brother.

Jacob's Dream

Jacob sets out from Beer-sheba on the heroic journey which forms the central structure of his story. The first episode on the journey takes place at an uninhabited spot where he stops to spend the night (the site is later said to be an existing Canaanite city, Luz). This is an instance of incubation, a practice that is found in the Hebrew Scriptures and in other religions, in which a person passes a night in a holy place in order to obtain a revelatory vision. In his dream, Jacob sees what is traditionally referred to as a "ladder" reaching to heaven with angels ascending and descending on it. The word for "ladder" can also be rendered "stairway" or "ramp." The description of it recalls the Tower of Babel and suggests a Mesopotamian ziggurat (see p.34 and p.37). In the Akkadian myth of Nergal and Ereshkigal, the gods are said to climb up "the long stairway of heaven," and the biblical account of "the gate of heaven" (Gen. 28.17) is paralleled in an Egyptian text that describes the ascent to heaven of a dead king by means of a ladder.

When Jacob wakes up, he erects a stone pillar and pours oil over it as an offering to the holy spirit of the place. Such sacred pillars, known as *masseboth*, were a regular feature of Canaanite shrines (see illustration on opposite page). Jacob names the place Bethel, which means "House of God" or "House of El." The Bible appears to be aware of the fact that the ancestors of Israel did not actually worship Yahweh. The story of Jacob's ladder, together with much other evidence, suggests that they venerated El, who was the head of the Canaanite pantheon, in his local manifestations at various Canaanite shrines. The episode is also one of several "sanctuary legends" in the Bible which justify the Israelite appropriation of Canaanite holy places in the Promised Land.

Jabob's Ladder, *by William Blake (1757–1828): "And he dreamed that there was a ladder set up on the earth, the top of it reaching to heaven; and the angels of God were ascending and descending on it" (Gen. 28.12).*

JACOB AND RACHEL

When Jacob reaches the land of his relatives (here called Paddan-aram), he finds himself in a situation similar to that of the servant who found a bride for Isaac in Genesis 24. Jacob's uncle Laban is still the head of the family in Haran, and Jacob meets his prospective wife, the beautiful Rachel, at a well. But from this point the course of the narrative is very different. Unlike Abraham's servant, Jacob has nothing to offer by way of a bride-price and, in order to win Rachel, he must work for Laban for seven years (Gen. 29.18). What follows is a lively folktale told with a degree of humor that would appeal to an audience. In the course of the story Laban cheats Jacob, Jacob cheats Laban, and Rachel cheats her father. Laban substitutes Leah for Rachel on the wedding night, justifying his action with the claim that custom forbade a younger daughter to be married before her elder sister. Such a custom is certainly widely attested, but the background may be a folklore motif known as the "false bride," in which another woman is substituted for the bride to foil the demons who were thought to threaten the marriage bed. A week after the wedding of Leah and Jacob, Laban permits Rachel and Jacob to marry, but in return Jacob has to work for Laban for another seven years.

Jacob increases his share of the family's flocks through quasi-scientific magic (Gen. 30.37–42), although the story attributes the outcome to God (Gen. 31.9). He becomes as rich as his grandfather Abraham. This arouses the jealousy of Laban's sons, and Jacob determines to return home secretly, with his family and possessions, to his father in Canaan. On this occasion, Rachel secretly steals her father's household gods or *teraphim* (Gen. 31.19), small idols similar to the *Lares* and *Penates* of the Romans. Her motive is unclear, but the theft has been related to a supposed custom of the people known as Hurrians, found in a legal document from the Mesopotamian city of Nuzi of ca. 1400 BCE. According to one interpretation of this document, the possession of the household gods conferred a right of inheritance. But the meaning of the Nuzi text is disputed, and the *teraphim* may simply have been thought to guarantee the safety, prosperity, and perhaps also the fertility, of their possessor.

Rachel's action leads to another example of the "jeopardy of the female ancestor" motif (see sidebar on p.38). When Laban catches up with Jacob and complains about the theft, Jacob, unaware that Rachel has taken the gods, pronounces the death penalty on the unknown thief (Gen. 31.32). Rachel averts the danger by another trick: she hides the idols in a camel bag and sits on it, pretending that she is menstruating and so cannot be approached: "Let not my lord be angry that I cannot rise before you, for the way of women is upon me" (Gen. 31.35). Laban's search proves fruitless and Jacob roundly rebukes him for what he sees as his false accusation (Gen. 31.36–42).

The sequel to the episode suggests what really lies behind this story of family hostility. Laban and Jacob make a treaty, which they ratify by erecting a pillar and a cairn of stones. These are given Aramaic and Hebrew names, meaning "cairn of witness." The deity enshrined in the stones, as at Bethel (see p.45), is the witness and guarantor of the pact. There is a parallel here to the well-known Mesopotamian *kudurrus* or boundary stones, which bore emblems of protecting deities. As the narrative brings out, the stones were to mark the frontier between Aramean and Israelite territory. The account reflects the politics of a much later period when, in spite of the old ethnic relationship between the two peoples, Aramean expansion posed a serious and long-lasting threat to the Israelite kingdoms.

WRESTLING JACOB

On his return from Haran to Canaan, at the ford of the Jabbok River, Jacob encounters a man who "wrestled with him until daybreak" (Gen. 32.24). It is now generally accepted that this "man" was the guardian spirit of the ford. To move from one region to another is often regarded as a dangerous procedure, and crossing places are often said to be haunted by dangerous spirits or demons. The story was originally a local legend that may have became attached to Jacob because of the assonance between the names Jacob and Jabbok.

As day breaks the man makes to leave: the idea that a spirit loses power at daybreak and must then depart is very common in myth and folklore. Before Jacob lets him go, the man gives Jacob a new name, Israel, which is explained as meaning "The one who strives with God" or "God strives." To know someone's name is a step toward controlling them, so Jacob's assailant refuses to say who he is. Jacob calls the site of his encounter Peniel, "The face of God;" in other words, it was originally a Canaanite holy place where worshipers could come into direct contact with the local manifestation of El. The spirit dislocates Jacob's hip by hitting him on "the hip socket at the thigh muscle." This is said to be the reason why the Israelites do not eat the thigh muscle on the hip socket, a taboo not otherwise attested in the Hebrew Scriptures. The "thigh" may be a euphemism for the genital area, as in Genesis 24, when Abraham's servant took an oath by putting a hand "under the thigh of Abraham."

The so-called "Face of Baal," a terracotta cult mask of ca. 1300 BCE discovered at Hazor in northern Israel. It has been thought by some to represent the important Canaanite god Baal.

The Twelve Tribes of Israel

The account of the birth of Jacob's twelve sons during his stay in Paddan-aram links together various originally separate groups, who federated together to make up the later nation of Israel. In the ancient Near East, the tie of blood was regarded as the closest bond of union, so the story gives the twelve tribes of Israel a common ancestor, Jacob. Benjamin, the twelfth son, is born after the family's return to the land of Canaan (Gen. 35.16–18), although his birth is probably anticipated in Rachel's prayer "may the Lord add to me another son" (Gen. 30.24).

The twelve-tribe system could not have originated in the age of the ancestors, whenever that may have been (the ancestors represent an individual family rather than groups of families or tribes), and probably not even in the early period of the Israelite settlement in Canaan. It is more likely to date from the time when Israel became an organized state. Even then, it always remained to a considerable extent a theoretical system. The tribal names were originally geographical names of parts of Palestine, but in Genesis they become the names of persons (similarly, the names of towns in northern Mesopotamia became those of members of Abraham's family; see p.37). The names of the tribal ancestors are all given popular etymologies, which in no way correspond to historical reality.

The sons are born to four different mothers, Leah, Rachel, and their respective maids, Zilpah and Bilhah (see family tree, below). Rachel is Jacob's first and greatest love, and Jacob shows most affection for her offspring, Joseph and Benjamin, who are his youngest sons. One of them, Joseph, subsequently achieves great fame (see pp.48–51). Jacob is recorded as having one daughter, Dinah, by Leah. The story of the rape of Dinah and her brothers' revenge is told in Genesis 34 (see sidebar on p.49).

The curious story of Reuben's mandrakes (Gen. 30.14–16) probably derives from a belief that mandrakes enabled Rachel to conceive her first child, Joseph, which happens shortly after the episode (Gen. 30.22). The mandrake root often resembles the human form, and it was once believed in many cultures to possess aphrodisiac powers and to be an aid to fertility. The Hebrew term for mandrake is connected with the verb "to love."

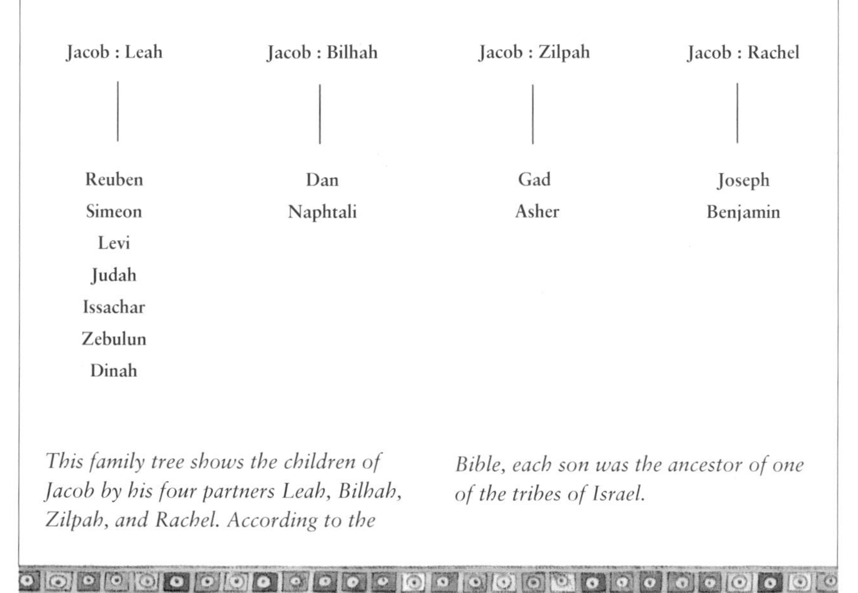

This family tree shows the children of Jacob by his four partners Leah, Bilhah, Zilpah, and Rachel. According to the Bible, each son was the ancestor of one of the tribes of Israel.

A clay fertility figurine that may represent the Canaanite goddess Astarte (ca. 1300 BCE). The household deities or teraphim *that the still childless Rachel stole in Genesis 31 may have been similar figures, the possession of which was thought to confer fertility (see main text).*

JOSEPH

The sagas of Abraham and Jacob are constructed from a variety of traditions, explanatory myths, local and sanctuary legends, which may have passed through successive stages of composition. However, the history of Joseph, which occupies the last fourteen chapters of Genesis, is the product of a single author and is best described as a short story or historical novelette. It displays considerable literary skill.

There have been frequent attempts to discover a historical setting for the story by asking whether the Egyptian customs and practices it mentions are characteristic of a certain period of Egypt's history. In particular, the great position achieved by Joseph (see pp.50–51) has often been explained with reference to the time of the Hyksos (ca. 1670–1562 BCE), foreign rulers of Egypt under whom a non-Egyptian could have risen to prominence. But the story of Joseph does not give the name of any particular pharaoh, and the general conservatism of ancient Egyptian society meant that the same beliefs and customs remained virtually unchanged over many centuries. However, it has recently been argued that some of the cultural features present in the Genesis story can hardly have existed before the late first millennium BCE. Beyond this, it is difficult to establish any unquestionable connection between the Joseph narrative and a specific period of Egyptian history.

While it is obviously fascinating to try to relate Joseph's story to historical events, this is a diversion from the true nature and purpose of the biblical narrative. At the heart of the tale is the common folklore theme of the insignificant character, in this case one of Jacob's two youngest sons, who rises to greatness. The same theme occurs again in the story of the anointing of David as future king of the Israelites (1 Sam. 16.1–13; see pp.80–81). The story of Joseph can also be seen as part of the Israelite Wisdom tradition (see pp.126–7). It illustrates its teachings about divine providence and retribution, and Joseph can be compared with a figure such as Daniel, who is a wise man and an interpreter of dreams (see pp.134–7). The narrative employs several themes which characterize the saga of the ancestors as a whole: sibling rivalry which is finally reconciled; the birth of sons; the superseding of the firstborn; and the theme of exile and return. This last theme is particularly expressed in the dying words of Joseph (Gen. 50.24), which foreshadow the following phase of the history of the Israelites.

DIVINATION

Not all dreams have a self-evident meaning, as Joseph's seem to have had (see box on opposite page). Often they contain strange images and apparent portents which the dreamer cannot understand and which cause acute anxiety, as happens with Pharaoh's cupbearer and baker (Gen. 40), and with Pharaoh himself (Gen. 41). Dream diviners were well-known and widely employed throughout the ancient Near East to interpret the significance of dreams. The pharaohs had a body of what were called "dream interpreters," and a papyrus of the New Kingdom period (ca. 1540–1070 BCE) gives a detailed key to dream interpretation.

Joseph has a silver cup "for divination" (Gen. 44.5). This indicates that he is also a professional diviner, a practitioner of leconomancy, divination by means of liquid in a vessel.

Above and opposite: Scenes from the Life of Joseph, *by Giovanni di Bartolommeo (1483–1511).* Left to right: *Jacob sends Joseph to his brothers (Gen. 37.14); Joseph is sold to the Midianites (37.38); Joseph and Potiphar's wife (39.12); Joseph in prison (39.20–21).*

Joseph and his brothers

The story of Joseph and his brothers unfolds alongside that of Joseph's rise to power in Egypt. The opening of the story relates that the seventeen-year-old Joseph is the favorite of Jacob, who has "a long robe with sleeves" – not "a coat of many colors," as it is traditionally rendered – made for him (Gen. 37.3). In 2 Samuel 13.18 "a long robe with sleeves" is said to be the dress of a princess, so Jacob can be seen as conferring a quasi-royal status on Joseph that sets him above his brothers. The coat introduces the motif of Joseph's special dress as an indication of his position, which the author follows through in the rest of his narration. Jacob's favoritism causes his other sons to be jealous of Joseph: "they hated him, and could not speak peaceably to him."

The same picture is presented in the following episode, which relates Joseph's two dreams (Gen. 37.5–11). (Such pairs of events are characteristic of the narrative: the dreams are paralleled later by the two dreams of Pharaoh's servants and the two of Pharaoh himself.) Joseph dreams that the brothers are in the fields binding sheaves: his sheaf stands upright and the others bow down to it. In his second dream, the sun, the moon, and eleven stars bow down to him. Dreams in the Hebrew Scriptures are not fantasies or illusions but realities that point to the dreamer's innermost nature. Joseph has a king's dreams, something which can only happen to someone who is in fact a king. What happens later, therefore, when Joseph is the virtual ruler of Egypt and his brothers prostrate themselves before him, comes as no surprise.

The dreams reveal clearly to his brothers that Joseph aims to be king and they resolve to kill him so as to prevent him from becoming their ruler. However, Judah urges mercy, "for he is our brother, our own flesh." The others agree and instead they sell Joseph to some Midianite traders bound for Egypt (Gen. 37.28). They present Jacob with the long-sleeved robe as evidence of Joseph's death. In Egypt, the Midianites sell Joseph to Potiphar, the captain of the royal guard, and he ultimately rises to a position of great power, second only to Pharaoh himself (see pp.50–51). The brothers' initial humiliation of Joseph brings about his ultimate elevation.

THE END OF JACOB'S TRAVELS

Jacob's return to Canaan (Gen. 33–37) rounds off his career and that of his brother Esau. Like Ishmael (see p.42), Esau is not viewed wholly unfavorably in Genesis. When he and Jacob meet again, he treats his brother magnanimously and the final territorial separation between them is amicable. Esau's importance as the ancestor of the Edomites is shown by the list of his eighty descendants.

Of the patriarchs, Jacob was connected primarily with central Palestine and Transjordan, while Abraham and Isaac represented a more southerly group. From Genesis 33.18, Jacob is associated with two important northern sanctuaries, first Shechem and then Bethel. There is further patriarchal altar-building at these originally Canaanite holy places, as well as two interesting additional stories. One, told in Genesis 34, relates a savage attack on Shechem by two of Jacob's sons, following the rape of Dinah, the daughter of Jacob and Leah. This seems to be an episode of conflict with the native population at a time when groups forming the later "Israel" were occupying the land. The other episode, in Genesis 35, describes Jacob's journey from Shechem to Bethel. He prepares for this by burying the people's "foreign gods," such as Rachel's *teraphim* (see p.46) under the sacred tree at Shechem. The biblical author believed that the worship of Yahweh had to be purged of foreign elements. But some ancient cult ceremony also lies behind the story: at various sanctuary sites, archaeologists have found figurines of deities that were deliberately buried, although it is unclear why.

Finally, Jacob and his family are depicted moving peacefully throughout the country until they settle at Hebron, the home of Isaac. Isaac dies at the age of one hundred eighty, and the narrative moves on to the story of Jacob's sons.

(Continued from opposite page:) *The brothers before Joseph, the high official of Pharaoh (Gen. 15); the brothers' donkeys are loaded for the return journey (44.1); the discovery of Joseph's cup in Benjamin's sack (44.12).*

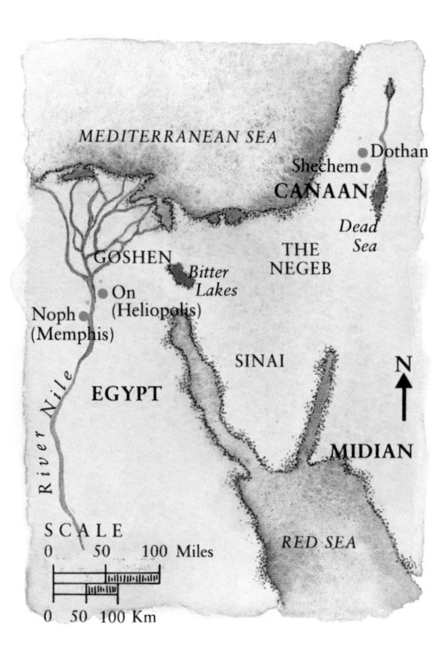

JOSEPH IN EGYPT

Just as Joseph's dreams led to his enslavement (see p.49), they were also a direct cause of his rise to a supreme position in Egypt. During his false imprisonment (see box, below), Joseph correctly explains the dreams of two of Pharaoh's servants (Gen. 40). His ability to interpret dreams comes to the attention of Pharaoh, who has had two vivid but puzzling dreams (Gen. 41). Joseph tells the monarch that his dreams presage seven years of plenty followed by seven of famine. He suggests measures to alleviate the effects of the famine and Pharaoh, impressed, puts him in charge of the relief effort. He grants him enormous powers "over all the land of Egypt" (Gen. 41.43). Although the annual flooding of the Nile usually guaranteed agricultural prosperity, famine was not unknown in Egyptian life. An inscription of ca. 100 BCE tells of a seven-year famine caused by the failure of the Nile floods in the reign of the pharaoh Djoser, who ruled some time around the twenty-eighth century BCE. The purpose of this inscription is to validate a claim by a group of priests for exemption from property tax. Similar priestly privilege is found in the Joseph story: when the Egyptians are compelled to surrender all their possessions to Pharaoh as a result of the famine, the priests are exempted (Gen. 47.22).

When the famine arrives, it is not confined to Egypt but is universal, and Jacob sends ten of his sons from Canaan to Egypt to get grain from the Egyptian storehouses. The narrative of Joseph's dealings with his brothers now becomes intertwined with the account of his rise to power. When they arrive in Egypt, the brothers do not realize that this great official is Joseph (he is at least twenty years older and speaks through an interpreter). But he recognizes them, and the intricate scenes by which the story is played out represent Joseph's plan to reunite his entire family, now much increased, in Egypt. He wants them to share in the good fortune that his wealth and position now make possible.

However, before this final reconciliation, Joseph plays a game with his

Judah and Tamar and Potiphar's wife

Two episodes in the story of Joseph stand apart from the main narrative of a hero's rise and his relations with his siblings. The first is the tale of Joseph's brother Judah and Judah's widowed daughter-in-law Tamar, which is related in Genesis 38, just after Joseph is taken to Egypt. The tale recounts how Judah mistakes Tamar, who is veiled, for a prostitute, and has intercourse with her. She becomes pregnant with twins and is denounced as a whore, until she reveals the identity of the father, who admits his fault. The story contains several of the recurring motifs in the history of the ancestors: the overcoming of childlessness, the birth of twins, the suppression of the firstborn, and problems with the local Canaanite population. The episode has often been seen as an insertion, misplaced from earlier in Genesis. However, just before the story Judah saves Joseph from death (Gen. 37.26–27), so perhaps the biblical author felt that it was appropriate to tell a story about Judah at this point.

The second episode is the attempted seduction of Joseph by the wife of Potiphar, the head of the royal guard and his first Egyptian employer (Gen. 39). The author may have used a well-known Egyptian story called "The Tale of the Two Brothers." In this tale, the elder brother is married and the younger works as his servant. One day, as the younger brother is going about his duties, his sister-in-law tries to seduce him. He rebuffs her, and when her husband returns she accuses his brother of rape. The biblical story has an almost identical plot, except that Joseph is falsely imprisoned until his sudden rise to power under Pharaoh. In the Egyptian tale, the accused man is able to convince the husband of his innocence and the two brothers are reconciled.

brothers that is a just retribution for their past behavior, and the narrator skillfully increases the suspense before the eventual happy ending. Finally Joseph reveals his identity to his brothers (Gen. 45.4) and, to allay their fears, he gives a theological explanation of their original misdeed. It was God who was responsible for his journey to Egypt, not Joseph's brothers. God sent Joseph ahead of his brothers "to preserve life," and "to keep alive for you many survivors." The area in Egypt where the Israelites are to settle, the land of Goshen, is allotted to them (Gen. 45.10). It remains their place of residence until the Exodus.

The last section of the story of Joseph (Gen. 48–50) centers on the deaths of Jacob (Israel) and of Joseph himself. On his deathbed, Jacob is once more shown as a great source of blessing. He blesses the two sons of Joseph, typically preferring the younger, Ephraim, over the elder, Manasseh. He is then credited with a lengthy poem in which he foretells the destiny, not invariably a happy one, of each of the twelve tribes in turn. When he dies, he is embalmed in the usual Egyptian manner and his corpse is taken by Joseph to Canaan to lie in the cave at Machpelah, which Abraham purchased as a family burial plot.

Joseph returns to Egypt and lives there for many years. The book of Genesis ends significantly with his last words, which look to the future of the Israelite nation: "Then Joseph said to his brothers: 'I am about to die; but God will surely come to you, and bring you up out of this land to the land that he swore to Abraham, to Isaac, and to Jacob.' "

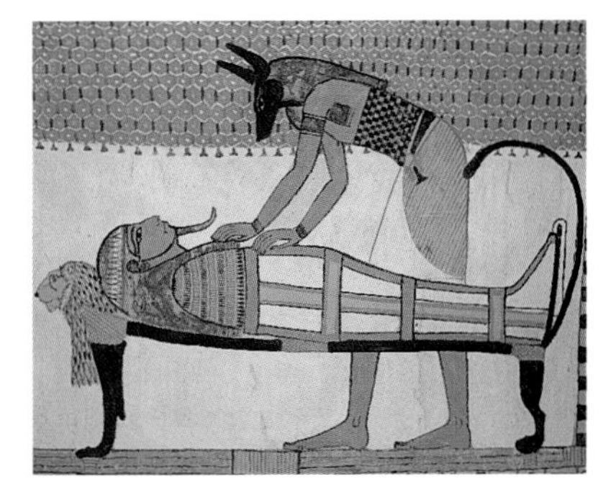

Anubis, the Egyptian god of embalming, prepares a body for burial, from a wall painting of ca. 1300 BCE. Embalming was usual Egyptian practice. The Bible says that after his death Joseph "was embalmed and placed in a coffin in Egypt" (Gen. 50.26).

Egyptian elements in the Joseph narrative

There are many apparently genuine Egyptian details in the narrative of Joseph's rise to power. Several Egyptian words are used, for example the shout "*Abrek!*" uttered before Joseph's chariot (Gen. 41.43). This is usually translated as something like "Bow the knee," although the exact meaning is uncertain. The new name given to Joseph, Zaphenath-paneah (Gen. 41.45), is genuinely Egyptian, as are the names of his bride, Asenath, her father Potiphera, and Potiphera's city of On, "City of the Pillar." On, better known by its later Greek name of Heliopolis, was a center of sun worship that was famous for its obelisks.

Pharaoh confers on Joseph authority "over my house." This phrase seems to be a direct translation of the title of the highest post in the Egyptian administration, which is accurately described in the biblical account as a position second only to that of Pharaoh himself. Pharaoh says to Joseph: "Without your consent no one shall lift up hand or foot in all the land of Egypt" (Gen. 41.44). Joseph is invested with Pharaoh's ring and a golden chain (Gen. 41.42): this was the regular practice for high Egyptian officials and is attested in a text from the reign of Ramesses II (ca. 1279–1212 BCE). Joseph later acquires all the land in Egypt for Pharaoh (Gen. 47.20). This seems to reflect the situation under the New Kingdom (ca. 1540–1070 BCE), when Egypt became a despotic state and everything was held to belong to the pharaoh.

These details, however, do not point unambiguously to any particular period of Egyptian history (see p.48). Not everything related about Egyptian customs in the Joseph story is accurate. For example, it is said that "the Egyptians could not eat with the Hebrews, for that is an abomination to the Egyptians" (Gen. 42.43) and that "all shepherds are abhorrent to the Egyptians" (Gen. 46.34). But there is no support for these statements in Egyptian sources.

III MOSES

The books of Exodus, Leviticus, Numbers, and Deuteronomy are dominated by the figure of one man, Moses. There is little doubt that Moses was a real person, though of limited status. He may originally have been a significant local figure, perhaps a member of the tribe of Levi operating especially in the region of Kadesh-barnea in the extreme south of Canaan. In the Bible he is transformed into the idealized leader of Israel and given an overriding theological significance. He is the agent of God, and through him God acts to determine for all time the fundamental character and structure of the nation of Israel. His story transcends any particular historical epoch, and Moses appears as the human source of the fundamental features and institutions in the life of Israel: the worship of the one God, Yahweh; the Covenant; the Law; prophecy; kingship; and priesthood.

THE FIGURE OF MOSES

Moses first rises to prominence during the oppression of the Israelites in Egypt, described at the beginning of the book of Exodus. This oppression is often said to have taken place under the pharaoh Ramesses II (ca. 1279–1212 BCE), with the Exodus itself occurring under his son and successor Merneptah (ca. 1212–1204 BCE; see table on p.58). Ramesses was a great builder, who founded the city of Rameses in the eastern Nile delta. According to Exodus 1.11, Rameses was constructed by Israelite forced labor. However, Moses' historical background is not the Bible's main concern. The unnamed pharaoh of the narrative is presented as a stereotypical oppressive ruler. His actions seem improbable, especially his desire to wipe out his large Hebrew slave workforce, as related in Exodus 1.15–22.

The account of the Egyptian oppression is a carefully constructed narrative in three stages of increasing severity. Its purpose is to pave the way for the emergence of the heroic figure of Moses. After the death of Joseph and his brothers (see pp.50–51), their descendants flourished in Egypt (Exod. 1.7). A new pharaoh succeeds who is alarmed at their size and power and enslaves them. But their numbers increase at an even greater rate, so Pharaoh instructs the Hebrew midwives, Shiphrah and Puah, to kill all newborn Israelite males (the narrative may be using the number two as a storytelling device rather than seriously implying that the entire Hebrew population had only two midwives). The midwives cleverly succeed in disobeying this command, but Pharaoh simply orders all Hebrew males to be thrown into the Nile at birth (Exod. 1.22).

At this point Moses is born to parents of the house of Levi, who are later named as Jochebed and Amram. His mother keeps his birth secret until she can no longer hide the baby, then lays him in a waterproof basket and places him among the papyrus reeds on the banks of the Nile. Moses' sister Miriam (not named until Exod. 15) watches as Pharaoh's daughter finds the child and sees that it is a Hebrew baby. Miriam goes to the princess and offers to find a Hebrew nurse for the child. This nurse is Moses' own mother, who brings the child up. He is then adopted by the princess, who names him Moses (Exod. 2.10). The name Moses is really Egyptian, meaning "son," a shortened form of names such as Thutmose. But the narrator did not know the true origin of the name and shaped the story so that it backed up his own explanation, as frequently occurs in the Bible. The writer derives the name (*Mosheh* in Hebrew) from the Hebrew *mashah*, "to draw out," referring to Moses being lifted out of the Nile.

Parallels to this story occur in many cultures. The closest is the legend of King Sargon of Assyria (ca. 2334–2279 BCE), whose mother, it was said, bore him secretly, placed him in a basket of rushes sealed with bitumen, and threw him into the river. He was rescued by a certain Akki, "the drawer of water," who adopted him as his son. In due course Sargon won divine favor and became a great ruler. The Moses story concentrates on legendary themes and does not address puzzling questions such as how Moses has an elder sister when he is apparently his parents' first child.

The pharaoh Ramesses II (reigned 1279–1212 BCE) dressed for war. He has been proposed as the despotic ruler who initiated the oppression of the Israelites, as related in Exodus.

Papyrus flags depicted on a piece of Egyptian faience of ca. 1350 BCE. Many European artists assumed that the "reeds" in which Moses was laid (Exod. 2.3) were bulrushes, but they would almost certainly have been papyrus reeds.

Moses in Midian

Moses is raised in the Egyptian royal household, but as a young adult he becomes aware of the plight of his own people. He kills an Egyptian for beating a Hebrew (Exod. 2.12), and then flees for his own safety to Midian, on the eastern side of the Gulf of Aqaba. Moses' stay in Midian illustrates the common folktale idea of the hero's withdrawal into the wilderness before his return to the land where his destiny lies. Moses' encounter at a well with the daughters of the Midianite priest (Exod. 2.15–17) is a scene found also in the stories of Isaac and Jacob (Gen. 24 and 29). As on those occasions, the encounter leads to the hero's marriage, and the birth and naming of a son.

The tradition of Moses' marriage into the family of the Midianite priest shows that the biblical author was aware of close ethnic ties between Midian and Israel. Moses' father-in-law (variously called Jethro, Reuel, and Hobab), is described in Judges 1.16 and 4.11 as a Kenite, a descendant of Cain (see pp.30–31). The "mark" that God put on Cain (Gen. 4.15) may have meant that the Kenites stood out from their neighbors as Yahweh worshipers. Jethro may be a type of high priest of Yahweh: in Exodus 18 he presides at a sacrificial feast attended by the Israelites, and then instructs Moses in the administration of justice. The mountain where the name of Yahweh is revealed to Moses (see p.54), appears to be in the Midianite or Kenite area.

Some have concluded that the Midianites or Kenites were the original devotees of Yahweh and that the Israelites took over the worship of Yahweh from them after they had left Egypt. But the theory creates many problems, because Yahweh is intimately involved in events before and during the Exodus. Perhaps the most that can be said is that the Midianites or Kenites had a tradition, reflected in the Bible, that they shared a common cult with the Hebrews.

THE BURNING BUSH

During his stay in Midian (see p.53), Moses goes up to Mount Horeb (Exod. 3), where he encounters God in the form of a burning bush. God reveals his divine name, Yahweh, which is usually rendered "the LORD" in English Bibles. He calls on Moses to return to Egypt and lead his people from servitude to the Promised Land of Canaan.

The revelation of the name of Yahweh probably refers to the adoption of a new deity, but the Bible emphasizes that the being who speaks to Moses is "the God of Abraham, the God of Isaac, the God of Jacob." When Moses asks God to give his name, he answers "I am who I am," apparently an explanation of the divine name (Exod. 3.14). There has been much debate about the real meaning of this enigmatic phrase, which can also be translated "I am what I am" or "I will be what I will be." Much of this discussion may be beside the point, as God, like many divine beings, is reluctant to say his true name (compare Jacob's encounter at the ford in Gen. 32), and "I am who I am" is simply a negative answer to Moses' question. The Hebrew verb "to be," *hayah*, has a similar sound to the word "Yahweh," and the narrative skillfully makes the connection (Exod. 3.14–15): God says "I am who I am," then "I am" (a single word in Hebrew), and finally reveals the name Yahweh itself. This passage may represent part of the liturgy of an ancient cult initiation ritual.

God assures Moses that he will remain with him and, using a well-known Near Eastern formula for the appointing of a messenger, commissions him as an envoy to Pharaoh (Exod. 3.18). This formula was also used when a person was called as a prophet (compare 1 Kings 18.1), and the Jewish prophetic tradition has probably influenced the tradition of Moses' call. God empowers Moses, as he did the later prophets, to demonstrate the divine origin of the mission by various miraculous signs. The objections that Moses raises to his mission in Exodus 4 parallel those often uttered by prophets when faced with the divine summons.

The episode on Mount Horeb is partly a sanctuary legend. It follows a common pattern in which a holy place is legitimized by a manifestation of the deity. "The angel of the Lord" that appears in the flaming bush means the visible presence of God himself, who was believed in Israel to manifest himself in a fiery flame (see picture on p.26).

Aaron

At the burning bush, Aaron appears for the first time as Moses' brother and his spokesman. In the earliest parts of the Pentateuch (the first five books of the Bible), he is depicted as Moses' helper and joint leader of the Exodus and the desert wanderings, and there is little sign that he has specific priestly functions. Aaron may once have played a distinct or even an independent role in these events, perhaps indicated by traditions which show him in an unfavorable light because of his opposition to Moses. But he is clearly subordinate to Moses as his agent and displays many features that double those of the greater leader: he too has a miraculous rod, with which he causes the Egyptian plagues; both he and Moses face hostility from the Israelites; both are denied entry to the Promised Land; and both die on a mountain outside it.

The later parts of the Pentateuch give Aaron much greater prominence as the founder of the Israelite priesthood and guarantor of the only true priestly line. In particular, Aaron represents the high priest and his position as the head of the Temple of Jerusalem after the Exile. In Exodus 28, the installation of Aaron with the vestments and regalia of a high priest is described in detail.

Moses and the Burning Bush, *from a French illuminated manuscript of ca. 1210. It illustrates Exodus 3.5: "Then he said, 'Come no closer, remove the sandals from your feet, for the place on which you are standing is holy ground.' "*

The bridegroom of blood

After Moses and his family set out to return to Egypt (Exod. 4.20), they camp for the night at a place where Yahweh, apparently, seeks to kill Moses. This episode (Exod. 4.24–26) is highly mysterious and difficult to interpret, no doubt because the biblical narrator no longer knew its real meaning. It seems to be a fragment of a once independent tradition and betrays archaic features, such as the representation of Yahweh as a kind of hostile night demon, and the use of a flint knife for circumcision. Although Moses is alluded to through the context, his name does not appear in the Hebrew text of this episode. Originally the threat was probably to his firstborn son, who escaped death by being circumcised. God's threat to kill Pharaoh's firstborn precedes the story, and in Israelite thought the life of the firstborn male belonged to God.

The narrative reveals the original significance of circumcision as a rite of puberty and preparation for marriage. Moses' wife Zipporah touches the feet (possibly a euphemism for the genitals) either of Moses or her son – it is not clear which – with the foreskin, saying "Truly you are a bridegroom of blood to me." The story explains the origin of the rite known as "bridegroom of blood by circumcision."

MOSES AND AARON AND THE TEN PLAGUES

On Mount Horeb (see p.54), God commissioned Moses to bring the Israelites out of Egypt into Canaan, the Promised Land. Acting in obedience to this commission, Moses and his brother Aaron confront Pharaoh in Exodus 5. However, at this stage they request only that the Israelites be allowed to make a three-day journey into the desert in order to celebrate a sacrificial feast to their god. This again suggests that there may have been a connection between the Israelites in Egypt and ethnically related groups, such as the Kenites or Midianites, who lived in the desert regions of the far south of Palestine (see p.53). It may be significant that the Bible calls the deity "the God of the Hebrews" (Exod. 5.3), a term with a wider sense than "Israelites."

The first meeting with Pharaoh is unsuccessful and only brings greater hardships to the Israelites (Exod. 5.4–21). Moses complains to God, who once again reveals his name Yahweh (Exod. 6.3). In this passage, which is usually ascribed to the priestly biblical author "P" (see p.26), God reassures Moses and Aaron that his promise of deliverance will prevail. He says that he did not appear to Abraham, Isaac, and Jacob as Yahweh, but as "God Almighty," in Hebrew *El Shaddai*. This title probably means "El of the mountains," which would confirm the view that the ancestors worshiped the supreme Canaanite deity, El (see p.72). The epithet *Shaddai* is attested in Aramaic inscriptions of the ninth or eighth century BCE from Deir Alla in Transjordan.

Moses and Aaron go back to Pharaoh and once more ask him to let the Israelites go. God commands Aaron to give a sign or portent to convince the monarch that their request comes from God. Aaron enters into a kind of magical competition with Pharaoh's own magicians. He turns his staff into a snake, a feat which is immediately duplicated by the Egyptians, but Aaron's staff swallows up those of his adversaries (Exod. 7.8–12; see also sidebar on opposite page). Aaron's feat does not move Pharaoh, so God commands Moses to turn the waters of Egypt into blood. Aaron does so by wielding his staff, but the Egyptian magicians repeat the miracle and Pharaoh is unimpressed. Through Moses and Aaron, God sends more plagues: frogs, gnats (or mosquitoes), flies, livestock pestilence, boils, hail, and locusts. In each case, Pharaoh either remains unmoved by the plague, or relents and promises to release the Israelites, but then goes back on his word. It is only after the tenth and most drastic plague, the death of all the firstborn of the Egyptians, that Pharaoh needs no further convincing of God's power. He summons Moses and Aaron and does not simply promise to free the Israelites but actually commands the brothers to conduct their people out of Egypt (Exod. 12.31).

This drawing from an ancient Egyptian relief depicts a figure called "Seth of Ramesses-Beloved-Of-Amon." Some scholars have suggested that it represents a magician turning his staff into a serpent, which appears behind him.

The narrative of the plagues is thought to represent a conflation of the work of the two biblical authors who are referred to as "J" and "P" (see pp.26–7), each of whom originally told of seven plagues, the number given in Psalm 105. According to one of the writers, it is Moses who wields his staff to bring about the plagues. The other writer gives much greater prominence to Aaron, who uses his own staff in the course of a contest with the Egyptian sorcerers. During the plague of boils (which was originally the fourth and central plague), the Egyptian magicians are afflicted with boils and therefore become ritually unclean, so they must withdraw. The two traditions could be successfully woven together because they both served to demonstrate the irresistible power of the god of the Israelites, Yahweh.

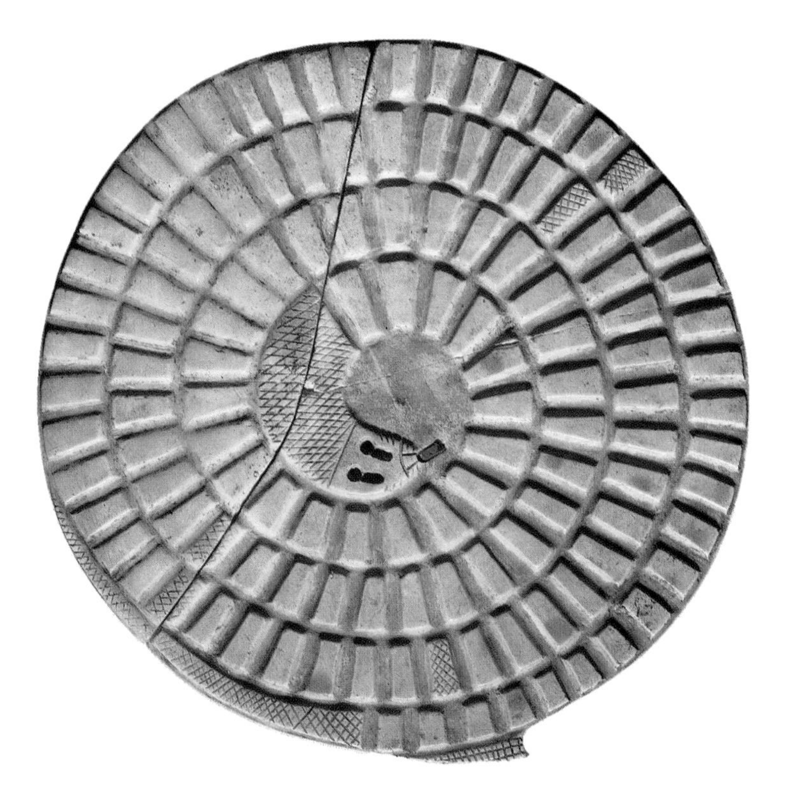

Serpents featured in all areas of Egyptian culture. This alabaster and jasper gaming board of ca. 250 BCE *is for a game called* Mehen.

AARON'S STAFF

In Israelite tradition, Aaron's staff (see main text on opposite page) was venerated as the sign of his superiority and symbolized the exclusive claim of Aaron and his descendants to the priesthood (see p.54). The book of Numbers (Num. 17–18) relates how God sought to end wrangling among the Israelites and asked Moses to obtain twelve staffs, one from each of the leaders of the tribes of Israel. The one bearing Aaron's name, for the tribe of Levi, produced buds, blossoms, and "ripe almonds." (Such miraculous sproutings are common in folktales.) As a result, God granted Aaron and his sons sole control of the priesthood. The staff was laid up in a holy place.

It is significant that God should first try to impress Pharaoh by turning Aaron's staff into a snake. Snakes were potent symbols in the ancient Near East (see also p.29). In Egypt, they were regarded as the embodiment of chaotic forces, and the magic which could control and harness them was greatly valued.

Natural phenomena

The plagues that God imposes on the Egyptians (see main text on opposite page) are often explained as natural phenomena that occur from time to time in Egypt. The first plague, in which the Nile turns to stinking blood, may refer to the occasional reddening of the river owing to volcanic deposits and algae from the White Nile and Blue Nile. In his account of the seventh plague, hail, the biblical author notes that the wheat and the spelt were not destroyed because they had not yet ripened (Exod. 9.31–32). This is probably an accurate reflection of the Egyptian annual planting schedule. The ninth plague, "darkness over the land," could refer to the thick dust clouds caused in March by the hot winds from the Sahara known as the *Khamsin*. Exceptionally severe infestations of frogs, gnats, flies, and locusts are characteristic of North Africa, and no doubt afflicted Egypt with some regularity.

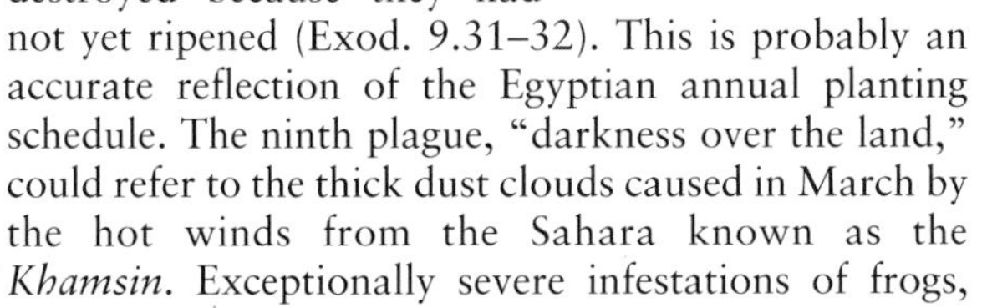

Moses Conjures up the Plague of Locusts *(Exod. 10.12–13), from the Nuremberg Bible of 1483.*

The biblical plagues may reflect ancient memories of real occurrences, but they are also of theological significance. The Nile waters do not simply become red but actually turn into blood, which possesses profound symbolism in the Bible. Goshen, where the Israelites lived, is spared from the plagues of hail and darkness: this seems more an instance of God's care for his people than something that really happened. The unique severity of the plagues is frequently stressed.

The plagues appear to represent a tradition of God's judgment taking the form of natural calamities. There are other examples in oracles that prophesy divine judgment, such as chapter 4 of the book of Amos.

CROSSING THE SEA

PASSOVER AND UNLEAVENED BREAD

Exodus 12 and 13 explains the origin of the ancient spring festival of Passover and Unleavened Bread. There were originally two separate celebrations. Passover was the ritual sacrifice of an animal in order to ward off evil from the home, and may have been connected with the migration of pastoral nomads to summer pastures. Unleavened Bread was more characteristic of a settled population. It was a communal festival at the start of the harvest of barley, the first crop to be gathered.

The Bible brings the festivals together as commemorations of the events of the Exodus. Passover (*Pesah* in Hebrew, whence the word "paschal;") is so called because Yahweh "passed over" the Israelite dwellings (Exod. 12.27), which were protected from the last of the ten plagues, the death of the firstborn (see p.56). The festival once began with the sacrifice at dusk of a kid or lamb, a practice still observed by the Samaritans (see p.125). Exodus explains that the blood of the sacrifice protected the Israelite firstborn from God's plague (Exod. 12.21–23). The Passover is a night of vigil, because God kept vigil to bring the people out of Egypt (Exod. 12.42). It is followed by a week in which no leavened bread is eaten; the lack of leavened bread is attributed to the Israelites' hasty departure (Exod. 12.39).

After their release from captivity (see pp.56–7), Moses and the Israelites head southwest from the city of Rameses (see map on opposite page), taking the bones of Joseph with them. God goes before them in the form of a pillar of cloud by day and a pillar of fire by night, which are symbols of the divine presence elsewhere in the Bible. The quickest route to Canaan is "by way of the land of the Philistines," but God chooses to lead the people "by the roundabout way of the wilderness towards the Red Sea." This will discourage the Israelites from turning back if they face war (Exod. 13.17), and will make Pharaoh believe that the former captives are "wandering aimlessly" in the wilderness (Exod. 14.3). God encourages the Egyptians to pursue the Israelites because, as he tells Moses, he plans to "gain glory for myself over Pharaoh and all his army." In doing so, God will show the Egyptians that he is "the Lord."

The Egyptian army catches up with the Israelites at the Red Sea. There is understandable fear among the fugitives, but Moses reassures them: "The Lord will fight for you" (Exod. 14.14). During the night, the two pillars move between the Israelites and the Egyptians. On God's instruction, Moses raises his staff and God sends a strong east wind which drives back the sea and lays bare the seabed. The Israelites cross the sea "on dry ground." The Egyptian army follows, but in the morning God throws the pursuers into a panic and clogs the wheels of their chariots. Moses raises his staff again and the sea flows back, drowning Pharaoh's army.

The events of the crossing of the sea are pictured in terms of the ancient creation myth (see p.18 and pp.26–7), with the dividing of the waters and the defeat of the hostile forces of chaos, represented by the pursuing Egyptian army. Like many other narratives in the Bible, the story derives from two different traditions, represented by the authors "J" and "P" (see pp.26–7). In the "J" tradition, Yahweh acts alone, while "P" gives Moses much greater prominence. The two traditions have been skillfully woven together to produce a coherent narrative, the purpose of which is to awaken Israel's faith in the divine warrior who fights for his people. After

KINGS OF THE NINETEENTH DYNASTY

The Exodus is often dated to the time of the Nineteenth Dynasty of Egyptian pharaohs (below), in particular the reigns of Ramesses II, who built the city of Rameses (see map), and Merneptah. All dates are approximate.

Ramesses I	1292–1290 BCE
Sethos I	1290–1279 BCE
Ramesses II	1279–1212 BCE
Merneptah	1212–1204 BCE
Sethos II	1204–1198 BCE
Siptah	1198–1193 BCE
Tewosret	1193–1190 BCE

Egyptian troops on the march, a sculpture of the period 1990–1786 BCE from a tomb at Assyut, Egypt.

The route of the Exodus

The Israelites are said to have escaped across the "Sea of Reeds," in Hebrew *Yam Suf*. There is no agreement on exactly where this is, although elsewhere in the Hebrew Scriptures *Yam Suf* refers to the Red Sea. But few scholars believe that it means Red Sea in the Exodus story, and alternative routes have been proposed (see map). For example: a northern route across Lake Sirbonis; two central routes across either the marshes of Lake Timsah or the Little Bitter Lake; and a southern route near the Gulf of Suez. The itinerary of the Israelites after the crossing is also much disputed. It is impossible to identify with certainty many of the place-names in the narrative and the Bible itself does not seem to be consistent about the route of the Exodus. Different biblical authors were probably influenced by different ancient trails across Sinai.

It appears likely that the book of Exodus includes elements of various different nomadic traditions in a story that originally related the movements of just one of the groups that made up the later nation of Israel: the Bible hints that other, related, groups also moved from Egypt and wandered around the Sinai peninsula. The wandering people of the Exodus story are thus made the representatives of the entire Israelite nation.

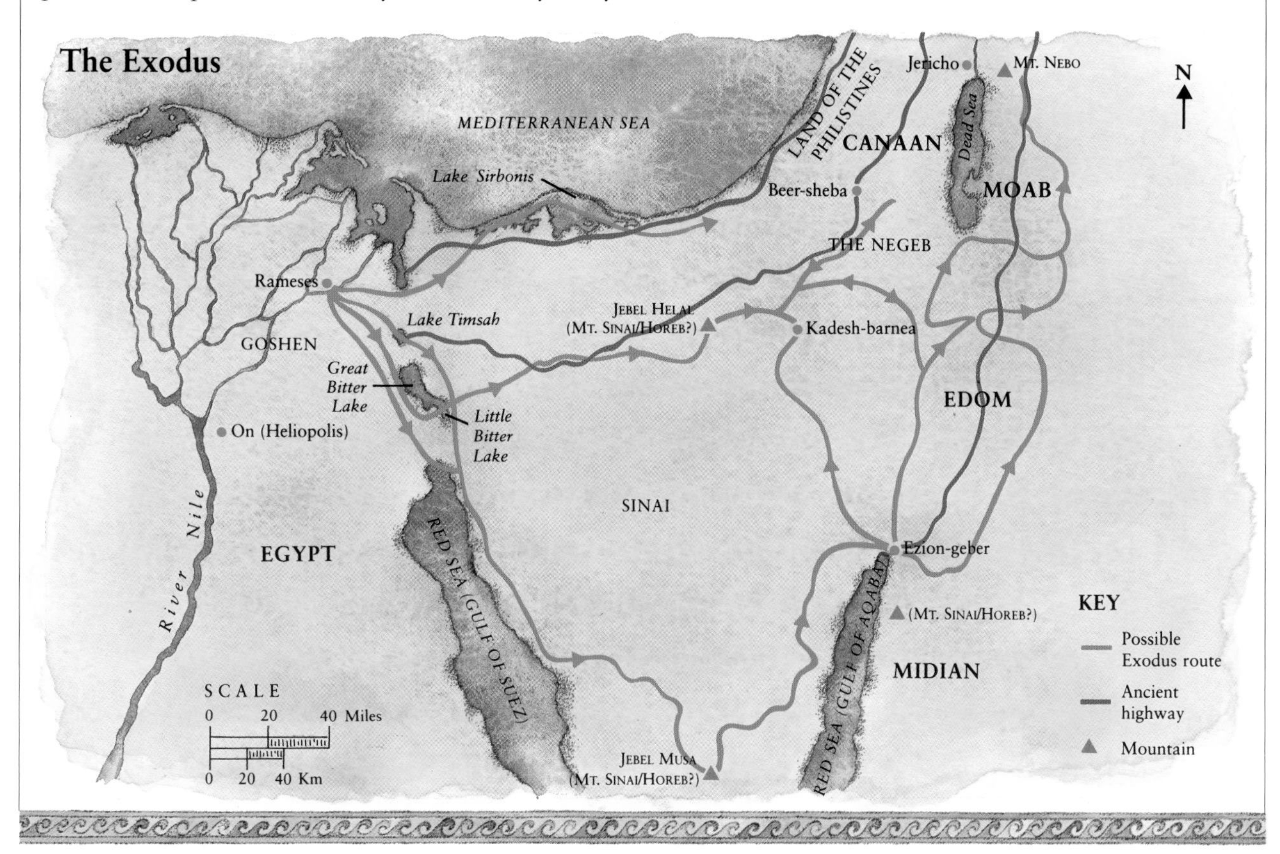

their deliverance from Pharaoh, the Israelites put their faith in Yahweh and his servant, Moses.

Both accounts agree on one essential point: something miraculous happened at the sea through divine agency, and that this was of fundamental importance for the nation of Israel. The episode is followed by a poem known as "The Song of the Sea" (Exod. 15.1–18). It is said to have been uttered by Moses, but in fact it is a later composition, possibly a liturgical hymn commemorating the Exodus. The first part of the text (verses 1–12) recounts the crossing of the sea and praises Yahweh's supremacy over all other deities. The second part (verses 13–18) shows how this event delivered Israel to become God's people, conquer the Canaanites, and establish the divine sanctuary on Mount Zion, Jerusalem.

MOSES ON MOUNT SINAI

After the crossing of the Red Sea (see pp.58–9), the Israelites journey into the wilderness until they come to Mount Sinai, where God transmits to Moses the divine laws by which the people of Israel will be governed. The giving of the law is the central element in the whole of the Pentateuch, as is implied by the amount of space devoted to it: twenty-one chapters of Exodus, the whole of Leviticus, and a considerable part of Numbers. The most important section, which determines all that follows, is the account beginning in Exodus 19 of the great theophany (manifestation of God) after the people arrive at the holy mountain (Exod. 19.2). This lengthy and complicated section is made up of a variety of components of different dates and backgrounds, but its significant themes are clear.

The narrative stresses the unique role of Moses as the mediator of the divine word. He alone is able to speak directly to God and to receive the divine instructions for transmission to the people. This is expressed by the repeated visits of Moses to the mountaintop to meet with the deity. The biblical picture of Moses includes features reminiscent of later kings, priests, and prophets who also appear as mediators. He sets a pattern for an important aspect of the future life of the Israelite nation. At Sinai,

Moses with the Tablets of the Law, *by Rembrandt (1659).*

THE TEN COMMANDMENTS

The basis of the covenant between God and Israel are the Ten Commandments, also called the Decalogue. They represent a type of law in the Hebrew Scriptures that consists of absolute ethical imperatives ("apodictic" law). There are two full versions of the commandments, in Exodus 20 and Deuteronomy 5, and they are summarized or quoted elsewhere in the Bible, especially in the prophets. They are also found in a papyrus of ca. 100 BCE and in fragments from the Dead Sea Scrolls (see p.145). The variations in these versions suggest that they all derive from an original catechism of ten brief sayings, or "Words," as the Hebrew describes them, which in some cases have been expanded. The basic commandments may go back to Moses, but their final form could have been influenced by Near Eastern treaties in which terms were imposed on a vassal by a sovereign. Alternatively, they may have originated as a cult liturgy of the sort also found in Psalms 15 and 24, setting out ethical conditions for participation in worship. Religious groups differ in their numbering of the commandments (see below).

THE COMMANDMENTS (EXOD. 20.1–17)

"I am the Lord your God who brought you out of the land of Egypt, out of the house of slavery." *[Prologue: the first commandment in Judaism.]*

"You shall have no other gods before me. You shall not make for yourself an idol." *[The second commandment in Judaism; the first commandment in the Roman Catholic and Lutheran churches; the first commandment – with the prologue – and second commandment in the Orthodox and Protestant Reformed churches.]*

"You shall not make wrongful use of the name of the LORD your God."
"Remember the sabbath day, and keep it holy."
"Honor your father and your mother."
"You shall not murder."
"You shall not commit adultery."
"You shall not steal."
"You shall not bear false witness against your neighbor."
"You shall not covet your neighbor's house; you shall not covet your neighbor's wife ... or anything that belongs to your neighbor." *[The tenth commandment in Judaism and in the Orthodox and Protestant Reformed churches; the ninth and tenth commandments in the Roman Catholic and Lutheran churches.]*

Moses is specifically the mediator of the Covenant, the formal agreement between God and the people that defines Israel as a nation. The terms of the Covenant are expressed as laws given by God that the Israelites must obey, and may well follow the general model of Near Eastern treaties between an overlord and his vassal. God promises that if the people abide by the terms of the Covenant, they will become "a priestly kingdom and a holy nation" (Exod. 19.6). Moses reports the divine promise and the people assent to it (Exod. 24.7). The narrative includes the giving of the Ten Commandments (see sidebar on opposite page) and other legal provisions. In fact, the Bible takes the view that all basic Israelite law was handed down by God to Moses on Mount Sinai (see also pp.118–19).

The Sinai episode appears to have been shaped by Israelite cult traditions, perhaps by a recurring festival of Covenant renewal. For example, the detailed purification rites before the theophany (Exod. 19.10–15) are a preparation for participation in worship. The theophany itself is accompanied by storm, fire, and earthquake (Exod. 19.16–18). These phenomena have sometimes prompted the view that Mount Sinai was a volcano, but they are in fact stereotypical features of a divine manifestation. The sounding of a trumpet, which is mentioned several times, represents the summons to a religious ceremony, and the cloud of smoke that hides Yahweh from the people (Exod. 19.18) may be based on an incense rite. Finally, the sacred meal and the sprinkling of sacrificial blood to seal the Covenant (Exod. 24.5–8) suggest a cult ritual.

The holy mountain

As far as the Bible is concerned, all that can be said for certain is that God's abode, the mountain where Moses received the Law, was outside the Promised Land, somewhere to the south of Canaan. Mount Sinai is traditionally identified with the mountain called, in Arabic, Jebel Musa ("Mount of Moses") in the south of the Sinai peninsula (see right and also map on p.59). But this identification seems originally to have been made by Christians and to go back hardly beyond the fourth century. It has been suggested more recently that the mountain was somewhere in northwestern Arabia, in what may have been a Midianite area, while some passages in the Bible locate Yahweh's dwelling in Edom (Seir), to the south of Palestine. There are various other terms for the holy mountain, notably Mount Horeb, but also "Mountain of God" or simply "the mountain." The concept of the holy mountain may have been applied to different sites in the interest of pilgrims after its exact location was forgotten.

In the Bible, there are several descriptions of Israel's history from the Exodus to the occupation of Canaan that make no mention of Mount Sinai. This is remarkable, given the focal role of Sinai in the Exodus narrative. In particular, there are strong indications of a tradition that Kadesh-barnea, an oasis in the desert region south of Beer-sheba, occupied the central place in the movement to Canaan and was the site of the giving of the Law. This tradition would have represented the experiences of one of the groups that made up the later nation of Israel. According to this theory, it was displaced by the tradition of another group, which located the lawgiving and Covenant-making on Mount Sinai.

Jebel Musa in southern Sinai, identified as Mount Sinai by Greek Christian monks in the 4th century CE.

THE BOOK OF THE COVENANT

Immediately after the giving of the Ten Commandments (see p.60), God instructs Moses to set a further, and much longer, series of laws before the Israelites (Exod. 20.22–23.33). When Moses reads them out, they are referred to as the "book of the Covenant" (Exod. 24.7), a term by which the collection of laws is usually known. The episode illustrates how all Israelite law came to be attributed to the work of Moses. It seems to be an insertion into the Sinai narrative, because it evidently refers to a time when Israel is a settled community, that is, after it has already moved into the land of Canaan: the ordinances refer to houses, cattle, fields and vineyards, slaves, silver, chattels, and the use of money. The book of the Covenant is thought to be the oldest of the biblical law codes and to reflect conditions soon after the occupation of Canaan, or perhaps even in the early days of the monarchy. Among the sacrificial laws is a prohibition on boiling a kid in its mother's milk (Exod. 23.19). There may be a parallel to this curious rule in a Canaanite text, so it is possible that the regulation is directed against a Canaanite rite. In fact, the book of the

Early Israelite religion

One section of the book of the Covenant (Exod. 20.22–26) provides some important insights into aspects of Israel's early religious practices. God states that he will come to bless the people "in every place where I cause my name to be remembered." This condones the existence of many sanctuaries, in contrast to the situation in the later Israel. It also appears to sanction the Israelite adoption of ancient Canaanite shrines, as seen in the ancestral legends of Genesis and elsewhere (see, for example, p.38 and p.45).

There are three divine regulations concerning the construction of an altar. The first is very ancient and states that altars must be made simply of earth (Exod. 20.24). This may reflect a preference for materials that were in their wholly natural state as they were created. A similar idea may underlie the second regulation, which allows altars of stone as long as hewn stones are not used, because a chisel would profane them (Exod. 20.25). This is probably directed against Canaanite altars, which were made of dressed stone. When Joshua builds the first Israelite altar in accordance with these prescriptions, the Bible states specifically that "no iron tool" was used on its stones (Josh. 8.31). There is no mention of iron in the book of the Covenant, so the law may go back to a time when ironworking was not yet known to the Israelites.

The third regulation prohibits the making of steps up to the altar (Exod. 20.26). It is likely that this law was originally aimed at the widespread ancient Near Eastern cult of the "high god," which involved the construction of altar steps. In the Bible, the reason for the prohibition is "so that your nakedness may not be exposed" on the altar, that is, so that the genitals will not be directly over any part of the altar. This reflects the Israelite abhorrence of any association between the sexual and the holy. A later regulation insisted that priests had to wear linen shorts beneath their robes to cover their genitalia.

Exodus 20 says "You shall not go up by steps to my altar," an ordinance probably designed to counter altars similar to this one at Petra, the center of the ancient Nabatean kingdom. Stepped altars were a characteristic of pagan worship.

Covenant as a whole may have been composed mainly to combat the religious and cultural influence of the indigenous Canaanites during the early period of Israelite settlement in Canaan (see p.72).

The book of the Covenant falls into two main sections that are distinguished by their literary style. The first section consists mostly of ordinances or case law, *mishpatim*, beginning with "if" or "when." For example: "When a slaveowner strikes the eye of a male or female slave, destroying it, the owner shall let the slave go, a free person, to compensate for the eye" (Exod. 21.26); and "If a thief is found breaking in, and is beaten to death, no bloodguilt is incurred; but if it happens after sunrise, bloodguilt is incurred" (Exod. 22.2–3). This is the most common form of ordinance found in other ancient Near Eastern law codes (see box, below). The second section is largely religious in content and combines humanitarian injunctions, standards for the administration of justice, directions for three great annual pilgrimage festivals, and sacrificial regulations. Like the Ten Commandments, this section is apodictic (imperative) in form, for example: "You shall not spread a false report. You shall not join hands with the wicked to act as a malicious witness ... when you bear witness in a lawsuit, you shall not side with the majority so as to pervert justice" (Exod. 23.1–2); and "Six days you shall do your work, but on the seventh day you shall rest, so that your ox and your donkey may have relief, and the homeborn slave and the resident alien be refreshed" (Exod. 23.12).

A Byzantine wall painting depicting Moses, from St Catherine's monastery on Jebel Musa, the traditional Mount Sinai (6th century CE). From the Middle Ages onward, artists portrayed Moses as an older, bearded figure (see picture on p.60).

Near Eastern law codes

The practice of codifying laws was common in the civilizations of the ancient Near East. In the last century, a number of law codes were discovered that display numerous points of contact with the legal texts of the Hebrew Scriptures, suggesting that they provided inspiration for the biblical codifications.

Seven ancient Near Eastern law codes are of particular importance. Four are royal codes, three of which date from the Sumerian period: the code of Ur-Nammu, the founder of the third dynasty of Ur (ca. 2100 BCE); the code of Lipit-Ishtar of Isin from the early nineteenth century BCE; and the code from the kingdom of Eshnunna of about the same period. The fourth is the famous law code of King Hammurapi of Babylon (ca. 1728–1686 BCE; see right). All except for the Eshnunna code have poetic prologues and epilogues praising the king and threatening divine retribution for anyone who seeks to alter his law. Three later codes, also containing parallels to biblical material, are not directly ascribed to monarchs. They are the Middle Assyrian Laws (twelfth century BCE or earlier), the Hittite laws from Asia Minor (fourteenth century BCE), and the laws of the Neo-Babylonian empire (ca. 625–539 BCE). The book of the Covenant resembles these codes in its largely secular subject-matter and in its businesslike tone, and there are many parallels. For example, the law that describes what happens when an ox is killed by another owner's animal (Exod. 21.35) is almost identical to one in the laws of Eshnunna.

These numerous parallels do not mean that the book of the Covenant depends directly on any particular Mesopotamian texts. It is more likely that the Israelites came to know the common Near Eastern pattern of setting out the law through their contact with the Canaanites, although no Canaanite law code has yet been discovered.

A Babylonian stela of ca. 1700 BCE inscribed with the law code of King Hammurapi.

THE TABERNACLE

A statue of king Gudea of Lagash. He was said to have had a dream in which the god Ningursu gave him details of a temple he had to build, together with a sketch of the structure.

Following the account of the giving of the Ten Commandments and the Book of the Covenant (see pp.60–63), the remainder of the book of Exodus is devoted mainly to the building of a sacred dwelling place for God among the Israelites. This sanctuary takes the form of a tent or tabernacle that will also house the tablets of divine law handed down to Moses on Sinai. The concept of the tent-shrine was not unique in the ancient Near East. The Canaanites designated the abode of El, their supreme god (see p.38 and p.45), and other deities as a tabernacle or tent, the same terms used in the Bible for the sanctuary of Yahweh. God showed Moses on Mount Sinai "the pattern of the tabernacle and of all its furniture" (Exod. 25.9). In the Near East, it was commonly thought that the design and dimensions of the earthly temple of a deity were divinely ordained, and corresponded to those of the deity's sanctuary in heaven. Gudea, the ruler of the Sumerian city of Lagash (ca. 2060 BCE), was said to have received divine instructions for the building of a temple.

Chapters 25–40 fall into two broad sections: God's detailed directions to Moses on the building of the tabernacle, followed by an account of how these instructions were carried out. The tabernacle consisted of a rectangular court around a central sanctuary structure. The court was enclosed by curtains supported on poles, and its dimensions have been estimated as one hundred forty-five feet long, seventy feet wide and seven feet high (44 x 21 x 2.1m). The exact length of a cubit, the biblical unit of measurement is uncertain, but it has sometimes been put at eighteen inches (45 cm). The furnishings were made from the finest and costliest materials by two expert craftsmen, who are named in Exodus 31 as Bezalel and Oholiab. The structure of the tabernacle is the focus of the narrative, but there are also details of its contents, such as the altars and, above all, the ark of the Covenant, the sacred chest in which the tablets of the Law will be laid up (see p.76). Instructions are also given for the consecration of Aaron and his sons to the priesthood (see p.54).

The tabernacle was designed to be portable, and when the Israelites moved about during their wanderings in the wilderness, the whole structure was dismantled by the Levites, who re-erected it wherever the tribes pitched camp. The constant movement of so large a structure is hard

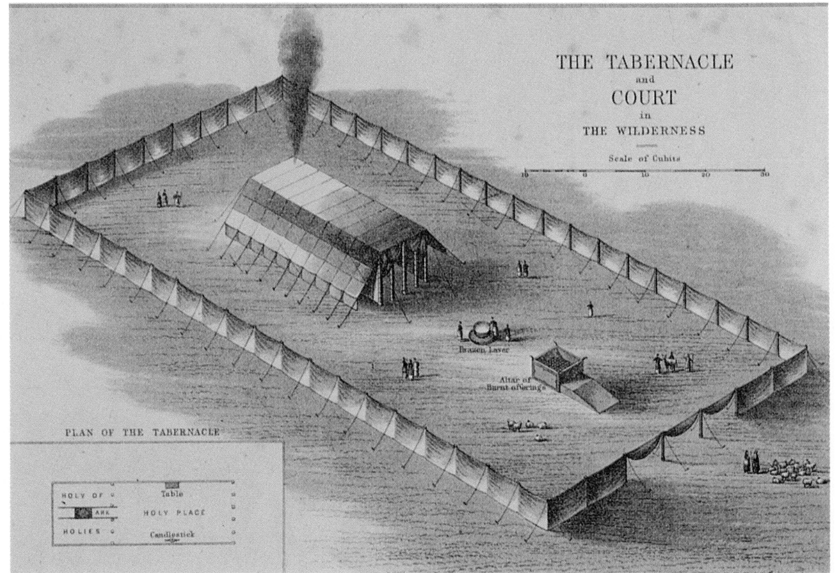

A 19th-century reconstruction of the Israelite tabernacle and court.

The Golden Calf

In between the two halves of the Tabernacle narrative is the episode of the Golden Calf (Exod. 32–33). Moses comes down from the mountain to discover that in his absence the people, led by his brother Aaron, have made an image of a calf in gold and erected an altar before it. When he sees them worshiping the idol, he smashes the tablets of the Law and destroys the calf.

The Tabernacle can only be built when the Israelites have been purged from this sin and the commandments have been reiterated. The text is essentially a mocking polemic against idolatry, the making of gods of silver or gold, which the book of the Covenant condemns as characteristic of Canaanite religion. The Israelites put their trust in a mere calf, and we are surely meant to laugh at Aaron's lame excuse that he only threw the gold into the fire and a calf emerged.

A bronze cult bull of the 12th century BCE from Samaria in northern Palestine.

However, the episode may represent what was originally a more favorable tradition of the fashioning of the calf. The story appears to be related to the account in chapter twelve of the first book of Kings of the setting up of two golden calves, at Bethel and Dan, by King Jeroboam I of Israel (see p.93). In Exodus the Israelites utter a ritual cry: "These are your gods, O Israel, who brought you up out of the land of Egypt" (Exod. 32.4). This cry is almost identical to Jeroboam's statement in the account in 1 Kings 12.28. The "calf" would have been a bull, a symbol of power that was commonly used to represent deities in the ancient Near East. Although Jeroboam's actions are condemned as sinful, his bulls would not have represented a foreign deity, but Yahweh. He may have been reviving a long-established cult which legend ascribed to Aaron.

to imagine in desert conditions, and it is unlikely that the Israelites, who were probably a relatively small group, then had the craftsmen, materials, or wealth to erect it. One explanation is that the account of the tabernacle stems from much later priestly sources, probably in the time of the Israelite monarchy, and in fact presents a description of Solomon's Temple at Jerusalem (see pp. 90–91) in the guise of a portable shrine. This is in accordance with the view that all Israel's religious institutions originated at the time of the Exodus.

The priestly writers were aware that no permanent sanctuary building existed before the occupation of Canaan, but this does not mean that the wilderness tabernacle was merely a fiction of the biblical narrators. As well as using the word "tabernacle," *mishkan*, the account also employs – some one hundred thirty times – the expression "tent of meeting," *ohel mo'ed*. This refers to a much simpler type of shrine (Exod. 33.7–11 and elsewhere) that was more likely to have existed in the wilderness. It was a real tent that one person could pitch, outside the camp, and functioned for the receiving of oracles. On these occasions, when Moses entered the tent to meet with God, the divine presence would descend in the form of a pillar of cloud. The priestly authors transferred the most significant features of the tent of meeting to their representation of the tabernacle, so the divine cloud descends on the tabernacle, and the meeting with God takes place over the ark within it.

The two traditions of desert tent and urban temple are brought together in the curious depiction of the "tent over the tabernacle" in Exodus 36.14. The materials from which it was made, tanned rams' skins and goats' hair, are in marked contrast to the rich fabric drapes of the rest of the tabernacle, and reflect a genuine desert tradition. The same materials are found in a Bedouin tent and, especially, in the Bedouin portable shrines or *qubbas*.

Il Nebi Musa, a Muslim shrine marking the spot on Mount Nebo where Moses is thought to have died (see box, below).

THE WILDERNESS YEARS

The account of the Israelites' wanderings in the wilderness, both before and after the stay at Mount Sinai, is made up of a number of separate episodes from different backgrounds and dates, some of which are repeated more than once. Their framework seems to have been the itinerary preserved in Numbers 33, which may well be an old pilgrimage route to and from Mount Sinai. Unfortunately, it is hard to identify with any certainty many of the names in the itinerary.

Certain themes characterize the narrative as a whole. The most prominent is what has been described as the "murmuring" motif, the repeated complaints of the Israelites against Moses, Aaron, and Yahweh himself (such as Num. 14.1–10). The complaints reflect internal dissent among the people as well as the hardships of desert life, such as the lack of food and water and the danger of attack from snakes and human enemies. The lack of trust which the murmurings betray often provoke Yahweh's anger and, in these stories, Moses appears as the supreme intercessor. For example, in the episode of the Golden Calf (see p.65), he averts God's wrath and persuades the deity to alleviate the condition of the people.

In some of the wilderness stories, Moses is also seen as a worker of divine wonders linked with particular localities or natural phenomena. He purifies the waters of Marah by casting a piece of wood into them (Exod. 15.23–25), and produces water from a rock at a place called Massah and Meribah (Exod. 17.6–7; Num. 20.11). Through Moses, God sends an abundance of quails and manna, which are phenomena characteristic of the Sinai peninsula (Exod. 16). In a battle with the Amalekites, victory is achieved through the power of Moses' uplifted hands (Exod. 17.8–13). When God sends a plague of snakes to punish the Israelites, Moses ends it by setting up a bronze serpent on a pole: anyone bitten by a snake could

The deaths of Moses and Aaron

Moses and Aaron both died outside the Promised Land of Canaan. This was an undeniable fact of Israelite tradition, but it was felt that some explanation was needed as to why these two great figures had not shared in the fulfillment of God's promise to the people. In the book of Deuteronomy, the explanation is that Moses was punished for the repeated sins of the people in the wilderness. In Numbers 20, Moses and Aaron repeat the miraculous provision of water from the rock at Meribah (Kadesh), after which God tells them: "Because you did not trust in me, to show my holiness before the eyes of the Israelites, therefore you shall not bring this assembly into the land that I have given them" (Num. 20.12). The reason for this is unclear. Perhaps Moses and Aaron were guilty of presumption by not giving God credit for the miracle (Num. 20.10). It may be that the story is deliberately vague about the offense so that Moses and Aaron are not incriminated too greatly. In any case, much care is taken to preserve the brothers' reputation. They install successors to carry on their work: Aaron's son Eleazar as the chief priest (Num. 20.28) and Joshua, son of Nun, as the national leader (Deut. 31.7). Both brothers die on mountains, Aaron on Mount Hor near Kadesh (Num. 20.28) and Moses on Mount Nebo in Moab (Deut. 34.5). When Moses ascends Mount Nebo (see picture, above left), God shows him the whole of the Promised Land (Deut. 34). It is now known that such a survey was a way of establishing a right of territorial possession, so that, in a sense, Moses does achieve his final goal.

At the end, Moses is presented like one of the ancient patriarchs. He dies, full of vigor, at a great age, and before his death he blesses all the Israelite tribes, as Jacob had done. Yet his stature is unique, as the final words of the Pentateuch make clear: "Never since has there arisen a prophet in Israel like Moses, whom the Lord knew face to face."

The Israelites gathering manna in the wilderness (Exod. 16.16–17), by the Master of the Manna, ca. 1470. The Bible derives the word manna from the question "Man hu?" *(Hebrew for* "What is it?"*), which the Israelites are said to have asked when they first saw the "fine flaky substance, as fine as frost" (Exod. 16.14). The true origin of the word, like the identity of the substance itself, is not known for certain.*

BALAAM

Balaam is described as coming from Pethor on the Euphrates River, and his divinatory activities (see main text) show him as a typical Mesopotamian seer-priest, the kind known as *baru*. In the Bible he acts under the direct inspiration of Yahweh, and what is recounted of him represents various conventions of Israelite prophetic legend.

Recent discoveries suggest that Balaam may have been a real person of the Transjordan region. Fragmentary Aramaic texts of the ninth century BCE from Deir Alla refer to a Balaam who, as in the Hebrew Scriptures, was the son of Beor. He is said to have a vision of a disaster that befalls his city, at which he weeps. This revelation is received from an assembly of divine beings described as *Shaddin*, which recalls the title *Shaddai*, "Almighty," an archaic name for God in the Hebrew Scriptures (see p.56). The name *Shaddai* occurs more than once in the biblical story of Balaam, for example in Numbers 24.4.

look at the serpent and be cured (Num. 21.6–9). This is the origin of the bronze serpent that stood in the Temple of Jerusalem (see pp.90–91) It was originally a symbol of Canaanite religion, but is here attributed to Moses, although its original significance as part of a cult involving serpent worship has been neutralized.

Moses' leadership does not go unchallenged, even by his sister Miriam and brother Aaron. They question Moses' unique role of mediator, but God reaffirms his position (Num. 12). Similarly, the stories of the rebellion and deaths of Korah, Dathon, and Abiram (Num. 16), and of Aaron's staff (Num. 17; see also p.57), reflect challenges by groups within the priesthood to the exclusive authority of the line of Aaron.

On their way to Canaan, the Israelites enter Moabite territory east of the Jordan River (Num. 22.1). Balak, the king of Moab, realizes the threat they pose and summons the seer Balaam to lay a curse on them. The account of Balaam's failure (Num. 22–24) is marked by folklore motifs and much ironic humor, as in the episode of the angel and the talking donkey (Num. 22.22–35), and in the way that Balak's schemes fail. Every time he asks Balaam to curse Israel, the seer obeys God's instruction and utters a blessing (see also sidebar, right).

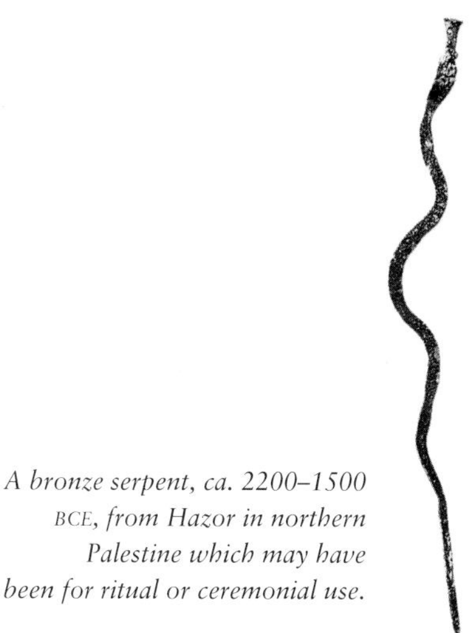

A bronze serpent, ca. 2200–1500 BCE, from Hazor in northern Palestine which may have been for ritual or ceremonial use.

IV THE PROMISED LAND

The Bible relates that, following the death of Moses (see p.66), the Israelites crossed the Jordan into Canaan. The book of Joshua gives an account of their first conquests and ends with a reaffirmation of the supremacy of Yahweh. But, according to the book of Judges, Israel repeatedly broke faith with God, for which they faced divine anger. Each time the people repented, God appointed a leader or "judge" to vanquish the nation's enemies. The greatest of them, Samuel, paved the way for a new institution: the monarchy.

The valley of the River Jordan.

THE INVASION OF CANAAN

According to the book of Joshua, after the death of Moses the Israelites conquered the Promised Land under his successor, Joshua, in three rapid campaigns: in the center, then in the south, and finally in the north. Archaeology has frequently been used to support and approximately date the biblical picture of a speedy and violent conquest. It has been claimed, for example, that four of the towns taken by Israel (Lachish, Debir, Bethel, and Hazor) suffered great destruction between ca. 1300 BCE and ca. 1200 BCE. This may be so, but there is no direct proof that the Israelites destroyed these particular towns or any others, because most cities in Palestine were devastated many times between 2000 BCE and 1000 BCE. Also, other cities said to have fallen to Israel could not, on the evidence, have been destroyed in the century 1300–1200 BCE. According to the Bible, Joshua and the Israelites captured and burned Jericho and Ai, but recent discoveries show that Jericho, while not uninhabited, was barely occupied during that century and certainly did not possess a city wall, while Ai had been abandoned hundreds of years before. Most of the territory conquered according to the book of Joshua is in the area later allotted to the tribe of Benjamin. The core of the book may represent how just one tribe came to possess its own particular territory.

Before the Israelites reach the land of Canaan, the book of Numbers gives an account of their conquest of Transjordan, the territory east of the Jordan. The narrative is confusing, because it is constructed from a variety of traditional elements and motifs. Transjordan is shown as a region contested by a number of peoples – the Edomites, Amorites, Moabites, and Midianites – who had settled in the area before the arrival of the Israelites. (It remained a much disputed area throughout the later history of the region.) The region is depicted as being somewhat strange and sinister. God's anger is provoked when some of the Israelites have relations with Moabite women and start to worship the local god, Baal of Peor (Num. 25). This apostasy is blamed in Numbers 31.16 on the seer-priest Balaam (see p.67), who is killed for it (Num. 31.8). In the book of Joshua, the Transjordanian Israelite groups build an altar, which the other tribes view as a departure from the true worship of Yahweh. According to Hebrew tradition, Transjordan had once been populated by Rephaim, a race of giants. The word is also found in Canaanite texts, where it refers to divine beings. King Og of Bashon, who was defeated and killed by the Israelites (Num. 21.33–35), was remembered as one of the Rephaim and his iron bedstead, nearly fourteen feet long and six feet wide (4.5m x 2m), was said to exist in the Ammonite city of Rabbah. It may have been a sarcophagus, or perhaps a basalt dolmen in the vicinity of Rabbah.

According to the Bible, Moses allotted the Transjordanian region to the tribes of Reuben and Gad, and to the half-tribe of Manasseh (Josh. 1.12–15). This probably points to what was in fact an independent movement of Israelite groups across the Jordan, perhaps originally from the west of the river.

The crossing of the Jordan

The crossing of the Jordan River into the Promised Land, recounted in Joshua 3, is of great symbolic and spiritual importance for the Israelites. In preparation for it, the people are told to undergo the regular ritual purification before a religious ceremony, and the passage over the river is led by the priests and the sacred ark. The story draws an explicit parallel between the crossing of the Jordan and the crossing of the sea during the Exodus (Josh. 3.15–17 and 4.23). Both events hark back to the language of the creation, when the waters of chaos were divided and the dry land emerged. The story is followed by an account of the origin of the sanctuary of Gilgal, which concludes with a circumcision rite at a spot called "the Hill of the Foreskins," and a celebration at Gilgal of the spring festival of Passover–Unleavened Bread (Josh. 5).

The character of the Gilgal legend indicates that it was a pre-Israelite holy place, probably the site of a Canaanite festival, which reenacted the victory of a deity over the forces of chaos, as in the stories of the gods Baal and Marduk (see p.27). The events at the Jordan and at Gilgal may well be the real source of the tradition of Israel's crossing of the sea.

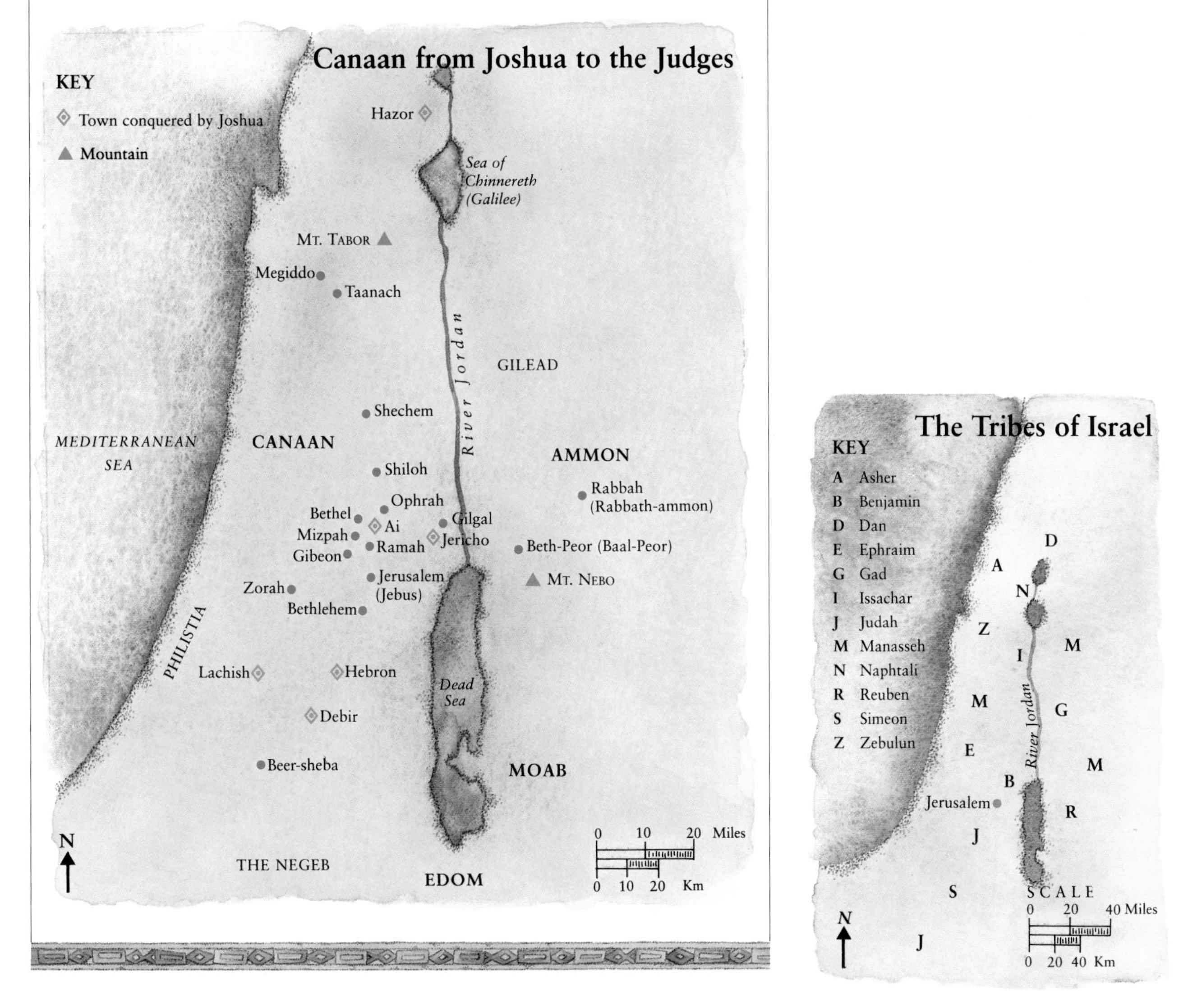

THE FALL OF JERICHO

The Fall of Jericho, *an illustration by Jean Fouquet (ca. 1425–1480) from* Antiques Judaïques, *a French translation of the work of the 1st-century Jewish historian Flavius Josephus.*

The book of Joshua records the capture of a chain of cities in the central plateau of Palestine. Jericho, situated by an important ford over the Jordan, is the first and most influential of these cities and becomes a symbol of the conquest. The Bible combines two stories of the city's fall. The first is told in chapter 2. Joshua sends two men on a spying mission to Jericho. They are hidden from the king of Jericho's men by a prostitute, Rahab. In gratitude, the spies promise her that if she ties a crimson cord in the window of her house, she and her family will be spared by the Israelites. The sending of spies is a common feature of military planning and implies that Jericho was captured by military attack. Rahab's house is said to be in the city wall, and it has been plausibly suggested that the crimson cord in fact indicated the point at which an army could breach the wall. The crimson cord is also reminiscent of the blood applied to the doors and windows of Israelite houses to protect them from God's plague during the first Passover (see p.57).

The walls of Jericho did not fall to a military onslaught but as the result of a great seven-day ritual, which is recounted mainly in Joshua 6. The second story of the capture of Jericho begins with a meeting between Joshua and the commander of the heavenly army, who promises to lead the attack (Josh. 5.14). On God's orders, a procession of priests and soldiers walk around Jericho for six days accompanying the ark of the Covenant. On the seventh day, the Israelites utter a great shout and the walls of Jericho collapse. The Israelite force charges into the city and the inhabitants are slaughtered. All the booty goes into Yahweh's treasury, a regular feature of "the wars of Yahweh," known as the *herem* or "ban."

The fall and capture of the city is represented as a liturgical act, a processional encirclement of Jericho of the ark and its attendants, accompanied by ritual blowing of trumpets, in a seven-day festival. The story may have its origins in a rite celebrated at the sanctuary of Gilgal, which means "circle of stones" (see p.69). The rite may have reenacted a legend that the Benjaminites captured Jericho, a city whose broken down walls were still to be seen nearby.

A stone tower of the Neolithic period (ca. 7000 BCE), discovered during excavations on the site of ancient Jericho, close to the modern town. Jericho is the oldest continuously inhabited city in the world; it was first settled ca. 8000 BCE.

After Jericho, the narrative of the occupation largely consists of a chain of local traditions from the area later assigned to the tribe of Benjamin. Following the story of the crime of Achan (Josh. 7), which explains the existence of a cairn of stones in the valley of Achor and the valley's name, there is an account of the capture of Ai (Josh. 8.1–29). This city was destroyed ca. 2400 BCE and remained a ruin until ca. 1200 BCE, so it was at most a small unwalled settlement during the likely period of the Israelite occupation of Canaan. In Hebrew, *ai* means "ruin" and the narrative may have served to explain that a prominent heap of ruins at Ai was the result of a destruction by Joshua.

The story of the Gibeonites accounts for the treaty relationship between Israel and four Canaanite cities, in the same way that the story of Rahab accounts for the friendly relations with a Canaanite group considered to be her descendants. A later tradition describes the Gibeonites as a group of temple servants, who were assigned to this position by Joshua. This, together with other indications of amicable relations between the newcomers and some of the indigenous inhabitants, suggests that the Israelite occupation of Canaan was not always accompanied by the wholesale conquest, destruction, and slaughter, which characterize the Bible narrative.

In fact, contrary to the biblical tradition, the "conquest" was probably a slow, gradual process, achieved by separate but ethnically related

Northern and southern campaigns

Compared with the detailed episodes relating to the central region, only highlights are given of the conquest of the rest of Palestine. These are the battle at Gibeon in the south and the destruction of Hazor in the north (see map on p.69).

One passage in the story of Gibeon refers to a series of unusual natural phenomena (Josh. 10.12–13). Part of this passage is a fragment of an old poem imploring the sun and moon to stand still. Gibeon was an ancient sanctuary, important in later Israelite history, and there is evidence that Shamash, the sun god, was worshiped there. The poem was originally addressed to Canaanite astral deities but was transferred to Yahweh by the Israelites.

Hazor was the leading city-state in Galilee, as the book of Joshua indicates (Josh. 11.10). Recent excavations have revealed the city's great size and strength. There seems to have been a partial destruction of Hazor by fire in the late thirteenth century BCE, but it is unlikely that the Israelite tribes would have had the military capacity to take such a well-fortified city. Also, the defeat of Jabin, king of Hazor, is ascribed in the book of Judges to a coalition of tribes, after Joshua's death. The story in Joshua probably represents a local tradition of the tribes that settled in Galilee, originally independently of the process of settlement in the rest of the country.

The site of ancient Hazor in northern Israel. According to the Bible, it was taken by Joshua (Josh. 11.10) and later by Deborah and Barak (Judg. 4.2 and 4.24). The city was rebuilt and fortified by Solomon (1 Kings 9.15) and Ahab.

After the conquest of the whole of Canaan, much of the remainder of the book of Joshua is taken up by a lengthy account of the division of the land among the twelve tribes. Many of the details of this account may derive from old materials, but it is generally agreed that the precise demarcation and assignment of Israelite territory would hardly have been of interest before the period of the unified kingdom of David and his successors.

groups over some hundreds of years (see p.74). At the time of the Israelite occupation of Palestine, the country was divided into a number of small city-kingdoms, inhabited by peoples known as the Canaanites and Amorites. This situation is reflected in letters written between ca. 1450 and 1350 BCE from the rulers of various Canaanite cities to the Egyptian pharaoh, their nominal overlord. Some of the letters, which were discovered at Amarna in Egypt, mention attacks by a people called *Hapiru*, a word that is probably closely connected with the name "Hebrew." The disturbances were obviously on a limited scale and it is too simplistic to equate the *Hapiru* attacks with the biblical record of the Israelite conquest. It seems likely that the movement of the *Hapiru* formed just one part of the long and piecemeal process which saw the settlement of various Israelite groups in Canaan.

The Bible unifies the traditions of a number of these groups, under the influence of the group that came out of Egypt. In Joshua 24, there may be a clue as to how this one tradition became dominant. Joshua gathers a great assembly of the Israelite tribes at the sanctuary of Shechem. Addressing them, he states clearly that their ancestors "served other gods." The aim of Joshua's speech is to win over the whole body of the Israelites, many of whom also worship other deities, to Yahweh, the one true God of Israel. The climax of the assembly is reached when all the people recognize Yahweh as the god who guided Abraham, Jacob, and Moses, and led the people into the Promised Land: "The Lord [Yahweh] our God we will serve, and him we will obey" (Josh. 24.24).

THE JUDGES

The book of Judges deals with the fortunes of the Israelite tribes in Palestine before the establishment of a unified state and kingdom. At Shechem, the tribes had pledged themselves to the worship only of Yahweh, but after Joshua's death the nation began to worship the Canaanite deities (see box, below). As a result, the Israelites incurred divine judgment and fell under the power of foreign enemies. When this occurred, the nation repented and turned back to Yahweh, in consequence of which Yahweh chose a "judge," who first led the people to victory over their oppressors and then governed Israel during his or her lifetime. But after each judge's death, Israel again turned from God and the same sequence of events recurred. This pattern was repeated many

The religion of the Canaanites

A basalt relief depicting Baal, the Canaanite god of storms. Baal was the guardian of the rains and therefore ensured the fertility of the soil. He is the central figure in an important cycle of myths that are probably intended to mark the end of the agricultural year and the coming of the autumn rains.

Before this century, what was known of Canaanite religion depended on the Hebrew Scriptures, some Phoenician inscriptions, and ancient Greek authors. This situation changed dramatically with the discovery, from 1929 onward, of a great number of written texts at Ras Shamra in northwest Syria, the site of the Canaanite city of Ugarit. These texts (often called "Ugaritic"), written in the cuneiform script widely used in the ancient Near East, give a first-hand picture of the Canaanite pantheon, mythology, and cult.

The Canaanites worshipped a group of deities, the most important of which were El (the creator god and head of the pantheon; see p.39 and p.45), Baal (the young god of storms and the most active deity), and Baal's consorts, the goddesses, Anath, Ashtaroth, and Asherah (see pp.22–3). There are references to all these gods and goddesses in the Bible, for example, in Judges 2.13 and 3.7. Canaanite religion was centered on fertility, on how to ensure an abundance of crops, animals, and children. Fertility is the underlying theme of most Ugaritic myths, which were probably recounted or reenacted in ritual form each year in order to guarantee the continued fruitfulness and prosperity of the land and its people.

It is not difficult to see why a religion of this kind appealed so powerfully to the Israelites, who had to survive in a strange environment and adapt to a new agricultural way of life. Agriculture was the central concern of the local Canaanite gods, so it was perhaps only natural for the Israelite farmer to turn to them or to transfer their functions to Yahweh.

The ruins of Ras Shamra, the ancient city of Ugarit. It was at its most prosperous in the 15th–13th centuries BCE, the period to which the most significant discoveries of Ugaritic mythological texts belong.

Through him, a new system of government, the monarchy, came into being.

As its title suggests, the book centers on the figure of the "judge," who was not primarily a judge in the modern legal sense. The word "leader" gives a more accurate impression of the role, and the Bible sometimes describes the judge as a "deliverer" or "savior." The regular and permanent leadership of the tribes was in the hands of "elders," senior men entrusted with the conduct of affairs on account of their wisdom and experience. The judges, however, emerged in response to a severe crisis that almost always took the form of a dangerous attack from outside. They were marked out for their mission when they became possessed by the "spirit of Yahweh," a violent seizure that enabled them to perform superhuman feats of strength and military prowess. The judge represents a form of leadership that characterizes tribal societies and is often described as "charismatic." The judges disappeared with the formation of the unified state, when overriding authority was transferred to the king.

The book of Judges portrays the judge as the ruler of the entire Israelite people, but this picture represents the outlook of a later editor, who almost certainly belonged to the school of writers who produced the book of Deuteronomy (see p.106). These writers always view Israel as a single entity, and constantly emphasize the point that disobedience to Yahweh automatically courted national disaster. On closer examination, the stories of the judges that the editor used in fact deal only with the popular heroes and heroines of individual tribes and particular localities. These heroic figures emerged to save their own group when it was threatened with attack (probably on a relatively small scale) by neighboring groups.

The walls of ancient Shechem, which features prominently in the Bible. The city occupied a strategic position in the pass between Mount Gerizim and Mount Ebal. It was destroyed by the Hasmonean ruler John Hyrcanus in 107 BCE (see p.142). The Roman town of Neapolis (modern Nablus) was later built nearby.

An alternative view of the Israelite occupation

The opening chapter of the book of Judges presents a markedly different account of the Israelite occupation of Palestine from that found in the book of Joshua. In Judges, individual tribes, or even smaller groups, act independently of one another, each moving separately into a particular area on its own initiative. For example, the towns of Hebron and Debir are said in Joshua to have been captured by a united nation under the leadership of Joshua. But, according to Judges, Hebron fell to the tribe of Judah and Debir to the small clan of Othniel, and there is no mention of Joshua at all (Judg. 1.9–13).

The account in Judges also reveals that in a number of places the Israelites settled alongside the indigenous inhabitants. The newcomers did not attempt to expel or exterminate the natives, and the Bible suggests that they were regularly used as a source of forced labor (Judg. 1.27–35). As this passage indicates, the region of Israelite occupation was restricted to the hilly districts and did not extend to the Canaanite city-states of the plains, which long retained their independence (see also Judg. 1.19). Relations between these cities and the incoming Israelites no doubt varied from place to place. Often there would be mutual hostility, but Judges indicates that the two groups frequently lived in a state of relatively peaceful cohabitation. There are even instances of actual confederations between them, recalling the alliance between Joshua and the Gibeonites in Joshua 9 and 10 (see p.70). The alliance between Gideon and Abimelech and the Canaanite city of Shechem, which is recounted in Judges 8 and 9, is one example. The story of the dealings between the tribe of Benjamin and the Canaanite city of Shiloh (Judg. 21) probably has the same explanation.

HEROES AND HEROINES

A Syrian-Hittite stele of ca. 850 BCE, thought to depict Gilgamesh (center), the greatest hero of ancient Mesopotamian literature. Tales of his exploits were popular throughout the Near East and may have influenced the story of Samson.

Mount Tabor, where Barak is said to have defeated Sisera (Judg. 4.14–15).

When the Israelites looked back on the time of the judges from the more settled days of the monarchy, they recognized a distant past very different from later times. It was an epoch marked by heroic individuals who took the initiative in great deeds of arms, a situation that is well summarized in the statement: "In those days there was no king in Israel; all the people did what was right in their own eyes" (Judg. 21.25). In terms of outlook and literary character, the closest parallel to the book of Judges is the *Ayyam al 'Arab*, "The Day of the Arabs," the collection of Arab tribal traditions of the pre-Islamic heroic age. Like Judges, the *Ayyam* is almost entirely concerned with warfare.

At the heart of Judges are five originally separate story cycles about individual heroes and heroines, once local figures who came to be looked on as national symbols. In most of these, three basic details are given: the person of the heroic figure; his or her tribe, which remembered the hero or heroine in its traditions; and a sanctuary with which the figure is linked, where those traditions would have been preserved and recounted. In each case, this bare outline has been filled out with local legends, folklore, and cultic features, to produce the existing biblical form of the story.

The first cycle concerns Ehud (Judg. 3.15–30), who assassinates King Eglon of Moab, the oppressor of the Israelites. This story began as a typical folktale of a tribal hero, told with some broad black humor (such as the gruesome details of the killing of the "very fat" king in 3.22). But, in its final form, Ehud has become a national symbol, a "deliverer" of the Israelites who ends the divine punishment of foreign oppression.

The next narrative deals with the military campaign of the prophetess Deborah, the only woman judge, and Barak (Judg. 4–5). Mount Tabor, the border sanctuary of the two tribes primarily involved in the battle, features prominently, and the great poem sung by Deborah (Judg. 5) is probably a psalm of thanksgiving for victory that was sung at a solemn festival there. The poem begins with a vivid description of a manifestation of God, and victory is gained when even the stars and the torrential waters of the Kishon join the fight against Israel's foes (Judg. 5.20–21).

The third narrative tells the story of Gideon (see box, below). It is

Gideon

The third great story cycle in Judges is long and elaborate and centers on Gideon (Judg. 6.11–8.32). He seems to have been originally a simple folk-hero of a small clan group, who was remembered as one who upheld the fundamental social institution of the blood-feud by slaughtering the two kings of Midian who had killed his brothers (Judg. 8.18–21).

However, in the bulk of the scriptural traditions about him, Gideon is a national hero who rises from obscurity to greatness as the result of divine choice and inspiration. He leads his tribe in battle and, as often happens in such legends, he earns a new name, Jerubbaal (Judg. 6.32). As a reward for his exploits, the Israelites ask Gideon to be king (Judg. 8.22), an offer that he rejects, according to the narrative. But the rivalry for the throne between two of his sons following his death implies that he did in fact accept the kingship and rule Israel.

Interwoven with the account of Gideon's call by Yahweh is a legend about Gideon's foundation of the Manassehite sanctuary at Ophrah (Judg. 6.24), which plays a significant role in this saga. He is also said to have erected a golden ephod (in this case the word means a cult image rather than a priestly garment) at Ophrah, and his story ends with the mention of his tomb there. It seems fairly clear that the story of Gideon's life originated at the sanctuary, where it was probably recited on ritual occasions.

Samson

The fifth extended saga in Judges is the famous story of Samson (Judg. 13–16). He is a remarkable and complex figure. At one level, he is a simple clan folk hero, the son of Manoah of the tribe of Dan. But Samson's birth recalls that of several of the great ancestors (see pp.36–51): his barren mother is given a divine promise that she will conceive (Judg. 13.3). The announcement of the conception takes place at Zorah and is interwoven with the legend of the foundation of the sanctuary there (Judg. 13.15–24).

Unlike the other judges, he is not elected at a moment of crisis to lead the Israelite armies to victory over their oppressors. Instead, Samson's greatness is divinely ordained and foretold (Judg. 13.5), and he inflicts defeat on the Philistines (see p.77) through extraordinary feats of strength, such as the killing of a thousand men with the jawbone of a donkey (Judg. 15.15). Like the "labors" of the Greek hero Herakles (who has been compared to Samson since at least the fourth century CE), these feats are derived from a variety of sources. Some are local legends, while others derive from religious rituals and common mythological themes of the ancient Near East, such as the killing of a lion (Judg. 14.6). The source of Samson's amazing strength is his long hair, in accordance with a widespread popular belief, particularly among Semitic peoples. He is a trickster, as is seen in his riddle (Judg. 14.14), and in his repeated deception of his treacherous wife Delilah when she pesters him to reveal the secret of his strength (Judg. 16.6–14). Samson also embodies another well-known character of folklore, the man who constantly gets involved with women and is regularly deceived by them.

God temporarily abandons the hero to his fate after he finally tells Delilah his secret: Samson's hair is cut off as he sleeps and he is captured and blinded by the Philistines. When they parade their prize during a great festival at the Philistine temple, Samson makes an appeal to God (Judg. 16.28) to grant him a final victory. He recovers his strength and pulls down the two pillars of the Philistine temple, killing everyone inside, including himself. This heroic act of suicide is a sign that Samson is received back by God. It resembles the apotheosis of Hercules (the Roman name for Herakles) in the work of the same name by the Roman writer Seneca: Hercules mounts his own funeral pyre, speaks to his father in heaven, and hears his reply. Samson's story comes to an end with his burial in the family grave at Zorah.

Samson carrying off the gates of Gaza (Judg. 16.2–3), from an early 15th-century Biblia pauperum *("Paupers' Bible"). Such Bibles were largely pictorial and aimed at the illiterate poor.*

followed by an account of Jephthah (Judg. 11.1–12.7), which appears to represent a collection of tales from the sanctuary of Mizpah. It contains several common mythological and folklore themes, such as the hero's lowly birth (Jephthah's mother was a prostitute) and the vow with tragic consequences (Judg. 11.30–31). This vow results in Jephthah's sacrifice of his virgin daughter, which is said to have initiated an annual women's festival of lament for Jephthah's daughter. In fact, the festival marked the death of the spirit of fertility during the dry season.

The fifth, and probably most famous, of the heroic narratives is the story of Samson (see box, above).

SAMUEL

The focal point of the pre-Israelite shrine at Shiloh may have been similar to this round altar in the Canaanite temple at Megiddo in northern Palestine, dating from ca. 2000 BCE.

The last and greatest of the line of judges was the prophet Samuel. The announcement of his birth (1 Sam. 1) follows the common biblical pattern of the promise of a son to a barren woman, but there are indications that this story originally referred to the future King Saul (see pp.78–81). Hannah, Samuel's mother, names her child because she "asked" the Lord for him (1 Sam. 1.20), and says that the boy is "lent" to the deity (1 Sam. 1.28): both verbs play on the name Saul. Hannah's thanksgiving prayer (1 Sam. 2.1–10) has nothing to do with the birth of a child but is in fact a royal psalm. A biblical editor of a later period may have sought to exalt the office of prophet, which Samuel represented, above that of king. This may be why Samuel is credited with the subjugation of the Philistines, which is unlikely to be historical. Like his predecessors, Samuel is linked with a particular sanctuary, in this case Shiloh, a former Canaanite holy place. Its chief celebration was a great autumnal pilgrimage feast, during which women performed a ritual dance to enhance the fertility of the vineyards (Judg. 21.21). It is on this occasion that Hannah is granted her prayer for a child. Shiloh had a temple to Yahweh (1 Sam. 1.9), the Bible's first mention of such a permanent structure.

The temple housed the ark of the Covenant (see also p.64), from which oracles could be obtained. The Hebrew word usually translated as "ark" properly means "box" or "chest," and it is likely that various sanctuaries possessed such an object. Its most essential feature was that it could be carried about from place to place: portable shrines are known from ancient Egypt, Mesopotamia, and Canaan, and a portable tent-shrine, the *qabba*, is a feature of Bedouin and pre-Islamic Arab culture. In particular, the Shiloh ark was a sort of protective war standard, leading the Israelite army into battle, as in 1 Samuel 4. It was a manifestation of Yahweh's presence and was virtually identified with him: when the ark is brought to the Israelite army, the Philistines cry: "Gods have come into the camp" (1 Sam. 4.7). All these elements occur in a section of 1 Samuel known as the "history of the ark," that tells how the Philistines captured the ark and laid it up in the temple of their god Dagon at Ashdod. But Yahweh, in the shape of the ark, destroyed Dagon's image (1 Sam. 5.4), and the Philistines sent the ark away (1 Sam. 6.11–12). It halted at two Israelite shrines, Beth-shemesh and Kiriath-jearim, at the second of which it remained under a custodian, Eleazar. This episode may reflect memories of a time when the ark went on circuit to neighboring sanctuaries.

There was a hereditary priesthood at Shiloh, which probably came to an end through the Philistine destruction of the city, although the Bible attributes it to God's displeasure at certain ritual practices, such as cultic prostitution (1 Sam. 2.22). God was known at Shiloh by a title, "Lord of hosts" (as at 1 Sam. 1.3), that was possibly also Canaanite in origin. "Hosts" refers primarily to the heavenly powers that came to Israel's aid in its battles, although it may include the earthly Israelite forces as well.

The figure of Samuel has been much inflated in the biblical tradition. The most authentic information about him indicates that he was a local seer-prophet at the shrine of Ramah, where he is said to have built the altar and presided at sacrificial feasts, and where he was buried. From there he went on an annual circuit, dispensing justice at three sanctuaries (Bethel, Gilgal, and Mizpah) within a comparatively small area of central Palestine (1 Sam. 7.16–17). In the Bible, Samuel is transformed into a judge over the whole nation, a great military leader, and a prophet, as that office was understood in later Israelite tradition (see p.94). As the Bible sees it, it fell to Samuel to inaugurate the next stage in Israel's history, the monarchy.

The Philistines

The Philistines, who gave their name to the land of Palestine, formed part of a great ethnic upheaval that took place in the Aegean area in the last decades of the thirteenth century BCE. This upheaval is known as the movement of the "Sea Peoples," groups (probably Indo-European) that descended on the eastern Mediterranean coasts and destroyed the Hittite empire and the Canaanite city of Ugarit (see p.72), and menaced Egypt itself. They were finally checked, ca. 1190 BCE, by the pharaoh Ramesses III, who settled the Philistines along the southern coast of Canaan.

A decorated Philistine beer jug of the 12th century BCE. Philistine ceramics show Mycenaean influences.

In Canaan, they organized themselves into a league of city-states, made up of Ashdod, Ash-kelon, Ekron, Gath, and Gaza, where a rich and complex culture developed. They rapidly sought to penetrate into the hinterland, where they encountered the Israelites. They were a grave threat to Israel because they presented a unified military force against the less well-organized individual Israelite tribes. As a result, the Israelites fell under Philistine domination for a long period, and it was only when Israel became a unified state under King David that the Philistines were defeated and confined within their own territory. After this, the Philistines no longer posed a serious threat, but intermittent warfare continued for many years between them and the Israelite kingdoms, until the Philistine rulers and people were deported by King Nebuchadrezzar II of Babylon at the same time as the exile of the Israelites in the early sixth century BCE.

The Israelites were aware of the Aegean origin of the Philistines (Amos 9.7 says they came from "Caphtor," that is, Crete) and that they were a non-Semitic people: in contrast to the Israelites, the Philistines are regularly described as "uncircumcised." But the Philistines assimilated the divinities of local Canaanite religion: of the three Philistine deities that are referred to in the Hebrew Scriptures, Dagon and Ashtaroth are found in Ugaritic texts, and Baal-zebub, a name whose meaning is disputed, is clearly an epithet of the great Canaanite god Baal (see pp.22–3 and p.72).

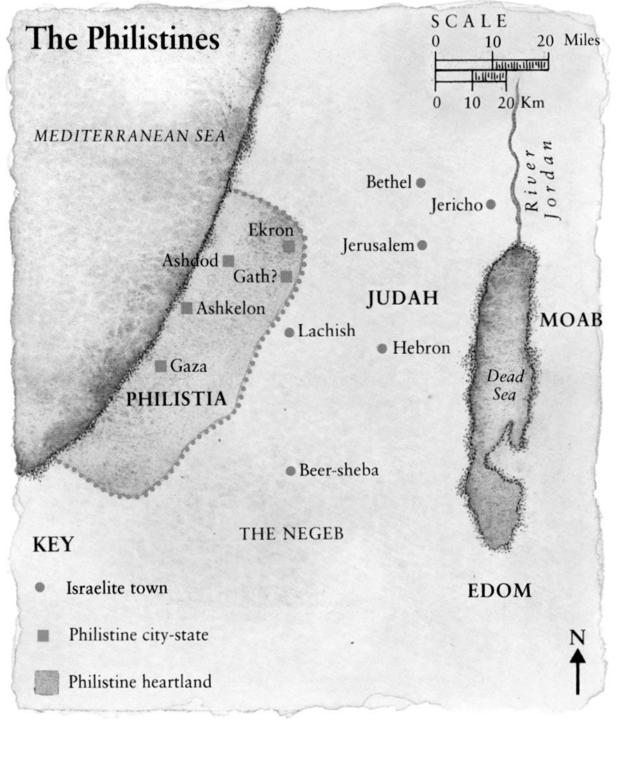

Philistine prisoners being led into captivity, from the temple of the pharaoh Ramesses III at Madinet Habu, Egypt.

V THE FIRST KINGS

It is widely believed that the monarchy was established in Israel as a response to Philistine pressure, and certainly Samuel was promised that Saul's calling would deliver God's people from the Philistines. But it seems more likely that the Philistine threat only brought to fruition an internal development that was already under way. Israel was undergoing the same evolution from a tribal society to a unified state that was occurring among neighboring peoples. The rule of Saul was the prelude to the reigns of David and Solomon, under whom the united kingdom of the Israelites reached the zenith of its power and prestige. During this period, Jerusalem and its Temple became the focal point of Israelite civilization and belief.

SAUL

The rise of the first king of the Israelites, Saul, is told at considerable length (1 Sam. 8–11). It is generally agreed that the narrative combines three different versions of how events unfolded. One account views kingship in a positive light, seeing it as an institution introduced to Israel on the initiative of God. It is essentially a folktale, in which the young Saul is described in idealized terms (1 Sam. 9.2). He is guided by divine providence to the home of the prophet Samuel at Ramah and, when he appears, Yahweh instructs Samuel to anoint him as king. The seer obeys (1 Sam. 10.1) and promises Saul three signs to authenticate his unexpected election, including possession by the divine spirit. All of these signs duly occur, but at this point Saul keeps secret what has happened.

Saul's reign

The reign of Saul has generally been considered to be of minor importance in Israel's history. Although the Bible stresses the novel character of monarchy (see box on opposite page), in some ways Saul's rule only continued the old system of charismatic leaders or judges (see pp.72–3), confined to only a small group of tribes and lacking any concept of territorial expansion. Unlike other ancient Near Eastern monarchies, including the later Israelite kingdoms, Saul had neither a great royal city nor a civil service. He failed to defeat the Philistines or to establish a dynasty.

Nevertheless, it can be argued that Saul established the main features of future monarchical rule. The Bible sees him as inaugurating the idea of the "kingdom" as a distinct entity that thereafter passed from one successor to another. The "ways of the king," which Samuel explains to the people in 1 Samuel 8, correspond to the legal rights of Canaanite monarchs, and Saul was indeed responsible for most of the things listed by Samuel. He drafted young Israelites into a standing army; he appropriated and redistributed his subjects' property; he imposed taxes; and, like other kings of his time, he had a harem. His residence was "Gibeah of Saul" (1 Sam. 11.4), not exactly a royal capital like David's Jerusalem (see box on p.82), but Saul did build a large fortress there. Saul attacked four Gibeonite cities, traditional allies of Israel, because he aimed to create a consolidated state: the cities formed a block separating the tribes of Ephraim and Benjamin from Judah. He wore a diadem and bracelet, the characteristic insignia of Near Eastern kings.

It was Saul's sacred character that most distinguished him from the judges. The ritual of anointing endowed him with God's spirit, and rendered his person inviolable and holy: he became "the Lord's anointed."

The second account views kingship negatively. The Israelite elders go to Samuel and request a king (1 Sam. 8.5), but God tells Samuel that in doing so the elders have in effect "rejected me from being king over them." He instructs Samuel to warn the people of the oppressive nature of monarchical rule, which he does in some detail (1 Sam. 8.11–18). But the Israelites persist, and Yahweh agrees reluctantly to their request. Some time later (the episode of Saul's anointing comes in between), the Israelite tribes assemble at the sanctuary of Mizpah. Saul, who has hidden himself among the baggage, is picked out by sacred lot and made king by a ritual acclamation. In this account, no mention is made of anointing (1 Sam. 10.17–24).

An Ammonite attack on the town of Jabesh-gilead is the starting point of the third account of Saul's accession. The news is conveyed to Saul, who is apparently living as a simple farmer at his home in Gibeah (1 Sam. 11.5). Like one of the judges of Israel (see p.72), he is at once seized by "the spirit of God," as a result of which "his anger was greatly kindled." Saul musters the Israelite militia and breaks the siege. In gratitude, the people go to the sanctuary of Gilgal and invest him as king. Samuel's call to the people to "go to Gilgal and there renew the kingship" (1 Sam. 12.14) was probably inserted by the biblical editor to harmonize the three accounts of the rise of Saul.

The negative view of kingship in 1 Samuel 8 has often been attributed to a later Israelite school that blamed the monarchy for the disasters which befell Israel after the reign of Solomon. But it may well reflect contemporary debate at the time of the introduction of kingship, together with different traditions from the sanctuaries of Ramah, Mizpah, and Gilgal, all of which could lay claim to Saul. These and other elements have been woven together to present the formal process of accession which characterized Israel's developed monarchy. The accession took place in three stages: God's designation of the king through anointing, accompanied by a divine oracle assuring the king of divine favor and setting out his rights and duties; a victory over the king's enemies, whether actual or staged as part of a ritual; and, finally, public recognition and acceptance of him as Yahweh's chosen ruler.

The significance of the monarchy

Before the accession of Saul, the Bible provides evidence that the old system of judges was breaking down and proving unsatisfactory. Samuel appointed his sons to succeed him as "judges over Israel," but they "turned aside after gain; they took bribes and perverted justice" (1 Sam. 8.1–3). The Bible recounts that the Israelite elders confronted the elderly Samuel and asked him to appoint a new type of ruler, in imitation of foreign practice: a king (1 Sam. 8.5). The institution of monarchy had already existed for many centuries in the Near East and, although kingship differed significantly in character from nation to nation, many features were common to all the monarchies of the region. Above all, monarchy was a sacred institution, everywhere closely linked with a nation's deities. The king, it was believed, was chosen by the gods and goddesses, and represented his people before them. The deities, in turn, regulated all national affairs through the sovereign. It was this fundamental characteristic that made the establishment of monarchy such a new and revolutionary step for the Israelites to take.

Samuel anoints Saul, king of the Israelites (1 Sam. 10.1), from the Nuremberg Bible of 1483.

THE FATE OF SAUL

The story of Saul's life is very like that of a hero in an ancient Greek tragedy. In the course of the drama, there is often a crucial moment when the hero's previous good fortune is reversed. In the case of Saul, this occurs when God, through Samuel, abruptly rejects him as king after he fails to carry out a divine command (1 Sam. 15.26). Yahweh withdraws his benevolent spirit from Saul, and instead sends an "evil spirit" to torment him (1 Sam. 16.14). Mental and physical illnesses are often attributed to evil spirits, so the Bible is probably referring to a state of paranoia or depression. Saul is driven to a succession of violent actions, such as the wholesale massacre of the priesthood at Nob (1 Sam. 22.18), and ultimately to his doom. The narrator tells Saul's tale with a degree of sympathy for his plight and depicts him as a heroic figure, albeit a tragic one.

Saul is enmeshed in three conflicts. The first is the break with Samuel, the event which determines Saul's fate. The rupture occurs in two incidents which center on the authority claimed by the new institution of monarchy. In the first incident, Saul displeases Samuel when he offers a public sacrifice (1 Sam. 12.9). His fault is not that he, a layman, has usurped a priestly function, but that the king has taken control of the national religion from the figure of the judge, represented by Samuel. In the second incident, Saul does not complete the destruction of the Amalekites, as God has demanded (1 Sam. 15.3). In sparing the Amalekite king and all the valuable booty, Saul in effect claims the right to change an absolute religious decree, and for this he is condemned. His actions also display political expediency, something which would be less shocking to us than perhaps it would have been to Saul's contemporaries.

Saul's second conflict is with David (see sidebar, left). This is caused by Saul's obsessive jealousy of the younger man, which is provoked by the victory song of the Israelite women after the slaying of the Philistine giant Goliath: "Saul has killed his thousands, and David his tens of thousands" (1 Sam. 18.7). Saul interprets the song as a wish for David to replace him as king, and from this point onward Saul's dealings with David take the form of a succession of desperate attempts to kill him.

Lastly, there is conflict within Saul's own family. His daughter Michal falls in love with David and marries him. When Saul makes an attempt on David's life Michal foils it by means of a trick (1 Sam. 19.13). More

THE RISE OF DAVID

David's early career and character are depicted in various ways. The account of his secret anointing (1 Sam. 16.13) reflects a common folklore theme, in which the youngest and apparently least worthy family member is unexpectedly preferred over his or her siblings. The folktale quality pervades the whole story of David's rise to a high position at Saul's court.

The young David's most famous exploit is his killing of the Philistine giant Goliath (1 Sam. 17.48–51). Elsewhere (2 Sam. 21.19), Goliath is said to have been killed by Elhanan, another Bethlehemite, suggesting that the feat was only later attributed to David. David is portrayed as the ideal of chivalry, magnanimous toward Saul and twice sparing his life, at En-gedi (1 Sam. 24.7) and at Hachilah (1 Sam. 26.9). He arouses universal adulation and is capable of deep affection, especially in his feelings for Saul's son Jonathan (1 Sam. 20 and elsewhere). David is a skilled musician and composer of songs, whose lyre playing has an almost magical power (1 Sam. 16.23).

There is another, probably more historical, aspect of David. He appears as the leader of a band of outlaws in the Judean desert, astutely planning for his future by presenting gifts to the leading families of the region and making marriage alliances with them (1 Sam. 25.39–43). In pursuit of his aims David is even prepared to become a vassal of the Philistine king of Gath. However, in the story as it stands, he avoids actually giving help to the Philistines against the Israelites (1 Sam. 27.8–12).

Below: David and Goliath, *from an English manuscript of ca. 1020.*

Below, right: *The heights of En-gedi, where Saul hid from David (1 Sam. 24).*

The Death of Saul (1562), *by Pieter Bruegel the Elder (ca. 1515–1569). Saul (left), already badly wounded and facing defeat, falls on his sword (1 Sam. 31.4).*

poignantly, Saul's son and heir, Jonathan, develops an intimate friendship with David and protects him from his father's plots (1 Sam. 19.1–7 and elsewhere). Jonathan recognizes that it is David, rather than himself, who will succeed his father as king.

Before his final battle with the Philistines, Saul follows the regular custom of ancient warfare and seeks divine guidance on its outcome. During his reign he has carried out a religious reformation, as later Israelite monarchs were to do, by eradicating "mediums and wizards" who claimed that they could call up ghosts and spirits, a practice always opposed by the cult of Yahweh (1 Sam. 28.3; compare Lev. 20.6 and 20.27). But after Saul's rejection, Yahweh refuses to speak to him through the normal legitimate means of divination: dreams, the Urim and Thummim (sacred lots), and prophetic oracles. Saul is forced to resort to a practitioner at Endor, who has apparently escaped his ban (1 Sam. 28.7). She has an *'ob*, a word usually translated as "familiar spirit," that is, she is a medium. The appearance of Samuel's shade (1 Sam. 28.14) reveals how the ancient Israelites saw the fate of the dead: they retained the physical state, and even the clothes, in which they had died, and led a shadowy existence under the earth, cut off from real life. Living in this mysterious realm, they were considered to have a supernatural power to foretell the future.

Saul is overcome with terror at Samuel's dreadful prediction of his defeat and death (1 Sam. 28.19). But the narrative tells how he recovers himself, and the reader is left to assume that he accepts his fate and faces it bravely. When three of his sons, including Jonathan, have died fighting against the Philistines, and all is finally lost, Saul meets death in a typically heroic act of suicide on Mount Gilboa (1 Sam. 31.4), although it is later said that he enlisted a foreigner to end his agony as he lay dying (2 Sam. 1.9). David's great lament (2 Sam. 1.17–27) is in praise of two men he saw as heroes: "Saul and Jonathan, beloved and lovely! In life and in death they were not divided; they were swifter than eagles, they were stronger than lions."

DAVID

The pool of Gibeon (modern El Jib), where the armies of Ishbaal and David confronted one another (2. Sam. 2).

After Saul's death (see p.81), his kingdom continued under one of his sons, Ishbaal or Ish-bosheth, in the northern part of the country, while the royal seat was removed to Mahanaim, east of the Jordan, outside the area of Philistine occupation. Meanwhile, in the south, the clans whose support David had secured (see sidebar on p.80) anointed him as king over the house of Judah (2 Sam. 2.4). War broke out between Ishbaal and David, and at this time the really powerful figures were the formidable commanders-in-chief, Abner for Ishbaal and Joab for David. A vivid story tells how their two armies met at the pool of Gibeon (see picture, left), where, following a common custom of ancient warfare, each side produced twelve champions to engage in single combat to the death. The outcome was inconclusive (2 Sam. 2.16), and a full-scale battle ensued, in which Abner killed Asahel, one of Joab's brothers, and thereby incurred a blood feud.

Later, Ishbaal rebuked Abner for allegedly having intercourse with Rizpah, one of Saul's concubines. (According to a Canaanite text from Ugarit, the taking of a dead king's wife represented a claim to the throne; compare 1 Kings 2.22.) Abner was angry at the rebuke and negotiated with David to bring the northern confederation onto his side (2 Sam. 3.12). Abner visited David to finalize the compact, but he was killed by

Jerusalem

David's acquisition of Jerusalem (see main text, opposite) had far-reaching implications for the nation. It became a royal capital city, an essential feature of any Near Eastern monarchy, and something which Saul had lacked (see box on p.78). It was a strategic bridge between the northern and southern Israelite tribes, but was not attached to any of them.

Until its capture, Jerusalem was a Canaanite enclave, inhabited by a people known as the Jebusites. David brought the city within the Israelite religious tradition by making it the permanent home of the ark of Shiloh (see p.76). Behind the detailed account of the transfer of the ark and the accompanying celebrations (2 Sam. 6–19), there probably lies a coronation rite. Through this rite, David assumed the role of the former Jebusite ruler, the heir of the priest-king Melchizedek who blessed Abraham in Genesis 18 (see p.39). A procession led by the ark was probably the climax of a great annual festival which celebrated God's choice of Jerusalem as his abode and of David's dynasty.

Samuel anoints David, a wall painting from the synagogue of Dura Europos, Syria, ca. 245 CE. The Roman dress reflects the cosmopolitan character of this Greco-Roman frontier town on the Euphrates.

David did not exterminate the Jebusite population, and many of their religious traditions directly influenced Israel's faith. Yahweh was assimilated with the local deity of Jerusalem, El-elyon, and, significantly, the Israelite god was considered to have made a special covenant with the king that paralleled his covenant with the nation. The reign of David therefore marked a profound change in the whole nature and understanding of monarchy in Israel, which assumed the general character of Near Eastern kingship. In particular, this meant that each succeeding monarch was regarded as the adopted son of the deity, who guaranteed the continuation of the dynasty (see, for example, 2 Sam. 7; Pss. 2 and 89).

David's empire

Every ancient ruler was expected to be a war leader who sought to expand the territory of his state, and David was no exception. During his reign (ca. 1000–961 BCE), it was feasible for him to adopt an expansionist program, because there was something of a political vacuum in the Near East. The empire of the Hittites had disappeared ca. 1200 BCE, and, following the death of the great emperor Tiglath-pileser I in ca. 1078 BCE, the Assyrian state entered a period of weakness that lasted for some two centuries. Since ca. 1065 BCE, Egypt, too, had been impotent under the twenty-first dynasty. The field was free for the Israelites and other peoples to establish themselves as powerful states.

Following the final removal of the Philistine menace (2 Sam. 8.1), the Bible tells of a series of conquests by David of important neighboring states. He defeated the Ammonites and became their king (2 Sam. 12.30). Moab was made a client state (2 Sam. 8.2), and Edom was garrisoned as a province under direct Israelite rule (2 Sam. 8.14). David is said to have stationed garrisons in two of the rising Aramean kingdoms, although they were to remain a thorn in Israel's side for many years.

David also forged important diplomatic ties. He made an alliance with Hiram, the king of Tyre, one of the important group of Phoenician trading cities that had recently gained independence from Egypt and were entering on a golden age. The Phoenician alliance became a cornerstone of subsequent Israelite foreign policy (see p.89).

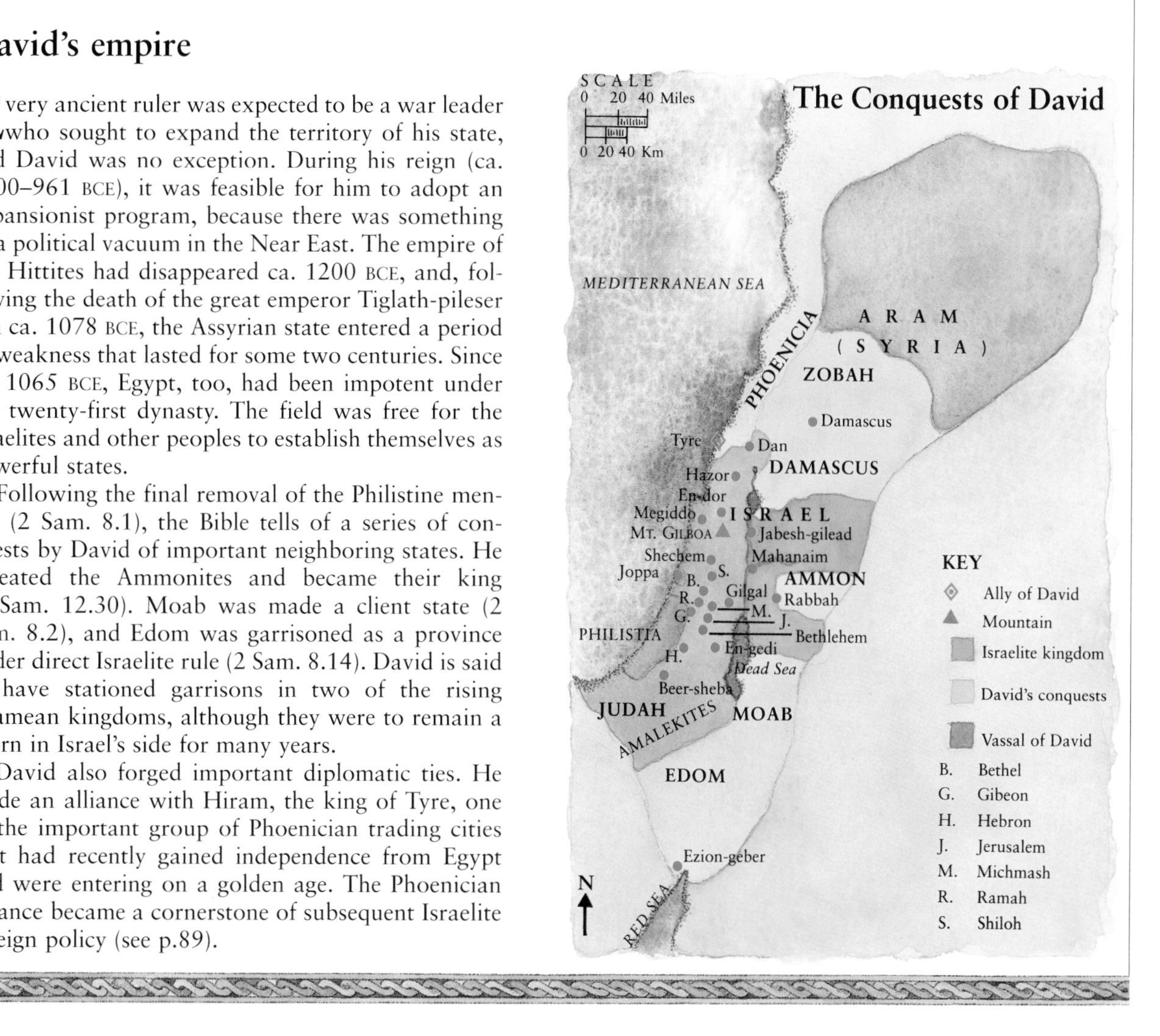

Joab in revenge for Asahel's death and the plan collapsed. With Abner gone, Ishbaal's entourage realized that his breakaway kingdom was no longer viable, and two of his officers assassinated him (2 Sam. 4.6–7).

Ishbaal's death opened the way for the second step in David's rise. David had always been careful to show respect and honor to Saul and his supporters. He sent a message of thanks to the inhabitants of Jabesh-gilead for giving Saul's corpse an honorable burial, laid a solemn curse on Joab for Abner's death, and executed Ishbaal's murderers. It was perhaps no surprise that the northern tribes now made an alliance with David and anointed him as their king (2 Sam. 5.3).

As the leader of a united kingdom, David was able, in two campaigns, to expel the Philistines from the central hill country and open the way for the third step in his rise to power, the taking of Jerusalem. Its capture is only briefly recorded, but the Bible indicates that it was taken by a stratagem, presumably because the Israelites had neither the experience nor the means for siege operations (2 Sam. 5.6–8). What is important is that Jerusalem was captured by David's personal troops and therefore became a private royal possession, held by David and his descendants by right of conquest. As if to emphasize this point, he renamed it the "City of David" and built a fortress there for himself (see also box on opposite page).

A clay figurine of a woman bathing, found at Achzib in Palestine and dating from ca. 1300–900 BCE. Bathsheba was bathing when David first saw her (2 Sam. 11.2).

DAVID'S REIGN

After David had established himself in Jerusalem (see p.83), he proceeded to organize his state. It was modeled on that of neighboring countries, particularly Egypt, and administered by a bureaucracy of court officials known as "servants of the king" after a Canaanite term. Its chief members were Joab, the commander of the native Israelite troops; Benaiah, the commander of the foreign mercenary forces (mainly Cretans and Philistines); two chief priests, Zadok and Abiathar; a royal herald, who announced the king's commands; and a secretary of state, who recorded the commands and other matters.

The internal history of David's kingdom is recorded in a number of chapters in 2 Samuel and the opening chapters of 1 Kings. These form a distinct section that may have been written by someone who was alive during David's reign. Its tone is markedly secular: God does not intervene directly in events and is not often mentioned, and the divine purpose is seen to operate through the interplay of human ambitions and motives. This section has been called "the court history of David" and "a narrative of the succession to the throne," and although it is unlikely that the narrative is as coherent and unified as is often supposed, both titles can claim some justification. It consists of tales of rivalry at court within the royal family, which is exacerbated by that invariable feature of ancient Near Eastern monarchy, the harem. A king and his several wives produced a large crop of ambitious siblings and half-siblings, all juggling for

Opposition to David's rule

Later ages idealized the reign of King David, but the Bible shows that in fact it was subject to severe problems and tensions. David never succeeded in reconciling the supporters of Saul to his rule. One very serious rebellion was instigated by Sheba, from Saul's tribe of Benjamin, who roused the northern tribes with a war cry: "We have no portion in David!" (2 Sam. 20.1). Sheba was eventually betrayed by the citizens of Abel (2 Sam. 20.22).

The Death of Absalom, *a 19th-century drawing of part of the floor of the 13th-century cathedral of Siena, Italy.*

Sheba's revolt followed the gravest crisis of the reign, the rebellion of Absalom, David's own son, and, after the death of Amnon (see main text, opposite), his expected heir. Absalom is a typical folk hero, renowned for his beauty and fabulously long hair (2 Sam. 25–26). By this time, David was growing old and losing his tight grip on affairs, and Absalom won the people's affection by promising speedy redress of their grievances (2 Sam. 15.2–6). He proclaimed himself king at Hebron, cutting off David from his traditional heartland. His uprising was serious, because he was followed not only by the southern Israelite tribes but also by the adherents of Saul's dynasty. The revolt drove David from his capital for a time, and he prevailed only with the support of mercenaries and other foreigners.

Before the final battle, David ordered his generals to "deal gently" with his son. But the commander-in-chief, Joab, felt no such fatherly indulgence, and killed Absalom when he was caught by his hair in a tree (2 Sam. 18.9–15). The old king was grief stricken and uttered a hauntingly brief, poetic lament: "O my son Absalom, my son, my son Absalom! Would I had died instead of you, O Absalom, my son, my son!" (2 Sam. 18.33).

David with the prophet Nathan (left) and kneeling in repentance (right), from a Byzantine Psalter of ca. 950 CE.

power among themselves. As an account of the succession, the narrative shows how those of David's sons who might have succeeded to the throne are gradually eliminated, until only Solomon remains.

The first story in this section tells of David's adultery with Bathsheba, the wife of Uriah, a Hittite officer in the royal army (2 Sam. 11.2–12.25). To get Uriah out of the way, David orders Joab to place him "in the forefront of the hardest fighting" during a siege. Uriah is killed and David marries Bathsheba, who bears a son. But the king's actions anger God, and the prophet Nathan succeeds, by means of a parable, in getting David to condemn himself (2 Sam. 12.1–6). David acknowledges his sin, but Nathan says that although God will not strike David down, his house will be troubled and he will lose the child born to Bathsheba. The child dies, and the story ends with the birth of a second child, Solomon, who is assured of Yahweh's love. David's despotic behavior in taking Bathsheba for himself and arranging her husband's death would not have been unusual for a Near Eastern monarch. But it is sharply contrasted with the condemnation of his actions by Nathan, who speaks for older and simpler Israelite values.

The story that follows, told in 2 Samuel 13, centers on Amnon, David's eldest son. Abetted in a typical harem intrigue by his "very crafty" cousin Jonadab, Amnon lured his half-sister Tamar to his bedside and raped her. In revenge, Tamar's brother, Absalom, killed Amnon, thereby removing the man who would normally be expected to succeed to the throne. Absalom's banishment for his deed provoked a resentment in him which threatened David's throne but ultimately brought about his own downfall (see box on opposite page).

THE ACCESSION OF SOLOMON

When it became clear that David's powers were waning, a struggle to succeed him broke out between two parties, each with clearly defined interests and backing one of David's sons (1 Kings 1–2). The two factions embodied the tension that existed between the population of Judah and the royal city-state of Jerusalem, and which persisted until the fall of the kingdom.

On one side was a group with a background in Judah. Its candidate was Adonijah, the king's eldest surviving son, who was born during David's time in Hebron (2 Sam. 3.4). Behind him stood Joab, David's military commander, and the priest Abiathar, the sole survivor of Saul's massacre of the priesthood at Nob (see p.80), who took refuge with David during his years in the wilderness. The opposing faction was based in Jerusalem and backed Solomon, the son of David and the Jerusalemite Bathsheba (see main text). Behind Solomon were figures who had become prominent only after David established himself in Jerusalem: the priest Zadok, the prophet Nathan, and Benaiah, the captain of the foreign palace guard.

After the death of David (1 Kings 2.10) Solomon prevailed, and he took rapid and ruthless steps to dispose of the opposition. Adonijah was executed on a trivial pretext. (1 Kings 2.13–25). It would have been politically dangerous to kill a priest, so Abiathar was simply deposed and banished (1 Kings 2.27). Joab sought sanctuary by seizing the horns of the altar (see below) in "the tent of the Lord" but was slain by Benaiah (1 Kings 2.34), who also killed Shimei, a leading figure from Saul's tribe of Benjamin (1 Kings 2.46).

A horned altar of ca. 900 BCE from Megiddo (compare 1 Kings 1.50; 2.28).

SOLOMON

After succeeding to the throne (see p.85), Solomon did not embark on any expansionist campaigns, and his reign (ca. 961–922 BCE) was recalled as a time of peace and prosperity, when "Judah and Israel lived in safety, from Dan to Beer-sheba" (1 Kings 4.25). This description was used in later prophetic books to describe the perfection of the future Messianic age. Above all, Solomon was remembered as the embodiment of wisdom. In Israelite thought, wisdom was a practical quality, something that all people needed in order to evaluate the circumstances of their existence and to make a success of life (see pp.126–7). As the person responsible for the welfare of the nation, the king required wisdom more that anyone else. In this respect, Solomon was the ideal ruler (see box, below).

At the beginning of his reign, Solomon visited Gibeon, described as "the principal high place" (1 Kings 3.4): it had not yet been replaced by Jerusalem as the chief shrine. At Gibeon the king had a dream in which God appeared to him. Solomon's dream closely parallels the experience of the pharaoh Thutmose IV, to whom the Sphinx of Giza appeared in a dream. The Sphinx told Thutmose of his divine election in childhood and of the legitimacy of his succession to the throne, and on waking, the

Solomon's writings

King Solomon's reputation for wisdom was so great that several later writings of a type known as "Wisdom" (see pp.126–7) came to be ascribed to him. The Wisdom of Solomon, one of the apocryphal books of the Bible, claims to be by him, and several sections of the book of Proverbs (see p.126). Solomon is said to have propounded some three thousand proverbs dealing with trees and animals (1 Kings 2.32). The Hebrew for "proverb" is *mashal*, from a verb which means "to be like" or "to compare." Solomon's "proverbs" were therefore similes and metaphors, pithy sayings (as seen in many examples in the book of Proverbs) in which an observation taken from the natural or animal world served to illuminate human behavior. To see the connection was evidence of a person's wisdom, and the ability to compose such proverbs was important for a monarch, because they were regularly used in diplomatic exchanges. The common stock of ancient Near Eastern wisdom consisted of such maxims. Since this wisdom was largely secular in character, it could be appropriated and adapted by different nations. This process was facilitated by the wide international contacts that were initiated by Solomon.

The traditions of Solomon's wealth and his love of women no doubt explain why the author of Ecclesiastes (see p.126–7) assumed his persona, and why Solomon was credited with the highly erotic Song of Songs (see p.131). The Bible claims that "his songs numbered a thousand and five" (1 Kings 4.33). Similarly, because of the tradition that Solomon wrote poetry, two of the Psalms (see p.121) were attributed to him.

Solomon Dictating the Proverbs, *from the French* Bible Historiale *of 1357.*

The Queen of Sheba

The Queen of Sheba Arriving at the Court of King Solomon *by the Master of the Cassoni. Italian, late 15th century.*

Another story praising Solomon's wisdom is the famous episode of the visit of the Queen of Sheba (1 Kings 10.1–13). Ceremonial visits between royal houses, accompanied by the exchange of rich presents, were a regular feature of the ancient Near Eastern political scene. According to the Bible, the Queen had heard of Solomon's renowned wisdom and comes "to test him with hard questions"(1 Kings 10.1). The ability of one sovereign to demonstrate his or her wisdom and knowledge to another in a riddle contest was a valuable diplomatic tool. A similar example occurs in the story of Samson (Judges 14).

The stone head of a woman from Saba (Sheba), 1st millennium BCE.

Many fantastic and erotic features came to be attached to the story of the meeting of the two rulers in the Jewish, Islamic, and Ethiopic traditions, but the biblical account probably reflects sober political realities. The Queen's state was in southern Arabia, in the area of present-day Yemen. The countries in this region had a virtual monopoly of the immensely valuable spice trade in myrrh and frankincense, a vast quantity of which the Queen of Sheba presented to Solomon. Solomon had begun to break into the trade there by sending a fleet to the Red Sea, and an agreement to regulate relations would have been to the benefit of both parties. Solomon's empire embraced the north-south "King's Highway" in Transjordan and the east-west route across southern Palestine to Gaza, two arteries by which Arabian luxury goods were exported to consumer nations. The encounter between the monarchs may well have led to an agreement under which the Queen gave Solomon's fleet access to the ports she controlled, while Solomon guaranteed the security of the trade routes through his territory (see also p.89).

pharaoh offered sacrifices. All these elements are present in the story of Solomon's dream (1 Kings 3.5–15). The Bible's account is distinguished by the king's request for wisdom and Yahweh's promise to give him "a wise and discerning mind," which will bring him greater "riches and honor" than any monarch of his day.

Immediately afterward there is a very different story, a folktale to illustrate how the king's wisdom operated in the administration of justice. Two prostitutes came before Solomon with a baby boy, each claiming to be the child's mother. Solomon was faced with the two conflicting claims and no hard evidence, so he asked for a sword and proposed simply to cut the baby in two and "give half to the one, and half to the other" (1 Kings 3.25). One of the women pleaded for the boy's life, begging Solomon to give him to her rival rather than to kill him. The other woman said spitefully that the child "shall be neither mine nor yours" and told the king to go ahead and cut it in half. Through his understanding of psychology, Solomon saw in the first woman the compassion of the real mother and gave her the child. When the Israelites learned of the decision, they recognized that their new king truly possessed "the wisdom of God."

SOLOMON'S REIGN

Phoenician warships depicted on an Assyrian relief of the 7th century BCE. The alliance with the Phoenicians, the greatest maritime power of the region, was central to Solomon's foreign policy.

The kingdom under Solomon increasingly assumed the form and manner of a despotic ancient Near Eastern state. Striking evidence for this is the institution of the levy or *corvée*, by which the population could be drafted to labor on public works, as the Israelites had been forced to do in Egypt. Ugaritic texts reveal that this was also a Canaanite practice. One passage seems to indicate that forced labor only applied to the peoples that had been incorporated into David's kingdom (1 Chron. 22–1), but it is clear that native Israelites were also involved (compare 1 Kings 12.4 and 12.18). The probable explanation is that non-Israelites had the status of permanent serfs of the state, while Israelites were only drafted for labor in exceptional circumstances, such as Solomon's extensive building program. David may perhaps have considered introducing the levy, but it was in his son's reign that it became central to the administration of the state.

Although he did not wage any aggressive wars, Solomon was concerned to protect his realm from attack. There is mention of several cities where he carried out construction work and stationed garrisons of chariots and cavalry (1 Kings 9.15–19). Their positions are significant: Hazor protected the extreme northern boundary and Megiddo controlled a strategic pass from the coast to the central highlands. Gezer, which had been destroyed by an Egyptian pharaoh and then given to his daughter as a dowry on her marriage to Solomon, was rebuilt as a base from which to dominate the area of Philistine occupation. Excavations at all these sites have revealed extensive building during the Solomonic period.

The ruins of the Phoenician city of Byblos (biblical Gebal, modern Jebail, northern Lebanon). Solomon's extensive diplomatic and commercial connections enabled him to receive assistance from the inhabitants of Byblos, the Giblites, with the building of the Temple in Jerusalem (see pp.90–91).

The king also built a number of store-cities in connection with a wholesale reorganization of the state recounted in 1 Kings 4. His territory was divided into twelve administrative districts, each with a regional governor at its head. The division was to some extent based on the old tribal areas, but was not rigidly bound by them. The object of the reorganization was that each district should supply food to the royal household for one month in turn, so the twelve divisions were more probably determined by the twelve months of the year. Old Canaanite cities were also incorporated into the new regional districts. The monthly provisions, which each governor was responsible for organizing, were collected in the store-cities before being sent on to the court at Jerusalem.

Under Solomon, the Israelite empire became an important power on the international scene. The king was able to use his new prestige to extend his influence with neighboring states by diplomatic means. Solomon concluded a treaty with King Hiram of Tyre, continuing the alliance begun by David (1 Kings 5.12). One benefit of this treaty was that Hiram provided Solomon with timber from the famous cedars of Lebanon for his building operations, together with skilled craftsmen to fell the trees and transport them by sea.

Solomon's large harem is condemned in the Bible, which claims that it led the king to accommodate the deities worshiped by his foreign wives (1 Kings 11.1–8). The size of the harem is probably exaggerated to present the monarch, in typical Near Eastern fashion, as a paragon of wealth and virility. But the royal harem was an important part of diplomacy, because treaties between states were cemented by intermarriage of the respective royal families. Solomon almost certainly followed this practice: the Biblical narrative gives especial emphasis to his marriage with a daughter of the king of Egypt, who had her own palace in Jerusalem (1 Kings 7.8 and 9.16). In Solomon's day Egypt was in decline, but it was still recognized as a famous country. The fact that Pharaoh himself thought it worthwhile to seek Israel's friendship shows the considerable international prestige that Solomon's kingdom had acquired.

Solomon's trading activities

Most of the biblical references to Solomon's immense wealth possess a legendary or folktale quality, but they doubtless also have some historical basis. Some of the king's wealth would have come from the regular tribute paid by foreign parts of his kingdom, but the main source of the finance which paid for Solomon's many building projects was trade. Several of the region's most important trade routes crossed Solomon's empire, and he was able to exploit his control of the caravans that used them. In particular, he tapped into the lucrative commerce in spices and precious metals between southern Arabia and the north (see box on p.87).

Solomon's alliance with the Phoenician king Hiram of Tyre (see main text, above) proved vital, because the Israelites had no maritime experience. Solomon built a fleet to operate in the Red Sea, and Hiram, the head of a great sea-going power, supplied the Israelites with skilled sailors and ships of their own. The royal fleet had its headquarters at the port of Ezion-geber near Eloth (Eilat) at the head of the Red Sea, although its exact site is uncertain.

Trading activities on this scale led to the emergence of a new phenomenon in Israel, a class of merchant traders. A note in the Bible records how these merchants imported horses from Egypt and Kue (in present-day Turkey) for Solomon's army and sold horses on to the rulers of states north of Israel (1 Kings 10.28–29).

A cult tablet from southern Arabia, with an inscription in ancient Sabean script (1st millennium BCE).

THE TEMPLE

The most important of Solomon's building projects was the Temple in Jerusalem, which was to play a central role in the life and worship of the nation. In constructing it he was fulfilling one of the recognized duties of the ancient Near Eastern king, who was expected to build as splendid an edifice as possible in his capital city as a dwelling place for the national deity. The Jerusalem Temple was an innovation: when King David had planned to build a home for Yahweh, the prophet Nathan had forbidden him on the grounds that until then God had lived only in a moveable tent or tabernacle (2 Sam. 7.6 and 1 Chron. 17.4).

Since the Israelites had no experience of temple building, Solomon turned to King Hiram of Tyre, who supplied him with materials and labor, and in particular to one craftsman, also called Hiram, who made the Temple's cult furnishings (see box, below). The Temple was thus essentially a Phoenician or Canaanite construction: its tripartite ground plan is similar to the shrine of Tell Tainat in Syria and a temple at Hazor. The Temple was built to the north of the original city of David (see map on p.190), on an ancient Jebusite holy place traditionally connected with a threshing floor where David had built an altar (2 Sam. 24.18). This became the place for the open-air altar of the Temple. The Temple's walls, lined with Lebanon cedar, were adorned with cherubim, palms, and open flowers (perhaps lotuses), common religious symbols in the ancient Near East. Outside the vestibule were two free-standing pillars, a feature of Phoenician temples. The significance of these pillars is disputed but their names, Jachin ("Yahweh establishes") and Boaz ("In him [it] is strength"), suggest that they symbolized God's presence. They may have corresponded to the guardian figures of deities that stood at the entrance to Mesopotamian sanctuaries and palaces.

The Temple furnishings

The furnishings of Solomon's Temple had a cosmic symbolism. There was a large metal basin, called the "sea" (1 Kings 7.44), supported by twelve bull statues. One biblical text states that the basin was used for the priests' ablutions, but since it was over seven feet (2.2m) high, this is unlikely. The name "sea" parallels the Mesopotamian *apsu*, the name both of the subterranean ocean and of a temple basin that symbolized it. Bulls were a frequent symbol of fertility, and were associated with the Canaanite storm god Baal (see p.72).

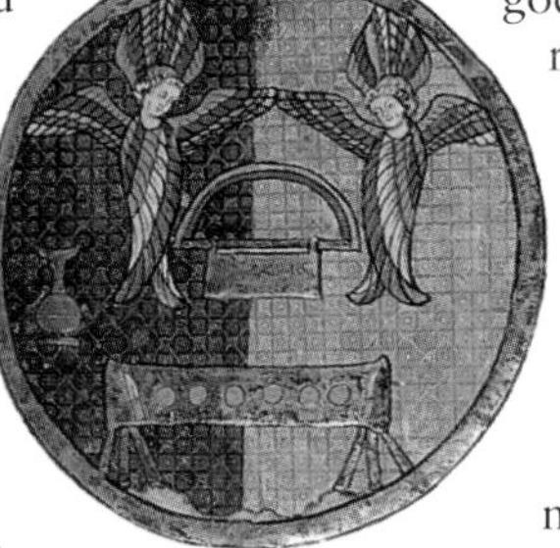

The ark of the Covenant and two angels, from a 13th-century Hebrew manuscript.

In the Temple there were also movable lavers (basins for ritual washing), decorated with wreaths, bulls, lions, and cherubim. The main sacrificial altar was described by a term indicating that it was considered to represent the mountain of the gods and the world center, with its base resting on "the bosom of the earth."

At the Temple dedication ceremony, the ark of the Covenant was placed in the central sanctuary, the Holy of Holies. It ceased to be a portable object and Yahweh, as the opening words of Solomon's dedicatory prayer express, was believed from this time to reside permanently in the most holy place: "The Lord has said that he would dwell in thick darkness. I have built you an exalted house, a place for you to dwell in forever." (1 Kings 8.12– 13). This belief led to the doctrine that Jerusalem and its Temple were inviolable. The ark was viewed as a throne on which God sat as an invisible deity. It is flanked by two large figures of cherubim, recalling a similar feature on the thrones of Phoenician rulers.

The building of the Temple (see 1 Kings 6) was accompanied by the construction nearby of Solomon's own (considerably larger) palace. This symbolized the Temple's role as both a national shrine and a royal chapel. The king was the head of the cult of Yahweh and led all important acts of worship. Solomon himself officiated at the consecration of the Temple, offering sacrifices and the dedicatory prayer.

In Canaanite religion the home of the gods was a mountain. The psalms used in worship at the Temple show that much of the mythology and symbolism of the divine mountain was transferred to Mount Zion, the dwelling place of Yahweh (see, for example, P. 48).

Later Temples of Jerusalem

Solomon's Temple was destroyed in the Babylonian sack of Jerusalem (587/6 BCE; see pp.110–11), but it was eventually restored by the returning exiles just over half a century later. The Second Temple seems to have been a more modest affair than the first and its significance was somewhat different (see pp.120–21). It was desecrated and despoiled under the Seleucid Greek empire (see pp.138–9) and later by the Romans (see p.142), and ca. 20 BCE Herod the Great, the client-king of Roman Judea (see p.142), embarked on the construction of an entirely new building. Descriptions of Herod's Temple, the bulk of which was completed in ten years, show that it reproduced the dimensions of Solomon's building. It followed contemporary Greco-Roman architectural fashions and its splendor was intended as a visual statement of the importance of Herod's kingdom to his Roman overlords. Herod was seen by many of his subjects as a usurper, and his lavish new Temple was an attempt to win them over.

There is considerable evidence, however, that Herod's motives were distrusted by the more devout and traditional elements of the population and that the new Temple did not win their affection and devotion. In any case, this last and grandest of the Jerusalem Temples had a comparatively short life: finally completed only in 64 CE, it was burned when the Roman emperor Titus took Jerusalem during a Jewish revolt in 70 CE and never rebuilt. However, the Temple site remains a holy one. Part of the ruins of the Temple complex, the massive Western Wall (the so-called "Wailing Wall"), is an important place of worship for Jews. On the site of the Temple's central sanctuary stands the Dome of the Rock, one of the holiest sites in Islam. It was built 685–691 CE to mark the spot where Abraham was said to have prepared to sacrifice Isaac (see p.43), and from where the Prophet Muhammad ascended to heaven.

Jews praying at the Western Wall, part of the few surviving remains of the Temple of Herod. The massive masonry is typical of Herodian architecture.

An artist's reconstruction of the southern façade of Herod's Temple (see also photograph on p.163).

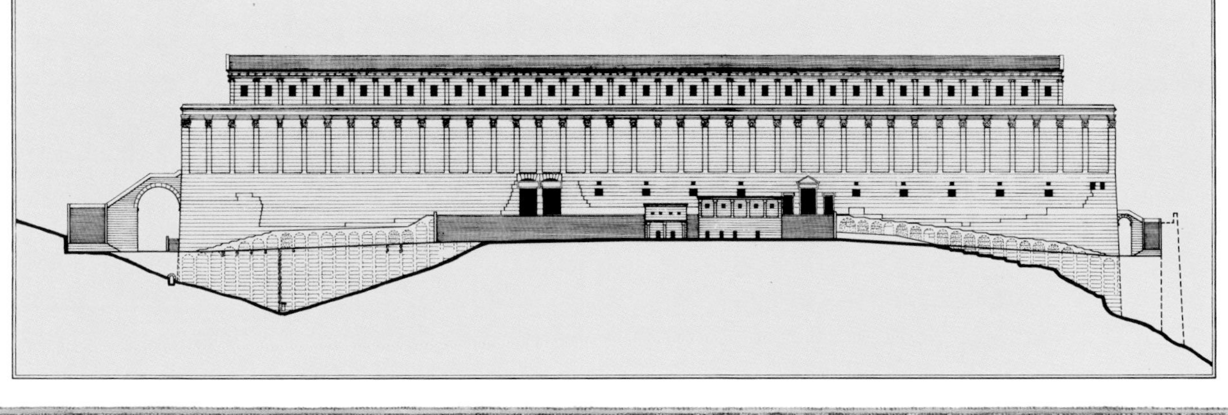

WEAKNESSES IN THE STATE

The Bible points to weaknesses and tensions in Solomon's kingdom which, while probably limited in their effects during his own lifetime, were to have serious consequences for the future. There was trouble in two dependent foreign territories, Edom and Aram (see map on p.83). The unrest there is represented as a divine punishment for Solomon's worship of foreign deities, as recounted in 1 Kings 11, and their significance may be exaggerated. One opponent, the Edomite prince Hadad, may have been no more than a minor irritant. Rezon, the Aramean rebel leader, certainly established an independent kingdom in Damascus, but this may have been in the period of Israelite weakness following Solomon's death. It has sometimes been claimed that Solomon faced financial difficulties, which is why he ceded twenty settlements in Galilee to his ally Hiram of Tyre for a large sum (1 Kings 9.11). But the story actually emphasizes Solomon's ability to bargain: Hiram is disappointed with his new property, and calls it "Cabul," which may mean "a land good for nothing."

More serious was the internal discontent, caused particularly by Solomon's institution of forced labor, the levy. Thirty thousand drafted male citizens worked abroad in shifts, spending one month in the Lebanon and two at home (1 Kings 5.14). The free Israelite farmers would have resented this state servitude, which must have brought serious disruption to farming. Thousands more were forced to work at quarrying and on Solomon's numerous building projects. Another source of internal pressure was the old tension between the north (Israel) and the south (Judah).

Egyptian influence on Israelite affairs

Egypt played an important role in the events of King Solomon's reign and immediately afterward. The country had long claimed hegemony over Palestine and, although an unnamed pharaoh of the weak twenty-first dynasty had thought it prudent to make an alliance with Solomon, the basic Egyptian policy did not change. The Egyptians were not averse to encouraging movements which might cripple Solomon's power. His opponents, Hadad and Jeroboam, received shelter and support from the pharaohs, and Hadad was even given the Egyptian king's sister-in-law in marriage.

Ruins of Megiddo, an important city under Solomon and his successors. It was captured by Sheshonk in the reign of Jeroboam I of Israel.

The Bible records that Jeroboam was welcomed by "King Shishak of Egypt" (1 Kings 11.40). This was the pharaoh Sheshonk I, who came to the throne ca. 945 BCE as the founder of a new Libyan dynasty. He was a strong ruler who revived his country's fortunes. In the reign of Rehoboam, Solomon's successor, Sheshonk sent an expedition into Palestine and reached Jerusalem. His campaign was on a considerable scale: a list of the cities he captured, carved on a wall of the temple of the god Amun-Re at Karnak, includes over one hundred fifty towns in Palestine. The list does not include Jerusalem, and the Bible's account, placed in the fifth year of Rehoboam's reign (see p.95), mentions only an attack on the city. Sheshonk was probably bought off by Rehoboam, who paid him tribute in the form of the treasures of the Temple and the royal palace. Further evidence of the extent of Sheshonk's raid are his victory column at Megiddo and a seated statue of the pharaoh found at Byblos in the Lebanon.

The two issues came together in the revolt of Jeroboam, a royal official from the northern tribe of Ephraim, who had first-hand experience of the effects of the levy: he was in charge of forced labor in his own tribal district. According to the Bible, Jeroboam received a divine promise from the prophet Ahijah that he would rule the ten northern tribes of Israel. The remaining two were the traditionally related tribes of Judah and Benjamin, which God reserved for David's dynasty (1 Kings 11.29–32). Ahijah's support for Jeroboam was significant, because Ahijah was associated with Shiloh, the original home of the national symbol, the ark (see p.76 and p.90): he represented the older Israelite religious tradition, which had been superseded by the new establishment centered on the Temple in Jerusalem.

Jeroboam's rebellion did not topple Solomon, because, as Ahijah said, God had promised that Solomon would be king "all the days of his life, for the sake of my servant David" (1 Kings 11.34). Jeroboam fled to Egypt to bide his time at the court of the pharaoh (see box, below).

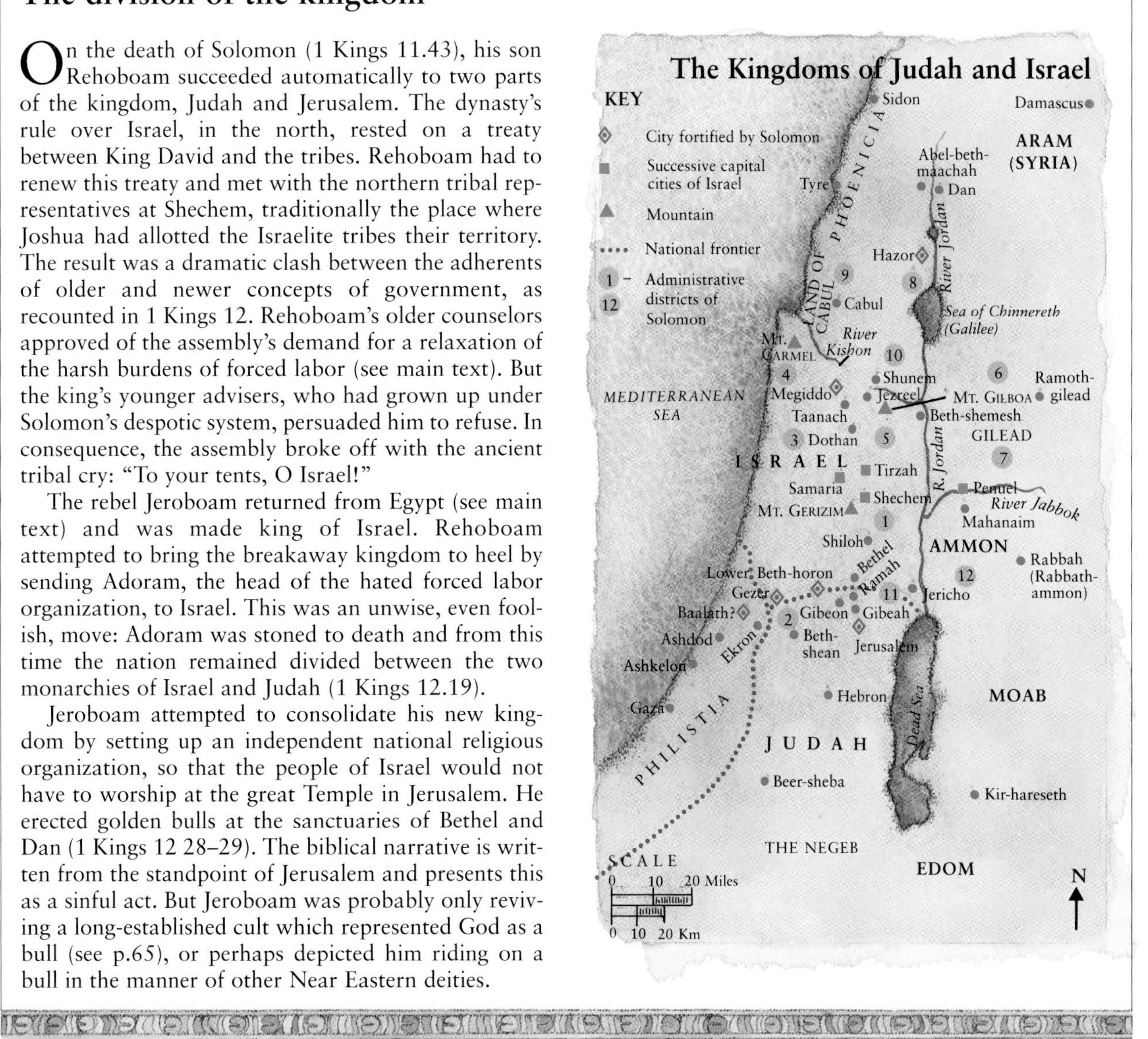

The division of the kingdom

On the death of Solomon (1 Kings 11.43), his son Rehoboam succeeded automatically to two parts of the kingdom, Judah and Jerusalem. The dynasty's rule over Israel, in the north, rested on a treaty between King David and the tribes. Rehoboam had to renew this treaty and met with the northern tribal representatives at Shechem, traditionally the place where Joshua had allotted the Israelite tribes their territory. The result was a dramatic clash between the adherents of older and newer concepts of government, as recounted in 1 Kings 12. Rehoboam's older counselors approved of the assembly's demand for a relaxation of the harsh burdens of forced labor (see main text). But the king's younger advisers, who had grown up under Solomon's despotic system, persuaded him to refuse. In consequence, the assembly broke off with the ancient tribal cry: "To your tents, O Israel!"

The rebel Jeroboam returned from Egypt (see main text) and was made king of Israel. Rehoboam attempted to bring the breakaway kingdom to heel by sending Adoram, the head of the hated forced labor organization, to Israel. This was an unwise, even foolish, move: Adoram was stoned to death and from this time the nation remained divided between the two monarchies of Israel and Judah (1 Kings 12.19).

Jeroboam attempted to consolidate his new kingdom by setting up an independent national religious organization, so that the people of Israel would not have to worship at the great Temple in Jerusalem. He erected golden bulls at the sanctuaries of Bethel and Dan (1 Kings 12 28–29). The biblical narrative is written from the standpoint of Jerusalem and presents this as a sinful act. But Jeroboam was probably only reviving a long-established cult which represented God as a bull (see p.65), or perhaps depicted him riding on a bull in the manner of other Near Eastern deities.

VI PROPHECY

Prophets and prophecy in the Israelite tradition emerged clearly on the biblical stage with Samuel's appointment of Saul as the first king of Israel (see p.78). Since prophets can arise in almost any society, they may well have been known to the Israelites before then. With the monarchy, Israelite society became more organized and prophets emerged as members of the recognized institutions, a process probably hastened by the existence of bands of prophets in Canaanite religion. It is not easy to define what makes a prophet or what distinguishes him or her clearly from other divinely-inspired figures, such as the judge or the priest. The prophets were individuals, not all of whom functioned in the same way, and the Bible provides considerable evidence of mutual hostility among different prophetic groups.

A Phoenician opal seal with various symbols representing the name YZBL, "Jezebel," the 9th-century queen of Israel.

THE PROPHET'S ROLE

The various terms used to describe prophets in the Bible indicate how they functioned to fulfill social needs, although they are to some degree stereotyped, to make it easier to recognize them. The general term "prophet" is a translation of the Hebrew word *nabi*, which probably means a speaker or spokesperson. Hence prophets were not primarily people who foretold the future, although they certainly had that power and exercised it, but proclaimers of the divine will, who spoke in God's name, with the regular introduction: "Thus Yahweh has said." Their essential function was to keep the nation in a correct relationship with God, and to strengthen that relationship by their words and actions. The verb derived from *nabi* means "to act as a prophet," and often denotes the uncontrolled behavior usually described as "ecstasy." Evidence of being possessed by the deity in some way, by Yahweh's "spirit" or "hand," was an essential characteristic of prophecy. It gave prophets an inner compulsion to deliver the divine message, even against their own will. Such behavior was not confined to the prophets of Yahweh: the eleventh-century BCE Egyptian story of Wen-Amon relates how a young man at the Phoenician city of Byblos was seized with prophetic frenzy during a sacrificial ceremony and delivered a message from the deity who had taken possession of him.

Prophetic figures were also called "seers," those who had the gift of second sight or who could discern omens from natural objects or phenomena. Another frequent appellation was "man of God," one who was in touch with the supernatural realm. Sometimes the prophet was said to have access to the court of heavenly counselors surrounding Yahweh. The prophet was thus a "messenger," and prophetic oracles often employed a formula by which an emissary spoke in the name of his or her superior.

Prophets functioned in a variety of social contexts. An individual prophet could serve the simple needs of a particular locality, usually centered on a shrine, as with Samuel at Ramah. But, particularly in the period of the early monarchy, there were also organized guilds of prophets, who moved about the country from sanctuary to sanctuary. Perhaps most significant were the court prophets, who were a common feature of ancient Near Eastern monarchies. They offered the king regular divine guidance in state affairs. Individual prophets are mentioned, such as Nathan and Gad at David's court, but the Israelite king maintained a large prophetic establishment, four hundred being mentioned in 1 Kings 22.6. This number closely parallels the four hundred fifty prophets of the Phoenician (Canaanite) god Baal in Jezebel's household (1 Kings 18.22), suggesting that the idea of the prophetic guild may have been borrowed by Israel from the practice of Canaanite monarchy. The Bible gives a vivid picture of the Baal prophets, on the occasion of their encounter with Elijah on Mount Carmel (1 Kings 18). In order to gain their god's attention, "they cried aloud and, as was their custom, they cut themselves with swords and lances until the blood gushed out of them. As midday passed, they raved on ... but there was no voice, no answer, no response." (1 Kings 18.28).

The divided kingdom after Solomon

After Solomon died, the Israelite monarchy was divided into the kingdoms of Israel, in the north, and Judah, in the south. There was more or less continual war between them, a situation that the new Aramean dynasty in Damascus was able to exploit under Ben-Hadad I (ca. 880–842 BCE). While Judah remained comparatively stable, continual disturbances wracked the north. Jeroboam I (922–901 BCE) was succeeded by his son, who was deposed by a military coup, three more of which followed. The Bible explains this instability as God's judgment for the sinfulness of Jeroboam (1 Kings 12.28–33 and 13.33–34), which his successors are seen as perpetuating. Two prophetic legends, full of vivid detail, recount how two prophets pronounced judgment on Jeroboam's house and ultimately on the northern kingdom itself (1 Kings 13).

The chaotic situation in the north was brought to an end with the accession of the final usurper, Omri (876–869 BCE), who was able to establish a powerful state and a stable dynasty centered on a new capital, Samaria. He checked the advance of Damascus and reconquered Moab, which had been lost during the years of Israel's weakness. Omri receives only a brief mention in the Bible, because the biblical editor regarded him as even more wicked than his predecessors. However, the great position he attained is indicated by the fact that Assyrian inscriptions refer to the kingdom of Israel as "the house of Omri," even long after Omri's dynasty had disappeared.

Rulers of Israel and Judah

Dates cannot be assigned to the reigns of the Israelite monarchs with absolute precision. This list follows the chronology of W. F. Albright, which is generally accepted as one of the most accurate. All dates are approximate and BCE.

THE UNITED KINGDOM

Saul	1020–1000
David	1000–961
Solomon	961–922

RULERS OF JUDAH

Rehoboam	922–915
Abijam	915–913
Asa	913–873
Jehoshaphat	873–849
J(eh)oram	849–842
Ahaziah	842
Athaliah	842–837
J(eh)oash	837–800
Amaziah	800–783
Uzziah	783–742
Jotham	742–735
Ahaz	735–715
Hezekiah	715–687
Manasseh	687–642
Amon	642–640
Josiah	640–609
Jehoahaz	609
Jehoiakim	609–598
Jehoiachin	598–597
Zedekiah	597–587/6

RULERS OF ISRAEL

Jeroboam	922–901
Nadab	901–900
Baasha	900–877
Elah	877–876
Zimri	876
Omri	876–869
Ahab	869–850
Ahaziah	850–849
J(eh)oram	849–842
Jehu	842–815
Jehoahaz	815–801
J(eh)oash	801–786
Jeroboam II	786–746
Zechariah	746–745
Shallum	745
Menahem	745–738
Pekahiah	738–737
Pekah	737–732
Hoshea	732–721

PROPHECY IN THE NEAR EAST

Some of the most interesting information about prophecy in the period of the two Israelite kingdoms comes from the West Semitic and Mesopotamian areas. A stele of King Zakir (ca. 780 BCE) of the Aramean state of Hamath records how he received an oracle of victory from his god through prophets who are called "seers" and "messengers." Letters from the eighteenth-century BCE city of Mari on the Euphrates contain several examples of oracles given to King Zimrilim by diviners whose words and actions resemble those of biblical prophets. These oracles include revelations through dreams, visions, and trances, and the employment of the messenger formula. Parallels have also been discerned in a number of texts containing oracles to various kings, from the period of the Assyrian empire (ca. 680–627 BCE).

Elijah Visited by an Angel *(1 Kings 19.5), from an altarpiece in Louvain, Belgium, by Dirck Bouts (ca. 1420–75).*

ELIJAH

The prophet Elijah was prominent in the reign of King Ahab of the northern kingdom of Israel (869–850 BCE). The northern Kingdom was much more subject to Canaanite (Pheonician) influence than Judah, the southern Israelite kingdom. The prophets of Israel therefore shaped the story of Elijah's career so as to present him as the ideal champion of Yahweh against the worship of the Canaanite deity Baal. The Elijah tradition greatly exaggerated his historical role, such as his reputed massacre of the prophets of Baal at Mount Carmel (1 Kings 18.40), which is more likely to have occurred under King Jehu (842–815 BCE). Elijah was credited with miraculous powers: according to the Bible, when he was sheltered by a widow of Zarephath, he replenished her supply of meal and oil, and raised her son from the dead (1 Kings 17). This story was probably transferred from Elijah's successor, Elisha (see 2 Kings 4), who was much more firmly rooted in the events of his day (see pp.98–9).

Elijah is presented as a new Moses. His confrontation with the prophets of Baal on Mount Carmel, where he challenges their beliefs (1 Kings 18), recalls the incident in Exodus 32 where the Israelites construct and worship a golden calf, and are punished by plague for their idolatry. But the Moses tradition is most obvious in the account of Elijah at Mount Horeb, the site of Moses' revelation. There, Elijah heard the voice of God,

Ahab

The Bible presents King Ahab, the son and successor of Omri (see p.95), in a highly unfavorable light. In the biblical view, he was essentially a pawn of his wife Jezebel, serving and worshiping her god Baal, and doing "more to provoke the anger of the Lord, the God of Israel, than had all the kings of Israel before him" (1 Kings 16.33). The heated antagonism between him and the prophet Elijah is developed in 1 Kings 17–19. Several narratives which probably refer to later kings have been transferred to him to derogatory effect. For example, the story of the battle of Ramoth-gilead (1 Kings 22) tells of the death of a monarch referred to only as "the king of Israel." This may be an inserted episode that was not originally about Ahab at all, whose name is not mentioned in the chapter until verses 39–40. Ahab is then said to have "slept with his ancestors," which in the Bible usually means that he died peacefully.

A more objective view of Ahab would show that he maintained a strong and stable government. He built cities and secured his state by renewing the Israelite alliance with the Phoenicians of Tyre (see p.89). He dominated the southern kingdom of Judah through the marriage of his daughter, Athaliah, to Jehoram, the son of the Judean king Jehoshaphat (873–849 BCE). Ahab's importance is strikingly shown in an inscription of Shalmaneser III, king of Assyria (ca. 859–824 BCE), the first Assyrian monument to include an Israelite king's name. Ahab allied with Israel's old foe, Damascus, against Shalmaneser, and the allies met the Assyrians at Qarqar in the Orontes Valley in 853 BCE. Although Shalmaneser claimed victory, his advance was checked. His inscription records that Ahab had two thousand chariots and ten thousand infantry. Ahab's contingent of chariots was the largest among the allied force, a clear indication of his wealth and power.

who told him to anoint Hazael and Jehu as new kings of Aram and Israel respectively, and a new prophet, Elisha, as his own successor (1 Kings 19).

Elijah seems to have spearheaded an attack on the worship of Baal, which was practiced by the Israelite king Ahab and his wife Jezebel, and to have initiated a new development in the worship of Yahweh. The ordeal on Mount Carmel shows that Baal is not just a usurper on Yahweh's territory, but that he is not a god at all: Yahweh alone is God. In Elijah's teaching, Yahweh acquires a universal aspect. He is the true lord of nature, a domain claimed by the Canaanite deities, and can control events in a foreign land. These insights are clearly expressed in the Mount Horeb episode, which also represents a genuine personal prophetic experience. At Horeb, Elijah had to recognize that his mission had failed and that Yahweh had decided to pass on his task to another.

Elijah's nemesis is Jezebel, who is depicted as a woman of great ability and determination. She exploited to the full the authority which her position allowed her, as in the episode of Naboth's vineyard (1 Kings 21), for which Elijah cursed her (21.23). It was accepted that foreign consorts worshiped their own deities, but Jezebel overstepped the mark by seeking to supplant Yahweh entirely with Baal, the god of her native Tyre. (This policy was also pursued in Judah by her daughter, Athaliah.) Her father Ethbaal, the Phoenician king of Tyre, was a priest of Astarte, and Jezebel introduced the worship of the goddess Asherah (1 Kings 18.19). Jezebel was accused of sorcery, and died a brutal death, pushed from her window by her own attendants during Jehu's coup d'état (see box, below), and, in accordance with Elijah's curse, devoured by dogs (2 Kings 9.30–37).

The Bible relates that before her death, Jezebel painted her face and looked out of the window in anticipation of Jehu's approach (2 Kings 9.30). Defiant in her religious beliefs to the end, she may have assumed the role of a goddess: "the woman at the window," possibly the goddess Astarte, is a motif of Assyrian art, as in the above ivory of ca. 800 BCE.

Palestine after Ahab

Ahab was succeeded in turn by his sons Ahaziah (850–849 BCE) and Jehoram of Israel (849–842 BCE), though their mother Jezebel (see main text) remained the power behind the throne. The policy of imposing the worship of Baal continued. Although Jehoram removed Ahab's sacred pillar of Baal in Samaria, the fact that it was later destroyed in Jehu's reformation may mean that he removed it only to replace it.

Jehu prostrated before the Assyrian king Shalmaneser III, from a monument known as the "Black Obelisk of Shalmaneser." It is the only known contemporary depiction of an Israelite ruler.

Omri's dynasty was finally overthrown by Jehu (842–815 BCE), a commander of King Jehoram's army. His brutal revolt, described in 2 Kings 9 and 10, took place in the context of an apparently unsuccessful attack led jointly by the kings of Israel and Judah on the Aramean city of Ramoth-gilead. Jehu was backed by the prophetic office of Elisha (see pp.98–9) and by the Rechabites, an Israelite group that stood out against the settled life of Canaan, abstaining from wine and agriculture and living in tents. But Jehu's usurpation seriously undermined the kingdom. He brutally extirpated the cult of Baal and murdered Jehoram of Israel, Jezebel, and all the royal family (Jehu may also have instigated the massacre of Baal prophets attributed in the Bible to Elijah; see main text). In doing so he broke Israel's alliances with the Phoenicians and Judeans, whose families had intermarried with that of Ahab. After the death of Ahab's and Jezebel's grandson, King Ahaziah of Judah, Ahaziah's mother Athaliah seized power. Athaliah, the widow of King Jehoram of Judah (849–842 BCE), imposed the same religious policies on Judah that her mother, Jezebel, had instigated in Israel.

As a consequence of his state's weakness, Jehu could not withstand the ever-increasing pressure of Assyria. He was forced to become the vassal of the Assyrian ruler Shalmaneser III (ca. 859–824 BCE).

The Ascent of Elijah *(2 Kings 11.1), a Russian painting of the 16th century. The prophet hands his mantle to Elisha as an angel carries him aloft in a chariot of fire.*

ELISHA

The Bible's stories about Elisha, the disciple and then successor of Elijah (see pp.96–7), form a narrative that is characterized by folklore and legendary motifs. Unlike Elijah, who was unswerving in his hostility toward the kings of Israel, Elisha at times advised the powerful figures of his day, in addition to helping ordinary people in distress. Also unlike Elijah, a solitary figure, Elisha acted in the context of groups of prophets who were probably responsible for shaping and recording his saga.

He is first mentioned in 1 Kings 19, when Elijah is commanded by Yahweh at Mount Horeb to anoint Elisha as his successor. In fact, Elisha was not anointed, but was chosen as a prophet while he was plowing a field, when Elijah approached him and threw his mantle over him (1 Kings 19.19). Elisha officially succeeded Elijah after accompanying him from Gilgal to Bethel, Jericho, and the Jordan, where they were separated by "a chariot of fire and horses of fire" and Elijah then "ascended in a whirlwind into heaven" (2 Kings 2.11). Elisha inherited the mantle of Elijah, representing the prophetic office, together with a double portion of Elijah's "spirit," his mysterious quasi-physical power. Elisha divided the waters of the Jordan with the mantle (2 Kings 2.14), as Elijah had done before him (2 Kings 2.8), reenacting the crossing of the Jordan by Joshua (see p.68). The "company of prophets" that had followed Elijah and Elisha from Jericho declared that "the spirit of Elijah rests on Elisha," and pledged themselves as his "servants" (2 Kings 2.15–16).

Many of the stories about Elisha in 2 Kings are essentially miracle tales, demonstrating the prophet's supernatural powers through magical acts. Thus, he purifies the spring at Jericho (2 Kings 2.19–22), brings a son to a Shunammite woman and resurrects him when he dies (2 Kings

The Arameans

The Arameans were a people who, in the last quarter of the second millennium BCE, spread out from the desert over the Fertile Crescent. In the Bible, the Hebrew patriarchs and matriarchs (see pp.36–51) are closely linked with the Arameans, who established a cluster of independent states west and north of the Euphrates River. The most important of these came to be Damascus, where King Rezon founded a new dynasty in Solomon's time (1 Kings 11.23–25). This began a three-hundred-year relationship, generally hostile but sometimes friendly, between Israel and Damascus.

In the disturbed conditions of the early years of Israel, Damascus, under Ben-hadad I, was able to dominate the northern kingdom. Under King Ahab of Israel, however, the Assyrian menace brought the two states into an alliance, until Jehu's revolt reversed Israel's policy (see p.97). At about the same time, Hazael of Aram, motivated, according to the Bible, by Elisha, usurped the throne of Damascus and was able to conquer all Israel's territory east of the Jordan. The Bible records warfare with Hazael's successor, Ben-hadad III, involving attacks on Samaria, which according to one legend was saved by Elisha.

But from this period onward, Damascus was weakened by repeated Assyrian assaults. As a result, Israel, under kings Jehoahaz (815–801 BCE) and Jehoash (801–786 BCE), recovered the initiative. The last scene of Elisha's life has the dying prophet ensuring victory over Damascus to Jehoash, albeit on a limited scale, by a vivid act of prophetic symbolism: Elisha commands the Israelite king to shoot an arrow eastward, representing "the arrow of victory over Aram," and then to strike the ground with his remaining arrows. But Joash strikes the ground only three times, ensuring him only three victories over Aram; had he struck the ground five or six times, he would have definitively ruined his rival (2 Kings 14–19). Damascus was finally overcome by Jeroboam II (786–746 BCE) and thereafter declined in importance, until it was finally brought to an end by Assyria in 732 BCE.

4.8–37), and makes an iron ax-head float (2 Kings 6.1–7). The heavenly chariot and horses of fire which carried Elijah into heaven almost certainly originally referred to Elisha, who was also known by the title "the chariots of Israel and its horsemen" (2 Kings 13.14). Elisha was seen as having the power to summon God's heavenly host in support of the nation: on one occasion, when the king of Aram surrounded Elisha's camp, Elisha was ringed by horses and chariots of fire that far outnumbered the Aramean forces (2 Kings 6.15–17).

These acts all occur in connection with prophetic groups, and it was in these circles that the Elisha saga took shape, when his disciples collected together various local traditions associated with him. Fundamental to the picture of Elisha is his relationship with the prophetic group called "the company of prophets," literally "the sons of the prophets" (2 Kings 2.5; 4.1; 4.38; 5.22; 6.1; 9.1). "The company of prophets" appears to be a technical expression for the followers of Elisha, who are described as his "servants," with him as their "master." In the Bible, almost all instances of the term are confined to the period ca. 860–840 BCE and to a relatively compact geographical area that contained the old Israelite sanctuaries of Gilgal, Bethel, and Jericho, where these prophets lived and operated just inside the northern kingdom of Israel. It seems to have been a movement organized in response to a specific historical situation, namely the religious policy of the dynasty of Omri that ruled the northern kingdom, and of Queen Jezebel in particular (see pp.96–7). "The company of prophets" ceased to exist when, with the overthrow of the dynasty by Jehu (see p.97), its work was done.

An ivory carving of a god, from the state bed of Hazael, ruler of the Aramean kingdom of Damascus, ca. 840 BCE (see box on opposite page).

Moab

Elisha was also involved in events concerning the kingdom of Moab, which had been conquered by King David (see map on p.83) but had regained its independence some time later. A contemporary account of events from the Moabite standpoint exists on the "Moabite Stone," a three-foot (1m) high basalt pillar of ca. 830 BCE celebrating the exploits of Mesha, king of Moab. It records that Omri of Israel had regained Moab, but that, after Omri's death, Mesha had thrown off the Israelite yoke, taken the offensive, and captured various Israelite cities in Transjordan. In language closely resembling the way Israel ascribed its victories to Yahweh, Mesha attributes his triumph to Moab's national god, Chemosh.

The inscription on the Moabite Stone is unclear as to whether Mesha's revolt occurred during the reign of Ahab or after his death, but the very first verse of 2 Kings states: "After the death of Ahab, Moab rebelled against Israel." Mesha is described as a sheep-breeder, who had to pay a huge annual tribute of lambs and wool to the king of Israel, but withheld it after Ahab's death (2 Kings 3.4–5). As a result, Jehoram of Israel (849–842 BCE), Ahab's son and successor, launched a campaign against him, but the expedition soon ran out of water for its men and animals, and Elisha was summoned to help. A musician was also summoned and, as he played, Yahweh's "hand" came upon Elisha, and, in God's name, promised both water and the defeat of Moab. Water duly arrived and the Israelites advanced on their target, but their conquest was frustrated when Mesha, seeing no other recourse, offered his eldest son and successor as a burnt offering on the city wall. The sacrifice threw the Israelites into terror (2 Kings 3.27) and they were forced to retreat. It is striking that Israel here implicitly recognizes the power of a foreign god, Moab's Chemosh.

The Moabite Stone, which was discovered in Transjordan in 1868.

ASSYRIA

King Ashurbanipal riding his chariot into battle, on a relief from the Assyrian royal palace at Nineveh, ca. 650 BCE.

Assyria, the heartland of which was a rich agricultural area along the plains of the Tigris River, had a long history. But its greatest era, and the time when it came into direct contact with the two Israelite kingdoms, was the period of the Neo-Assyrian empire (ca. 911–609 BCE). Around the beginning of this era, at least one of the campaigns of Shalmaneser III (ca. 859–824 BCE) involved the northern kingdom, Israel (see box on p.96).

After Shalmaneser's death, Assyria entered a period of decline. The removal of the Assyrian threat, together with the weakness of Damascus, enabled Israel and Judah under, respectively, Jeroboam II (786–746 BCE) and Uzziah (783–742 BCE), to make significant territorial gains. However, the accession of Tiglath-pileser III (745–727 BCE) ushered in the Assyrian golden age and, under a succession of able rulers, Assyrian policy became one of permanent conquest, creating an empire of provinces and vassal states (see map on p.107), marked by an increasing use of deportation to control the conquered peoples. To administer the state, a highly efficient and responsible bureaucracy was created and the imperial cities were enlarged and embellished, especially Nineveh, which was made into a magnificent royal capital by Sennacherib (705–681 BCE). The northern Israelite kingdom of Israel was eventually incorporated into the empire (see box on opposite page) and its southern neighbor, Judah, was forced into vassalage. But Assyria's resources were overstretched, and its eventual disappearance was rapid and dramatic. Nineveh fell to the Medes and Chaldeans in 612 BCE, and the final outpost of Assyrian resistance, Haran, was wiped out in 609 BCE (see p.108).

Assyria had a more civilized side. Assyrian art, as seen in reliefs on the walls of temples and palaces, reached a very high standard, and the great library of King Ashurbanipal (669–627 BCE) at Nineveh preserved the religious and other writings of the Sumerians, the main cultural influence

Attended by servants, Ashurbanipal reclines on a couch to dine with his queen, from a relief at Nineveh.

The end of the kingdom of Israel

King Jehu (842–815 BCE) founded a dynasty that lasted for about a century, bringing a period of stability to Israel, the northern Israelite kingdom, despite some profound weaknesses in the state (see p.97). The absence of Assyria meant that neighboring states were free to advance their own ambitions and rivalries (see Amos 1–2). Israel was primarily involved with the Aramean state of Damascus. Under Jehoahaz (815–801 BCE), Hazael of Damascus (see p.98) reduced Israel's army to fifty horses, ten chariots, and ten thousand infantry. However, Jehoash (801–786 BCE), the son of Jehoahaz, recovered a number of towns that his father had lost.

The long reign of Jeroboam II (786–746 BCE) was a period of respite for Israel. Like Omri, he is only briefly mentioned in the Bible, but he restored his kingdom's frontiers to their full extent and Amos indicates considerable prosperity during his reign. After Jeroboam's death, various usurpers ascended the throne, causing internal instability. When a resurgent Assyria invaded under Tiglath-pileser III, one of these usurpers, Menahem (745–738 BCE), was forced to buy off the Assyrians with a huge tribute of silver and Israel became an Assyrian vassal. After Tiglath-pileser's death, Israel's last king, Hoshea (732–721 BCE), revolted and, with Egyptian support, denied tribute to the Assyrians. Disaster ensued: Tiglath-pileser's son, Shalmaneser V, invaded Israel and the capital, Samaria, fell after a three-year siege. Under his successor, Sargon II, the former northern kingdom became an Assyrian province, and, according to his annals, he deported nearly thirty thousand members of Israel's upper classes to the northern Euphrates region. The deportees, who never saw their homeland again, were replaced with foreigners from other parts of the Assyrian empire (2 Kings 17).

Naked prisoners taken captive by King Sargon II of Assyria (722–705 BCE)

on Assyrian civilization. Nineveh's splendor was renowned. It was surrounded by extensive fortifications sheltering palaces and temples, including the magnificent shrine of the goddess Ishtar, the city's patron.

The Bible sometimes views Assyria as an instrument of God's purposes (Isa. 10.5–11). Nevertheless, smaller states could only see the Assyrians as oppressors, with brutal methods of warfare and a policy that might nowadays be called "ethnic cleansing": controlling conquered peoples through mass deportations and population exchanges. In addition, the regular tribute they had to pay was burdensome. Above all, it was the heavy tribute imposed on vassals that led small states like Judah to rebel, in spite of the odds against them. Several passages in the Bible proclaim a divine judgment on Assyria and predict its imminent fall. A few verses in the prophecy of Zephaniah foretell the devastation of Nineveh (Zeph. 2.13–3.5), but its actual destruction in 612 BCE is celebrated in the book of Nahum, one of the finest products of Israelite literary genius. Nahum's brief work consists of two interlinked poetic compositions and has the character of a prophetic curse. Nahum was probably a prophet attached to the Temple of Jerusalem, and his writing may be described as a liturgy. The fall of Nineveh is pictured as a cosmic event, described in motifs drawn from the ancient Mesopotamian myth of the defeat of primeval chaos. The poem expresses Judah's rapture at its release from the Assyrian yoke, which leaves it free to celebrate its religious festivals, "for never again shall the wicked invade you" (Nahum 1.15). The final words of the book show what a great and welcome impression the fall of Nineveh made in the Near East: "All who hear the news about you clap their hands over you. For who has ever escaped your endless cruelty?"

RULERS OF ASSYRIA, 934–609
All dates are approximate and BCE

Ashur-dan II	934–912
Adad-nirari II	912–891
Tukulti-Ninurta II	891–884
Ashurnasirpal II	884–859
Shalmaneser III	859–824
Shamshi-Adad V	824–811
Adad-nirari III	811–783
Shalmaneser IV	783–773
Ashur-dan III	773–755
Ashur-nirari V	755–745
Tiglath-pileser III	745–727
Shalmaneser V	727–722
Sargon II	722–705
Sennacherib	705–681
Esarhaddon	681–669
Ashurbanipal	669–627
Ashur-etel-ilani	627–624
Sin-shumu-lishir	624–623
Sin-shara-ishkun	623–612
Ashur-uballit II	612–609

AMOS

All the features of "classical" prophecy (see main text) are present in the book of Amos. He was from Judah, but prophesied in the northern kingdom of Israel in its last great days under Jeroboam II (786–746 BCE; see box on p.101). He was probably a substantial landowner and his message reflects a wide acquaintance with Israel's religious and cult traditions. The book of Amos begins with a traditional feature, a series of oracles of judgment against foreign nations (Amos 1–2.3). But then Amos switches abruptly (2.4) to a proclamation of similar judgment against the Israelites. Israel has flouted the kind of divine laws found in the book of the Covenant (2.4), and is guilty of the oppression of the poor through corrupt commercial practices (2.6), the perversion of justice (6.14), and ostentatious luxury (6.4–7). The prophet also denounces worship at famous sanctuaries, such as Bethel and Gilgal (5.5), not on the grounds of idolatry, but because those who went there believed that zealous ritual observance absolved them from moral and ethical obligations (5.21–24).

Amos refers to the moment of his calling (7.14), an event that was an essential feature of the prophetic office, when God summoned him from his normal life to prophecy. However, he denies that he is a prophet in the sense of belonging to the regular prophetic guilds. In a string of dramatic visions, he foresees the end of Israel, yet the final words of the book, which may not be by Amos himself, are a promise of restoration. The Davidic empire will regain its full glory, and God assures his people that "they shall never again be plucked up out of the land I have given them" (Amos 9.15).

THE CLASSICAL PROPHETS

The book of Amos (ca. 760 BCE; see sidebar, left) marks the advent of what may be called "classical" prophecy, which was to be of central and enduring significance for both Judaism and Christianity. Instead of collections of stories in which prophets (such as Elijah and Elisha; see pp.96– 9) play a leading role, "classical" prophecy is represented in the Bible by fifteen books that consist largely of oracles attributed to a particular prophet. In most cases, only a few biographical details of the prophet are given. The three longest books are known as the "Major Prophets": Isaiah, Jeremiah, and Ezekiel. The other twelve are all much shorter and are hence known as the "Minor Prophets": Hosea, Joel, Amos, Obadiah, Jonah, Micah, Nahum, Habakkuk, Zephaniah, Haggai, Zechariah, and Malachi. It was once common to speak of the so-called "writing" prophets, but in fact the classical prophets did not write down their messages themselves. The books that go under their names represent the work of their followers, who collected and transmitted their sayings, and also adapted them to meet changing circumstances.

Behind the books, however, it is possible to glimpse individual figures and their particular teachings. The line between them and their prophetic predecessors and contemporaries is not clear-cut: the source of their inspiration and the way they operated were essentially the same. Nevertheless, the classical prophets were aware of a fundamental conflict between them and other prophets, whom they denounced as "false." The "false prophets," as successors of the earlier style of prophecy, saw their function as ensuring the welfare of the nation. In contrast, the great individual prophets, such as Isaiah and Jeremiah, proclaimed God's impending destruction of Israel, because of the people's failure to observe the terms of the Covenant (see pp.62–3). While announcing the nation's doom, though, they also promised that, through this destruction, God would bring his people to new life. (The term "Israel" is at times ambiguous in this period, because it may denote either the entire nation of Israel, covenanted with God, or the northern Israelite kingdom, to distinguish it from the southern kingdom of Judah.)

The Prophet Amos, *by Juan de Borgoña (ca. 1470–ca. 1535).*

Micah and Hosea

Micah was from Judah, but he prophesied against Samaria, the capital of the northern kingdom of Israel: this means that he must have been active before the fall of that city in 721 BCE. The original collection of his oracles has been much expanded with later material, in particular the work of another contemporary prophet from the north. There is no agreement as to the precise extent of the additional material.

Micah's own message consists of both optimism and doom, but all of it reflects the language and liturgy of the Jerusalem Temple. It begins with a traditional description of a manifestation of God, with Yahweh coming from his shrine (Mic. 1.3). But his judgment is directed not against the enemy, but against the Israelite kingdoms of Judah and Israel. Among those denounced are priests, official prophets, and traders, together with those who rely on military power, soothsayers, or idols. To do justice and walk humbly is worth more than innumerable sacrifices.

However, after Jerusalem has been destroyed on account of its sins, Micah claims that there will be a reversal of fortune. Micah takes up one of the themes of the Temple cult, the attack of the nations on Zion and their annihilation. He provides the first picture of a coming savior or Messiah, an idea which reflects royalist ideology: a future ruler will be a second King David. This ruler will be struck on the cheek, a rite that was part of the installation of the Israelite monarch, and he is described in terms of the first man, another motif of Israelite kingship (Mic. 5.1–5).

The prophet Hosea also prophesied in the northern kingdom, probably a little later than Amos (see sidebar on opposite page), during the confused period following the death of Jeroboam II in 746 BCE, which saw a rapid succession of six kings in fourteen years (see table on p.95). His primary concern was to combat the popular religion of Israel, in which Yahweh was associated with Baal and the sexual rites of Canaanite religion: recent archaeological discoveries have confirmed the truth of this situation. The rites of the Canaanite cult provide the prophet with the material for launching his attacks. In a biographical section at the beginning of the book, Hosea is commanded by God to "Go, take for yourself a wife of whoredom" (Hos. 1.2).

The prostitute becomes a symbol of Israel's unfaithfulness to Yahweh. Israel's adoption of Baal, who in Canaanite religion was seen as the husband of the land and the source of its fertility, represents Israel's desertion of its true husband, Yahweh. As a result, national life is poisoned, and most of the book of Hosea consists of a sweeping condemnation of all the institutions of the northern kingdom, which it sees as doomed to destruction.

However, there is another side to the idea of the ancient Covenant between God and Israel being seen as a marriage. Hosea is again told to love a woman, as "the Lord loves the people of Israel, though they turn to other gods" (Hos. 3.1). In this case, the woman is a sign of God's desire to bring the people that he has chosen above all others to repentance. The prophet's distinctive message is that Yahweh is a god of love: God has always loved his chosen people, Israel, and he will never desert the nation, however hard it is punished for its sins.

Far left: The Prophet Micah, *by André Beauneveu (ca. 1335–1410).*

Left: The Prophet Hosea, *from the Worms Bible (German, ca. 1148).*

VII THE END OF THE TWO KINGDOMS

After the fall of Israel, the remaining Israelite kingdom, Judah, had a checkered history. In spite of the efforts of Hezekiah, Manasseh, and Josiah, the kingdom was ultimately unable to withstand the pressure of its imperial neighbors. It fell to the Babylonian king Nebuchadrezzar II in 587 or 586 BCE, and much of its population was forced into exile.

HEZEKIAH

After a period of turbulence following the fall of the house of Omri (see p.99), Judah was able to reassert its independence from the northern kingdom of Israel and take advantage of the weakness of the Aramean kingdom of Damascus. Edom was recovered, and under the long reign of Uzziah (also called Azariah, 783–742 BCE), Judah enjoyed a similar period of stability and prosperity to Israel, which also revived under Uzziah's contemporary, Jeroboam II (786–746 BCE).

This era came to an end with the revived expansion of Assyria under Tiglath-pileser III. To meet the Assyrian threat, King Pekah of Israel (737–732 BCE) and King Rezin of Aram renewed the alliance of King Ahab's time and apparently tried to persuade King Ahaz (735–715 BCE), Uzziah's grandson, to join them. When Ahaz refused, they attacked him and, to counter this crisis, Ahaz became a vassal of Assyria, which removed the pressure on him by defeating the northern allies (2 Kings 16.9). Ahaz may not have been a wholly unwilling vassal because the Bible depicts him disapprovingly as a man who sought to assimilate the cult of Yahweh to the general Near Eastern religious pattern (2 Kings 17; 2 Chron. 18). In particular, when Ahaz met Tiglath-pileser at Damascus, Ahaz saw an altar, probably a Syrian one, which he ordered to be copied (2 Kings 16.10–11). He subsequently introduced it into the Temple at Jerusalem as his own cult object.

By contrast, Ahaz's son and successor, Hezekiah (715–687 BCE), is regarded very favorably in the Bible, which declares that "there was no one like him among all the kings of Judah after him, or among those who were before him" (2 Kings 18.5). He reversed the religious policy of Ahaz and instituted a reformation to remove Canaanite practices from the religion of Judah, notably the worship in the Temple of a bronze serpent that was traditionally said to have been erected by Moses (2 Kings 18.4; 2 Chron. 29). This reformation represented a reassertion of the national religion and identity. Through it Hezekiah attempted to win over the inhabitants of the former kingdom of Israel, which ceased to exist after its conquest by the Assyrians (2 Kings 18.9–10 and 2 Chron. 30; see p.101).

Hezekiah pursued a consistent anti-Assyrian policy throughout his reign. He is reported (2 Kings 20.12–19) to have received a delegation

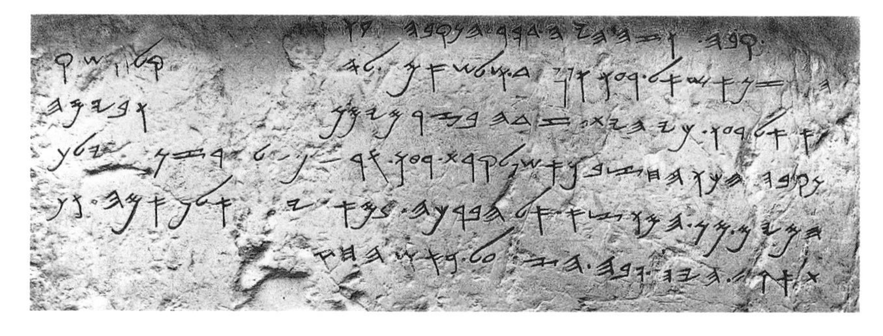

This Hebrew inscription of ca. 700 BCE commemorates the moment when two gangs of miners digging from either end of King Hezekiah's new water conduit broke through in the middle. The tunnel runs for 1750 feet (533m) from the Gihon Spring outside the city wall to the Pool of Siloam in the old City of David (see map on p.190). The inscription was installed at the point where the breach was made.

from the king of Babylon, Marduk-apal-iddina (Merodach-baladan in the Bible), who had struggled, with some success, to free his own kingdom from Assyrian overlordship. The Judean king no doubt intended to coordinate a joint resistance against Assyria. To safeguard Jerusalem's water supply in the event of a siege, Hezekiah constructed a conduit that still exists (2 Kings 20.20; 2 Chron. 32.30). Eventually, soon after the accession of Sennacherib to the throne of Assyria in 705 BCE, Hezekiah joined an anti-Assyrian revolt by a broad coalition including Phoenicia, some Philistine cities, and probably also Moab, Edom, and Ammon. The revolt was supported by Egypt, where a new and vigorous Ethiopian dynasty had come to power ca. 715 BCE. The pharaohs saw that, if all Palestine fell, Egypt would be Assyria's next target. From this time onward, Egypt was regularly active against Assyrian expansionism in the region.

In response to Hezekiah's rebellion, King Sennacherib of Assyria invaded Judah in 702 BCE. He has left an account of the campaign in his annals, according to which he took forty-six Judean cities, including the strategic town of Lachish. The siege and capture of Lachish were depicted in reliefs at his palace in Nineveh. His annals relate that he then laid siege to Jerusalem, shutting up Hezekiah "like a bird in a cage." To prevent the sack of the city Hezekiah was forced to submit and pay a heavy tribute. The biblical narrative confirms the payment of tribute, but then says that the Assyrians demanded the handing over of Jerusalem (2 Kings 18–19; Isa. 36–38). Hezekiah conveyed the news to the prophet Isaiah (see sidebar, right), who delivered an oracle reassuring him that the city would not be attacked. According to the Bible, the enemy withdrew after the angel of God had struck down several thousand men in the Assyrian camp (2 Kings 18.35–36).

It is difficult to reconcile the account in the annals of Sennacherib with that in the Bible. The leading figure in the biblical account is Isaiah, and the text bears all the marks of a prophetic legend. The fact that Jerusalem was spared and Hezekiah remained on the throne may have seemed a miracle to the Jerusalemites. They may have interpreted events in the light of the official theology, which stated that God would protect his dwelling place (see Isa. 35, 37, and sidebar, right). However, the harsh reality seems to have been that Hezekiah became a vassal of the Assyrian king. Sennacherib distributed most of Judah's territory among those Philistine rulers who had remained loyal to him.

Israelite prisoners being led into captivity after the fall of Lachish to the Assyrians; from a relief in the palace of King Sennacherib of Assyria at Nineveh.

ISAIAH

The prophet Isaiah was an important figure in the reigns of Uzziah, Jotham (742–735 BCE), Ahaz, and Hezekiah of Judah. He was the confidant of kings, closely involved in the political movements of the day, and profoundly associated with the worship and theology of the Temple at Jerusalem. It was there that he received his prophetic call (Isa. 6). The book of Isaiah is a complex document containing material of widely different dates, but it is agreed that the voice of the prophet himself can be heard in the first thirty-nine chapters of the book. Even these chapters probably incorporate later material, although much less than has sometimes been supposed.

Like other prophets, Isaiah promises a mix of doom and prosperity. He strongly condemns the social abuses for which the nation's leaders are responsible, and announces a coming judgment. But, compared with earlier prophets, his note of reassurance is even stronger. He reasserts the doctrine that the home of the Lord is inviolable (Isa. 37, 35), as long as the nation repents of its offenses and puts faith in Yahweh's promises (Isa. 1–2). If the people place absolute trust in God, they will always be secure, so power-politics and the entanglements of foreign alliances are unnecessary. This message lies at the heart of his address to Ahaz when the king is faced with the coalition between Israel and Damascus (Isa. 7.3–15), and of his warnings against reliance on the new power of Egypt (Isa. 19–20, 31). His vision for the future is of a holy Jerusalem, where an ideal king of David's line will rule in peace and righteousness (Isa. 9, 11, 32).

Parts of the book of Isaiah are attributed to two different prophetic authors known as "Second Isaiah" or "Deutero-Isaiah" (see p.112) and "Third Isaiah"(see sidebar on p.117).

DEUTERONOMY

The "book of the law", found in the Temple in the reign of Josiah (see main text on opposite page), is an important landmark in the history both of the religion of Israel and of the Bible. Its appearance is the first mention of a written document containing essential regulations for life and worship. It represents the beginning of the development of the written religious Law (*Torah*), which became a fundamental feature of Judaism. The book also marks the first step in the compilation of a canon of Scripture.

The "book of the law" is generally thought to be related to the book of Deuteronomy in the Bible. Deuteronomy stresses the duty to eradicate all foreign elements in the worship of Yahweh (Yahwism), a message that may well have provided the impetus for Josiah's reform program. However, the biblical book as it now stands is the product of a complex process of composition that took place over many years. Whether Josiah's book really lay hidden in the Temple until it was "discovered" is questionable. Some features of Deuteronomy suggest that it originated among northern exiles who had fled to Jerusalem, the remaining center of Yahwism, after the fall of the northern kingdom of Israel in 721 BCE. Their work was a program for the nation's future faith and practice, and demanded a complete purification of Yahwism.

It is also possible that the work originated during the reign of Josiah's predecessor, Manasseh, as a response to his religious policies (see main text, right). It was with the accession of Josiah that the ideal political and religious climate arrived in which the book could be made public. The book "found" in the Temple probably consisted only of the "statutes and ordinances" that make up chapters 12–26. This may have largely been the work of prominent Judeans of Josiah's circle, such as Hilkiah, the chief priest, and the prophetess Huldah. They are closely connected with the book's appearance (see 2 Kings 22; 2 Chron. 34).

MANASSEH AND JOSIAH

Following Hezekiah (see pp.104–5), the Bible presents two sharply contrasted kings of Judah. Hezekiah's son and successor, Manasseh (687–642 BCE), is presented as the worst of all the Judean rulers, because of his wholesale reintroduction of pagan practices. The Bible sees this as the cause of the kingdom's fall (2 Kings 21.12–14; see p.110). However, Manasseh can be seen in a more favorable light as someone who sought to reverse his father's religious and political policies, which had led Judah to the brink of disaster (see p.105). He probably felt that the only hope of maintaining any kind of national existence lay in vassalage to the Assyrian empire (see pp.100–101). Assyria reached its height in the early part of Manasseh's reign with the overthrow in 671 BCE of the Egyptian pharaoh Tirhakah by King Esarhaddon (681–669 BCE). Assyrian texts show that Manasseh was in fact a loyal subject, at least for the majority of his long reign. As a result, he appears to have recovered territory lost by his father, because he was able to garrison "all the fortified cities in Judah" (2 Chron. 33.14). His religious policy aimed to restore the traditional situation that existed before Hezekiah's destruction of local shrines, which may have been unpopular and divisive.

According to 2 Chronicles, the Assyrians "took Manasseh captive in manacles, bound him in fetters, and brought him to Babylon" (2 Chron. 33.11). During his imprisonment, Manasseh apparently repented, and after his release he revoked his religious measures. Much of the biblical account is clearly legendary, but the episode of his capture and imprisonment suggests that he was not always a subservient vassal of Assyria. It may be that he rebelled when, in the later part of his reign, Assyria lost control of Egypt under Esarhaddon's successor, Ashurbanipal (669–627 BCE).

In contrast to Manasseh, his grandson and successor, Josiah (640–609 BCE), is portrayed as an ideal monarch, who reestablished the exclusive worship of Yahweh and restored the true kingdom of David. Josiah's father, Amon, "abandoned the Lord" and was assassinated in a palace conspiracy after ruling for only two years (640–642 BCE). But the conspirators were killed, and the eight-year-old Josiah was placed on the throne by "the people of the land," probably the free property-owning classes who were committed to Judean independence (2 Kings 21.23–24).

The angel of God striking down the Assyrians (2 Kings 19.35–37), from the Nuremberg Bible of 1483.

National independence seems to have been Josiah's consistent aim during his reign. He was able to pursue such a policy because in this period the Assyrian empire became too weak to intervene in the affairs of Palestine. The Bible concentrates on Josiah's religious reforms, which were part of this reassertion of national identity. According to 2 Kings, Josiah carried out an exhaustive purge of Judah's religion in response to the teaching of "the book of the law," which was found in the course of repair work on the Temple (2 Kings 22.8; see sidebar on opposite page). But 2 Chronicles states that the destruction of heathen cult objects and practices took place six years *before* the "discovery" of the lawbook (2 Chron. 34.14). This is probably closer to what actually happened. Once the nation's worship had been purified, the "finding" of the book was the occasion for the formal renewal of God's Covenant with his people. In particular, the Passover was properly celebrated for the first time since before the period of the monarchy (2 Chron. 35.18; 2 Kings 23.22; but see 2. Chron. 30).

Josiah's reformation has often been interpreted as the removal of the Assyrian deities and idols. The Bible does mention what are said to be typical Mesopotamian astral cult practices, such as the worship of heavenly bodies. However, there is no real evidence that Assyria imposed its own deities on vassal states. Besides, the divinities and cults that Josiah suppressed existed among neighboring peoples, such as the Canaanites, Sid-onians, Moabites, and Ammonites. Josiah's aim was to establish a cult of Yahweh, free from foreign taint. His reforms extended to the old northern kingdom of Israel. It was technically an Assyrian province, but the enfeebled Assyrian state was unable to resist Josiah's moves toward a restoration of the united kingdom of David and Solomon.

A letter of King Esarhaddon of Assyria about the Ethiopian dynasty of pharaohs. Esarhaddon ousted the dynasty in 671 BCE.

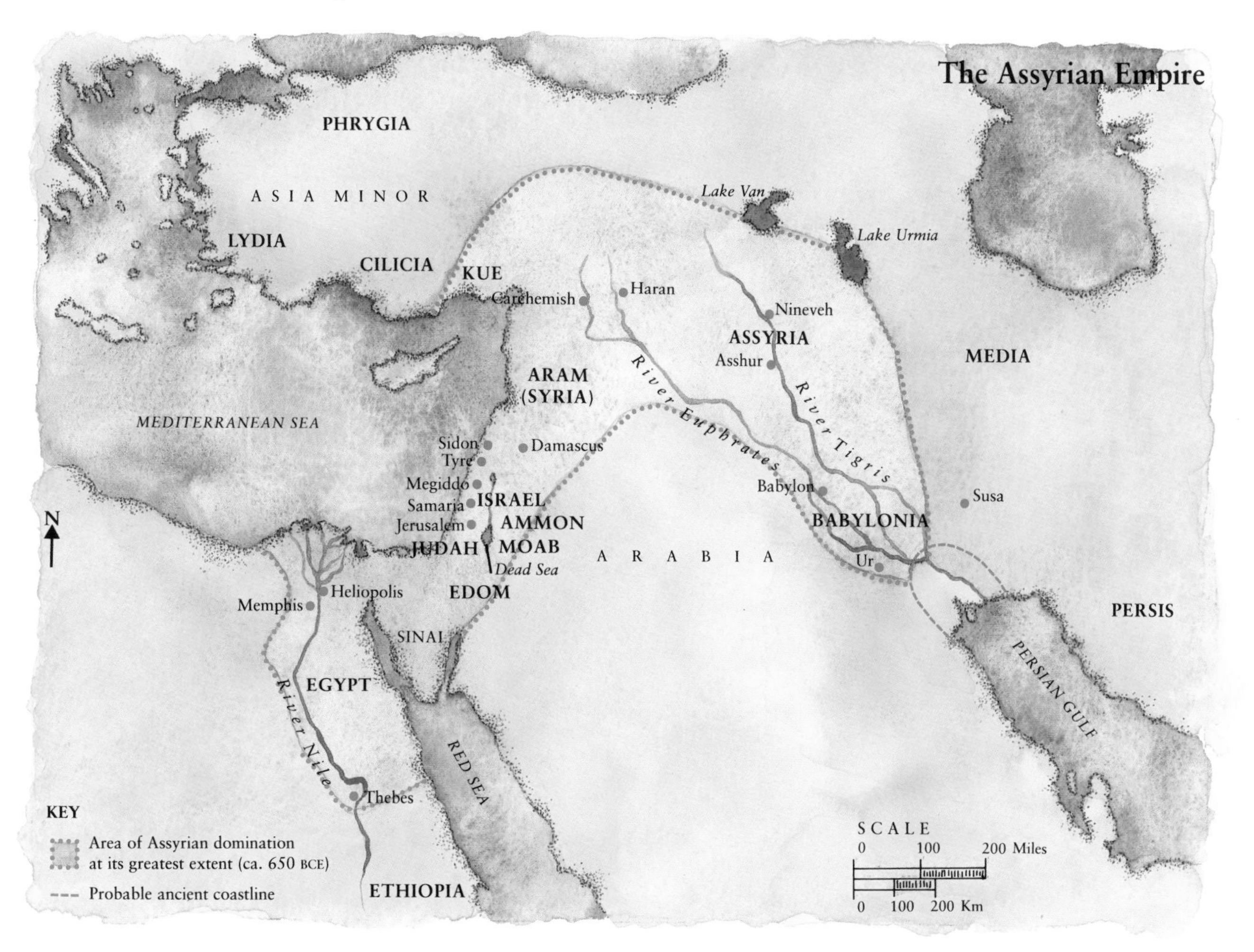

THE DECLINE OF JUDAH

RULERS OF THE CHALDEAN (NEO-BABYLONIAN) EMPIRE
Alternative spellings in round brackets; biblical spellings in square brackets. All dates are approximate and BCE.

Nabopolassar	625–605
Nabu-kudurri-usur II [Nebuchadrezzar, Nebuchadnezzar]	605–562
Amel-Marduk [Evil-merodach]	561–60
Nergal-shar-usur (Neriglissar)	559–56
Labashi-Marduk	556
Nabu-naid (Nabonidus)	555–39
Bel-sharra-usur [Belshazzar, Baltasar], regent	552–42

The achievements of King Josiah of Judah (see pp.106–7) did not last for long. He died in 609 BCE in battle against the pharaoh Necho II of Egypt, which had regained its independence from Assyria by 663 BCE under Necho's predecessor, Psammetichus I of the new twenty-sixth dynasty. In 612 BCE the Assyrian capital, Nineveh, fell to a coalition of the Neo-Babylonians and the Medes (see box, below), and Egypt changed its anti-Assyrian policy, realizing that a weak Assyria was less of a threat than the victorious Neo-Babylonian empire which had supplanted it. The Egyptians sought to bolster the Assyrian state of Asshur-uballit II (612– 609 BCE), centered on Haran in northern Mesopotamia. To aid Haran, Egypt established a base at Carchemish on the Euphrates, and Necho was passing through Palestine with his army on his way there when Josiah confronted him. Josiah probably feared that, unless Necho was stopped, Assyrian domination of Judah would only give way to Egyptian rule. The Judean king attempted to check the pharaoh's advance at Megiddo, but fell in the subsequent battle (2 Kings 23.29; 2 Chron. 35.23–24).

Necho was unable to prevent the fall of Haran, but, for a time at least, he was free to assert control over Syria and Palestine. After Josiah's death, "the people of the land" anointed his younger son, Jehoahaz, as king in preference to his elder brother Eliakim, probably because they believed Jehoahaz would continue his father's policies in support of the national religion and political independence. Necho seems to have understood

The Neo-Babylonians and the Medes

The Ishtar Gate, one of the city gates built as part of the lavish redevelopment of Babylon by Nebuchadrezzar II. It is faced with glazed bricks including reliefs of guardian bulls and serpent-dragons.

During the first millennium BCE a number of tribal groups, known collectively as Chaldeans, infiltrated the extreme south of Mesopotamia and gradually gained control of the ancient towns of Babylonia. Here, Chaldean kings such as Merodach-baladan championed national independence from Assyria. In 2 Kings 20 and Isaiah 39, Merodach-baladan is reported to have sent a delegation to King Hezekiah of Judah (see pp.104–5). The Chaldean period of greatness began with Nabopolassar, who defeated Assyria, took over its empire, and founded a dynasty referred to as "Neo-Babylonian" to distinguish it from earlier Babylonian civilizations. Under his son, Nebuchadrezzar II (605–562 BCE), the state attained the height of its power and splendor, but Nebuchadrezzar's three successors had short and troubled reigns. The last Neo-Babylonian ruler, Nabonidus, was not a member of the royal family. In 539 BCE the Neo-Babylonian empire fell to the Persians. In the Bible, the Chaldeans also appear as magicians and soothsayers, professions for which they were renowned throughout the Near East.

Comparatively little is known about the Medes, an Indo-European people closely related to the Persians. Over several centuries, they occupied roughly what is now northwest Iran, the region known in ancient times as Media. Toward the end of the seventh century BCE, King Cyaxares established a strong Median empire, with its capital at Ecbatana (modern Hamadan, Iran), and joined Nabopolassar in his carve-up of Assyria. But Cyaxares' successor, Astyages, was defeated by the Persian king Cyrus in 550 BCE, and Media became a province of the Persian empire.

this, because after only three months, he deposed Jehoahaz and took him as a prisoner to Egypt, where he died (2 Kings 23.34). Necho replaced him with his brother, changing his name from Eliakim to Jehoiakim (609–598 BCE). On his accession, a king often received a new name from God; the fact that the name was conferred by the pharaoh emphasized Judah's vassal status, which was confirmed by the imposition of a heavy tribute.

Egyptian hegemony was shortlived, because in 605 BCE the Babylonian king Nebuchadrezzar II defeated Necho at Carchemish. Necho had to withdraw to Egypt and effectively surrender the whole Syria-Palestine region to the Babylonians. In the following year, Nebuchadrezzar sacked the Philistine city of Ashkelon and moved against Jerusalem, forcing the submission of Jehoiakim.

Jeremiah

Like other prophetic works in the Bible, the book of Jeremiah has been extensively edited, in this case notably by authors reflecting the outlook of Deuteronomy (see sidebar on p.106). Nevertheless, Jeremiah emerges as more of a distinct personality than any other of Israel's prophets. He was apparently the son of Hilkiah, the high priest in King Josiah's time, and his career spanned the time from Josiah until the fall of Jerusalem and the exile to Babylon (see pp.110–11). There are anguished passages where he gives vivid expression to his compulsion to prophesy, knowing that his message will be rejected. The book contains lengthy biographical passages (such as ch. 26), from which we can gain a reasonably authentic picture of the prophet's life, and information about the last days of Judah. Jeremiah used a scribe, Baruch, to write down his words, and much of the existing text may reflect Baruch's work.

The book of Jeremiah follows the common pattern of a prophesied divine judgment on Judah, and its eventual restoration. Jeremiah tells Judah that before its fortunes are restored, it must first accept the punishment of Babylonian domination (Jer. 27). This message brought him into conflict with the advocates of rebellion, not least with the anti-Babylonian King Jehoiakim. In one famous episode, Jehoiakim publicly cut up and burned the scroll on which Baruch had transcribed the prophet's oracles (Jer. 36.23). Jeremiah was imprisoned for his views, and even King Zedekiah, who greatly respected him, could only consult him in secret and finally felt unable to accept his advice.

If Judah submitted to its fate, this would not mean the end of all hope. For Jeremiah, the destiny of the nation lay in the hands of those exiled to Babylon. He has a vision of two baskets, one of good figs, representing the exiles, and one of rotten figs, representing those left behind (Jer. 24). In a letter to the exiles (Jer. 29), he exhorts them to seek Babylon's welfare and to make a new life for themselves there. If they do so, he says, God will hear them and bring them back home.

Jeremiah, portrayed as a scholarly medieval churchman by the Master of the Annunciation of Aix (ca. 1440).

THE FALL OF JERUSALEM

Nebuchadnezzar II and his Advisers, *by Frans Francken the Younger (1581–1642).*

The accession of King Jehoiakim (see p.109) ushered in the final act in the history of the kingdom of Judah. In 601 BCE, Nebuchadrezzar II of Babylon launched an expedition to conquer Egypt, but was defeated by an Egyptian army and forced to withdraw. His vassal Jehoiakim, who is depicted in the book of Jeremiah as a brutal and arrogant ruler, exploited this weakness to break away from Nebuchadrezzar, probably by withholding the payment of tribute. It took some time for Nebuchadrezzar to regroup his forces, but while this was underway he apparently took steps to weaken Judah. The Bible records that the country was devastated by guerrilla bands of desert tribes sent by God (2 Kings 24.2), but they probably acted on the instigation of the Babylonian king.

Jehoiakim died in 598 BCE and was succeeded by his son Jehoiachin (598–597 BCE). The transfer of leadership came at a bad time, because in the following spring Nebuchadrezzar invaded. Jehoiachin surrendered Jerusalem without any marked resistance and averted the destruction of the city by paying a heavy tribute (2 Kings 24.13). More importantly, Jehoiachin, his family, and his court, together with seven thousand of the leading classes in Jerusalem and a thousand skilled craftsmen, were deported to Babylon. Nebuchadrezzar replaced Jehoiachin with his uncle Mattaniah, a son of Josiah, and renamed him Zedekiah (2 Kings 24.17). Under this name he is known as the last king of Judah.

Zedekiah found himself in an impossible situation. Many Judeans continued to regard the captive Jehoiachin as their rightful king, and looked on Zedekiah as little better than a usurper. Some of those who remained in Judah probably felt this way, and it was certainly the view of the exiles in Babylon. Administrative tablets from the time of Nebuchadrezzar list the rations of oil given by the royal household to Jehoiachin and his sons. The documents call Jehoiachin "king of Judah," suggesting that even the Babylonians continued to regard him as the rightful Judean ruler.

The final verses of both 2 Kings and the book of Jeremiah record how Nebuchadrezzar's successor, Evil-merodach (Amel-Marduk: see table on p.108), released Jehoiachin from prison and gave him an honored position at the imperial court. In the Bible, this act of clemency serves as an assurance that God will not forever abandon the house of David.

In Judah, the deportation of the kingdom's most senior officials had gravely weakened the whole structure of the state. The book of Jeremiah presents a vivid picture of how Zedekiah was torn between different groups of advisers, some advising rebellion and others appeasement and submission to Babylon (Jer. 37–38). The rebel faction relied on the support of Egypt, which still harbored ambitions in Palestine. The prophet Jeremiah, who was among the appeasers, repeatedly warned against the folly of trusting the Egyptians and antagonizing Nebuchadrezzar. But the hawkish faction prevailed and led Zedekiah finally, and fatally, to rebel.

The fall of Judah is recounted in the last chapter of 2 Kings, 2 Chronicles, and Jeremiah. In response to Zedekiah's revolt, Nebuchadrezzar laid siege to Jerusalem. The strength of the city's defenses enabled it to hold out for eighteen months, and there was apparently a brief respite for the defenders when the appearance of an Egyptian army caused the Babylonians to break off the siege temporarily (Jer. 37.5). Like the Assyrian king Sennacherib (see p.105), Nebuchadrezzar preceded his attack on the capital by taking various strategic Judean towns. According to Jeremiah 34.7, eventually only Lachish and Azekah held out, apart from Jerusalem. A number of letters, written by Judean officers in surrounding outposts to the commander of the garrison at Lachish, report

Judah after 586 BCE

The second book of Kings recounts the aftermath of the destruction of Jerusalem in 587/6 BCE much as it describes the misfortune which had befallen the city under King Jehoiachin a decade earlier (see main text). In both cases, the narrative claims, extravagantly, that most of the population was deported, leaving only "the poorest people of the land" (2 Kings 24.14 and 25.12). A more trustworthy source is preserved in the book of the prophet Jeremiah (see p.109), where it states (Jer. 52.28–30) that three thousand twenty-three Judeans were deported in 597 BCE, but only eight hundred thirty-two after Jerusalem's final fall and destruction. In fact, Babylon's policy toward its new province was relatively moderate, and life carried on in Judah.

The Babylonian king appointed a local man, Gedaliah, as governor: this name appears on a contemporary seal that identifies the owner as chief steward of the Judean royal palace. A number of prominent individuals and military commanders in the country rallied to Gedaliah at his headquarters at Mizpah. According to Jeremiah, Judean refugees returned home from the surrounding lands and were able to reap a good harvest (Jer. 40.12). The Bible also tells of a group of pilgrims from the north, who are murdered while taking sacrifices to "the temple of the Lord." This suggests that worship continued at the holy site of the ruined Temple in Jerusalem (Jer. 40.4–5).

Gedaliah was assassinated at Mizpah by Ishmael, a member of the deposed Judean royal family, who was also responsible for the slaughter of the pilgrims. Ishmael had returned from exile in Ammon, and probably intended to restore the deposed Davidic dynasty, with Ammonite support. Gedaliah's remaining followers were afraid of Babylonian reprisals in the wake of the murder and fled to Egypt, taking the prophet Jeremiah with them, against his will (a later legend claimed that the prophet had been stoned to death in Egypt).

The book of Jeremiah records that a second mass deportation to Babylon, involving over seven hundred Judeans, took place five years after the fall of Zedekiah, that is in 582/1 BCE (Jer. 52.30). This may represent the Babylonian response to the murder of Gedaliah. From this time, Judah was probably placed under the direct control of the Babylonian governor of Samaria, the province immediately to the north of Judah. But whatever happened, little further is known about conditions in the country for some forty years.

An Israelite seal inscribed "To Pedaiah son of the king." Pedaiah was a member of the Judean royal family in the last days of independent Judah (see 1 Chron. 3.16–17).

on Babylonian troop movements. One letter indicates that the town of Azekah has fallen, while another complains of the activities of the anti-war party in Jerusalem. There is also mention of a senior army officer being dispatched to Egypt, presumably to seek assistance.

Jerusalem finally fell in 587 BCE or 586 BCE, and the punishment for its lengthy resistance was harsh. Zedekiah was taken to the Babylonian headquarters at Riblah in the Lebanon. There, the Bible relates, his sons were killed before his eyes, his officers were executed, and Zedekiah himself was blinded and taken in chains to Babylon (2 Kings 25.7; Jer. 52.10–11). In Jerusalem, Nebuchadrezzar's commander-in-chief, Nebuzaradan, supervised the destruction of the Temple, the royal palace, and the houses of prominent citizens. The Temple furnishings and remaining treasures were carried off as booty. It was the end of autonomy for Judah, the end of ancient Israel as an independent organized state.

EXILE IN BABYLON

LAMENTS FOR JERUSALEM

The roads to Zion mourn,
for no one comes to the festivals;
all her gates are desolate,
her priests groan;
her young girls grieve,
and her lot is bitter.

Her foes have become the masters,
her enemies prosper,
because the LORD has made her suffer
for the multitude of her transgressions;
her children have gone away,
captives before the foe.

Lamentations 1.4–5

By the rivers of Babylon –
there we sat down and there we wept
when we remembered Zion
On the willows there
we hung up our harps.
For there our captors
asked us for songs,
and our tormentors asked for mirth,
saying,
"Sing us one of the songs of Zion!"

How could we sing the Lord's song
in a foreign land?
If I forget you, O Jerusalem,
let my right hand wither!
Let my tongue cling to the roof of my
mouth,
if I do not remember you,
if I do not set Jerusalem
above my highest joy.

Psalm 137.1–6

The shock of the destruction of Jerusalem (see pp.110–11) is powerfully reflected in the dirges of the book of Lamentations, an anonymous work traditionally attributed to the prophet Jeremiah. The end of Judah's independence raised profound spiritual questions for Israel. The most important issue was whether the fate of Jerusalem meant that Yahweh had been defeated by more powerful deities, who now had to be worshiped in his place. The Bible implies that some Israelites accepted this opinion. In Egypt, Jeremiah and his fellow exiles argue vehemently over the cause of the national disaster. In a notable diatribe, the prophet blames it on the nation's worship of foreign gods, but his hearers attribute their lot to the fact that they had failed to worship the "queen of heaven," the great Near Eastern mother goddess, as their ancestors had done (Jer. 44). In the period after the first deportation of 597 BCE, the prophet Ezekiel (see box, below) saw sun-worship and the rites of the Near Eastern vegetation deity Tammuz being carried out in the Temple (Ezek. 8.14).

Israel's faith and sense of nationhood seem to have been maintained and reinterpreted most effectively by the Judean exiles in Babylon.

"Second Isaiah"

Chapters 40–55 of the book of Isaiah form a distinct section that contains the oracles of an anonymous prophet, commonly referred to as "Second Isaiah" or "Deutero-Isaiah." He preached in Babylon toward the end of the Exile and his message is one of joyful hope. Israel, he declares, has paid the penalty for its sins, and the era of divine mercy and restoration is about to dawn (Isa. 40.1–5). He pictures the return to the Promised Land as a second Exodus, even more wonderful than the first. In his teaching, monotheism is clearly expressed for the first time: God is not merely the greatest deity, but the only one (Isa. 44.6), and Second Isaiah mocks foreign gods and goddesses as lifeless idols with no real existence (Isa. 44.9–20). The one true God, Yahweh, alone controls the world and determines its history.

Second Isaiah was aware of the political situation of his time. Babylon was under threat from the Persian king Cyrus (see p.115), and the prophet looks forward to its imminent fall. He sees Cyrus as the Lord's anointed ruler (Isa. 45.1), who will return the Jewish exiles to their homeland after God has led him to victory (Isa. 45.13).

Second Isaiah's oracles speak of a figure called the "servant," whose identity has been much disputed. Early Christians saw him as the Messiah, and identified him with Jesus, notably in Isaiah 53 (see p.225). The picture of the servant appears to be made up of a number of traditional images. Certain aspects of him recall Moses, a typical Davidic king, and one of the prophets. But he stands primarily for Israel, with which he is explicitly identified in places, such as Isaiah 44.1. Israel is the nation that has been chosen by God, but endures contempt and suffering. This marks the emergence of a new and remarkable concept: Israel's vocation as a sacrificial victim for the whole world. The divine purpose for God's chosen people is that it should atone, by its sufferings, for the sins of all humanity. "The righteous one, my servant, shall make many righteous, and he shall bear their iniquities" (Isa. 53.11). (See also pp.105 and 117.)

Conditions in Babylonia enabled them to preserve their identity. Governed by elders, the leading figures of the community, the exiles lived together in organized settlements, such as Tel-abib (Ezek. 3.15) and other places in the vicinity of the irrigation channel of Chebar, near the city of Nippur. One concern of the exiles was to preserve and record the traditions of Israel's past in writing, to create a body of Scripture upon which they could base their hopes for the future. It was during the exile that the Bible as we know it first began to take shape, in the form of the Pentateuch (the first five books); the history of Israel (Joshua to 2 Kings) by writers of the school that produced Deuteronomy (see p.106); and the prophetic writings. Living in a foreign environment, the exiles were anxious to stress the elements in Israel's religion which distinguished the Jews from other peoples. At this time, therefore, there was a new emphasis on such features as the Sabbath, circumcision, and the need for ritual purity.

The calamities which had befallen the Israelites came to be seen not as the result of Yahweh's powerlessness, but as his punishment for the people's repeated religious and social sins. This belief underlies the Deuteronomic school's interpretation of Israelite history. Similarly, the general message of the prophets, that a divine judgment would restore the nation by purging it, was presented as the norm. The Israelites began to see Yahweh as a truly universal deity, who controlled the destinies of all nations and remained with his people, even when they had been taken from his land. The exiles never abandoned the hope that God would eventually bring them back to their native country, and create Israel anew.

One of the most remarkable figures of the period of the Exile was the prophet Ezekiel. He was a priest who was taken to Babylon with King Jehoiachin in the first deportation of the Jews in 597 BCE (see p.110), and his whole message is couched in priestly terms and imagery. The book of Ezekiel paints a remarkable picture of the nature of prophetic inspiration, with visions, ecstatic experiences of levitation, and strange symbolic signs.

The book begins with Ezekiel's calling as a prophet, "among the exiles by the river Chebar" (generally identified with a canal that ran from the Euphrates near Babylon), in the fifth year since the first deportation (that is, ca. 592 BCE). It takes the form of a great vision of God, which became a central component of later Jewish mysticism. It has often been suggested that details of the vision, such as the winged cherubim (Ezek. 1.4–5), recall features of Mesopotamian religion. But they are perhaps more likely to reflect the cult of the Temple of Jerusalem, the main point being that Yahweh now manifests himself not only in Zion but also in a foreign land. Ezekiel's explanation of Jerusalem's destruction is that the actual presence of Yahweh, "the glory of the God of Israel," has left the polluted Temple, removing its divine protection (Ezek. 10.18–19). Restoration will only be achieved when the divine glory returns to the sanctuary.

Ezekiel's hope for the future is expressed in various ways. The people must accept responsibility for their own actions and no longer attempt to blame the disasters that have befallen them on the sins of their ancestors. He sees his defeated and exiled people as a heap of dry human bones, which is restored to life by the onrush of the divine breath or spirit. From this time the nation becomes a spirit-filled community (Ezek. 37.1–14). The book ends with a detailed description of a new and perfect Temple, the center and symbol of a newly revived Israel (Ezek. 40–48). The climax of the vision of the new Temple is the return of the divine presence (Ezek. 43.2–5), a sign of Israel's restoration. The divine voice tells the prophet that the Temple is the place "where I will reside among the people of Israel forever" (Ezek. 43.7).

Illustrations in the book of Ezekiel, from the Lambeth Bible, produced in England, ca. 1140–1150 (clockwise from top left): Ezekiel's calling (Ezek. 1.1–3); the vision of the scroll (2.9–3.3); six executioners, including "the man clothed in linen," receive God's instructions (9.1–5); the instructions are carried out (9.6–7).

A four-winged spirit from the palace of King Sargon II of Assyria (722–705 BCE) at Khorsabad. Such figures were common in Mesopotamian art and may have inspired the winged cherubim of Ezekiel's vision (compare Ezek. 1.5–6).

RETURN FROM EXILE

ZECHARIAH
Like Haggai (see main text), the prophet Zechariah was active during the reign of Darius I of Persia (522–486 BCE). The prophet also pins nationalistic hopes on Zerubbabel, who is promised royal dignity and lays the foundation of the new Temple (Zech. 2.6–10). He is crowned and given the royal title "Branch" (Zech. 6.12–13). In this passage, the name of the high priest, Joshua, has been substituted for that of Zerubbabel. Perhaps Darius, after securing his rule, deposed Zerubbabel, who disappears from history at this point. With the failure of this last attempt to reestablish the kingdom of Judah, much of the old royal ideology and ritual would have been adopted by the remaining senior figure of Judean society, the high priest. This would explain why the name Joshua has replaced Zerubbabel in Zechariah 6 and elsewhere.

The book of Ezra (see p.116) opens with a decree of King Cyrus the Great of Persia, following his capture of Babylon in 539 BCE. He attributes all his victories to Yahweh, "the God of heaven," and states that Yahweh "has charged me to build him a house at Jerusalem." He goes on to authorize the return of Judean exiles to undertake the work (Ezra 1.2–4). The authenticity of this decree has often been doubted, largely because it has been thought unlikely that a Persian king would credit his successes to the god of a foreign subject people. However, in one genuine decree, Cyrus ascribes his victories to Marduk, the national god of Babylon. This document also parallels the Bible in stating that, in various areas, Cyrus has rebuilt temples and returned former populations. In Ezra 6 there is another version, often thought to be more reliable, of the decree concerning the Judeans. The decree was apparently found at Ecbatana in western Persia (Ezra 6.2). It is really a memorandum to the royal treasurer there, authorizing the expenses for the building of the Temple. It does not mention the return of exiles, but they would not have been the treasurer's responsibility.

According to the book of Ezra, following Cyrus's decree, a caravan of deportees returned to Judah. The number given is very large – about fifty thousand – and probably represents several groups that in fact returned at different times under different leaders. The new Temple is said to have been founded by Sheshbazzar, a Persian official who was apparently made governor of Judah (Ezra 5.16–18). Sheshbazzar was probably accompanied by a small group of Judeans, who were able to set up the open-air altar of burnt offerings on the Temple site, and to celebrate the autumn feast of Tabernacles or Booths (Ezra 3.3–4).

Construction was halted owing to opposition from those who had remained behind in Judah and from the ethnically mixed inhabitants of the former northern kingdom of Israel. The opponents of the Temple were supported by other Persian provincial officials. The result, although there is a good deal of confusion in the narrative, was that work on the Temple was stopped until 520 BCE, the second year of the later king, Darius. Then, at the instigation of the prophets Haggai and Zechariah (see below and sidebar, left), work resumed under two leading Judeans, Zerubbabel and Jeshua (Ezra 5.1–2). The royal governors consulted the king as to whether the Judeans had obtained proper authorization. Darius sent an affirmative answer, and the new Temple was completed in late March 516 BCE. The narrative of the book of Ezra is clearly designed to emphasize the favorable attitude of the Persian empire toward the Judeans, and to show that the rebuilding of the Temple had royal approval. In general, it presents a neat and unified picture of events, but it is likely that the reality was much more complicated.

The monumental tomb of King Cyrus II of Persia at Pasargadae, Iran.

Like Ezra, Haggai refers to a delay in rebuilding the Temple, but attributes it to a lack of enthusiasm among the inhabitants of Judah ("these people," Hag. 1.2), and does not mention returned exiles. In response to the prophet's message, the governor of Judah, Zerubbabel (a grandson of King Jehoiachin; see p.110), and the high priest, Joshua (Jeshua in Ezra), began work on the Temple in the second year of Darius's reign. Haggai addresses an oracle to Zerubbabel, in which God tells him "I have chosen you," and promises "to overthrow the throne of kingdoms." This corresponds well with the period of turmoil in the Persian empire at the beginning of Darius's reign. Haggai suggests that the oracle to Zerubbabel and the rebuilding of the Temple, far from having Persian approval, represented an attempt to restore the kingdom of Judah.

Cyrus the Great and Darius I

The reign of King Cyrus II of Persia was a landmark in ancient Near Eastern history, and Cyrus probably deserves his epithet "the Great" more than many other rulers. He was the founder of an empire that was to last for over two centuries and bring the whole of the Near East, from India to the frontiers of Greece, under one ruler for the first time. Legends grew up around him, and ancient Greek writers, such as Herodotus and Xenophon, regarded him as a model ruler. This was largely because, unlike the Assyrians and Babylonians, he and his successors did not seek to eradicate long-established institutions, or to deport native populations. They preserved and restored the old legal and social systems of their subject territories, and paid particular attention to the upkeep of holy places. It is hardly surprising that Cyrus was welcomed as a deliverer, not only by the people of Judah, but also by the inhabitants of Babylon, who (according to his own annals) opened their gates to him without a struggle.

Cyrus's son, Cambyses, had a comparatively short reign, but he succeeded in incorporating Egypt into the empire. The end of Cambyses' reign was marked by political unrest, and his successor Darius, whose origins are obscure, had a hard fight to secure his accession. Under Darius, the Persian empire reached its greatest extent, expanding as far as northern India, and the magnificent palace that he built at his capital, Persepolis, testifies to the wealth and splendor of his state. The only shadow on his reign was the failure of his expedition against the Greeks, who defeated the Persians at the battle of Marathon in 490 BCE. This abortive campaign was of profound significance for the future: the Persian empire eventually fell to the Greeks under Alexander the Great in 330 BCE.

RULERS OF THE PERSIAN EMPIRE
All dates are approximate and BCE.

Cyrus II ("the Great")	550–529
Cambyses II	529–522
Darius I	522–486
Xerxes I	486–465
Artaxerxes I	465–425
Xerxes II	425–424
Darius II	423–404
Artaxerxes II	404–359
Artaxerxes III	359–338
Arses	338–336
Darius III	336–330

A relief from the palace of Darius I of Persia at Persepolis, ca. 500 BCE, which shows the enthroned Darius receiving a Median dignitary, behind whom stand two attendants.

VIII AFTER THE EXILE

The Exile in Babylon brought about a fundamental reappraisal of the spiritual life of Israel. This began among the exiles and was continued after their return to Judah, especially through the work of Ezra and Nehemiah. The life of the nation was now centered on the Temple, and Yahweh was increasingly seen not just as Israel's national god, but also as the one true and universal deity. In the postexilic period the biblical canon started to take shape. For most of this time, Palestine was dominated by the Persian empire and the Hellenistic empires which succeeded it. However, in the mid second century BCE, *more than four centuries after the fall of the last king of Judah, the Maccabees established an independent Jewish state.*

EZRA AND NEHEMIAH

According to the Bible, the leaders Ezra and Nehemiah brought about a fundamental reconstruction of the religion and society of postexilic Judah. Ezra's career is related in the last four chapters of the book which bears his name, together with two chapters now incorporated in the book of Nehemiah. The story of Nehemiah is told in the first person, but this does not mean it is an autobiography. The use of "I" throughout is a stylistic device modeled on Near Eastern royal inscriptions that were designed to glorify the national hero.

The traditions about the two men once existed separately, but they were brought together by the biblical author known as "the Chronicler" to form a sequel to the first and second book of Chronicles, the earlier retelling of Israel's history. It is not known for certain whether Ezra and Nehemiah were in fact contemporaries. In the Bible, both are dated to the time of a Persian king called Artaxerxes, who could be either the first or second ruler of that name (see table on p.115). It is generally agreed that Nehemiah was active under Artaxerxes I (465–425 BCE), but some have argued that Ezra came later, under Artaxerxes II (404–359) BCE. The current trend is to place both men in the time of Artaxerxes I and to see their careers as overlapping.

Ezra is described as a priest and as "the scribe of the law of the God of heaven" (Ezra 7.12), that is, he was the official in the Persian administration responsible for Israelite religious affairs. In accordance with the Persian imperial policy toward its subject peoples, Ezra was commissioned to investigate how Jewish law was being observed in Judah. On his arrival in Jerusalem, he at once put a stop to intermarriage between the returned exiles and foreigners. He saw this practice as polluting the land and "the holy seed" of Israel, and ordered that all foreign wives and their offspring be sent away in order to purify the nation and avert the wrath of God (Ezra 9–10). Later, as recounted in Nehemiah 8–10, he read out the law of Moses at a great ceremony that saw the revival of the feast of Tabernacles. This ceremony probably marked the first stage in the codification of the Pentateuch, the biblical books containing the divinely-given laws of Israel (see p.118). Ezra caused the leading inhabitants of Judah to bind themselves to the observation of the law by a solemn oath (Neh. 9.38–10.29).

Nehemiah's work was largely in the secular sphere. He was a cupbearer at the Persian royal court in Susa (modern Shush, Iran), when he received information that the fortifications of Jerusalem had been destroyed (Neh. 1.3). In 445 BCE, the twentieth year of Artaxerxes' reign, the king authorized Nehemiah to go to Judah (Neh. 2.6), apparently as imperial governor, a post he says he held for twelve years (Neh. 5.14). Behind this move may lie Persian concern to build a strong Judah in order to counter possible expansionist moves by the Egyptians.

Nehemiah's great achievement, recounted in a series of dramatic scenes (Neh. 3–6), was a rapid rebuilding of Jerusalem's walls, in the face of fierce opposition from neighboring leaders, led by Sanballat, the imperial governor of Samaria. There is an attempt to trick Nehemiah into seeking refuge from his enemies in the Temple (Neh. 6.10), but Nehemiah refuses, knowing that to do so would be sinful (Neh. 6.13). This episode has given rise to a theory that Nehemiah was a eunuch, and as such forbidden by the Jewish Law from entering the sanctuary (see Deut. 23.1).

Nehemiah also devised a new administrative structure for Judah and Jerusalem, tackled the issues of the proper observance of the Sabbath, and, like Ezra, sought to end mixed marriages (Ezra 10–13). Between them, Ezra and Nehemiah laid a firm religious and civil foundation for the future life of the Jewish community.

Ezra copying the Scriptures, from a 7th-century CE *Italian illuminated manuscript.*

The Jews of Elephantine

Between 1906 and 1908, a number of papyri, written in Aramaic and dating from the Persian period, were discovered on the island of Elephantine or Yeb in the Nile in southern Egypt. The documents came from a colony of Jewish mercenaries in the Egyptian army, who, like the Judeans who fled to Egypt with Jeremiah (see p.109), had probably migrated as the result of the Assyrian and Babylonian invasions of Judah.

The documents show that they had a temple dedicated to Yahweh, but their cult also involved the worship of other deities, including the goddess Anath (see p.72). This probably reflects popular Judean religious practices in the time before the Exile. This temple was destroyed at the instigation of Egyptian priests ca. 410 BCE, and one of the texts is a letter to Bagoas, the Persian governor in Jerusalem, seeking support for its restoration. The authors of the letter say that they have also approached the sons of Sanballat, the governor of Samaria and the chief opponent of the prophet Nehemiah (see main text). Bagoas's reply, which is also preserved, sanctions the restoration of the Elephantine temple. The letter suggests that the postexilic religious authorities in Judah were beginning to claim Jerusalem as the sole legitimate sanctuary. The Elephantine sacrifices, Bagoas says, must only be offerings of food and incense. Animal sacrifice, a central element of the sacrificial cult because it involves the shedding of blood, is to be permitted only in Jerusalem.

The ruins of the temple on Elephantine dedicated to the deities Khnum and Anuket.

MALACHI AND "THIRD ISAIAH"

Conditions in Judah after the return of the exiles, but before the arrival of Ezra and Nehemiah (see main text), can be reconstructed from two documents from this period, the book of Malachi and chapters 56–66 of the book of Isaiah (see p.105), whose author is often called "Third Isaiah." The prophet Malachi's main concern was the prevalence of intermarriage with women from neighboring nations, particularly among the upper classes and even the clergy. He believed such unions posed a threat both to Israel's national identity and to the uniqueness of its faith (Mal. 2.11–12). Malachi also attacked widespread abuses, such as the sacrifice of unclean animals and fraud in the payment of tithes.

Third Isaiah may have belonged to a faction that believed the destruction of the world was near, and was opposed to the established religion represented by the Temple and its clergy. He certainly paints a gloomy picture of the state of society. Its leaders are venal and drunken, there is social oppression, and the course of justice is perverted. In particular, he lists a whole catalog of religious abuses: neglect of fasting and the Sabbath; heathen religious practices; idolatry; child sacrifice; fertility cults; consulting the spirits of the dead; and eating the flesh of forbidden animals, such as the pig. Clearly, the expectations aroused by the ending of the Exile and the rebuilding of the Temple were not being realized.

THE LAW

The word "Law" is used to translate the Hebrew *Torah*, which means "instruction," "guidance," or "direction." It is used primarily to refer to the first five books of the Bible (Genesis through Deuteronomy), which are also called the Pentateuch, or "books of Moses." However, the Pentateuch contains much more than legal regulations. It includes the whole story of God's dealings with humanity from the Creation of the universe to the death of Moses (see pp.26–67).

The codification of the Pentateuch was central to how the Law of Israel came to be viewed in the postexilic period. The final version of the Pentateuch has generally been seen as a combination of three documents: a narrative from the period of the monarchy, the book of Deuteronomy (see p.106), and a work produced in priestly circles. Recently there has been a tendency to question whether any part of the Pentateuch as it stands dates from before the Exile, although the biblical authors certainly drew on many ancient traditions. In any case, the Pentateuch was written down in something approaching its present form during the period of Persian rule (539 BCE–330 BCE; see p.115).

The emergence of the written Pentateuch led to the codification of the other books of what Christians call the Old Testament. In this way there arose what is known as the "canon," a list of books which formed a body of authoritative Holy Scripture, in contrast to other religious writings. However, the Pentateuch was always regarded as supreme – the Samaritans, for example, have never recognized the other books in the Jewish canon (see p.125). The canon was built up in stages over several centuries and probably only reached its definitive shape toward the end of the first century CE.

The theory of an orally-transmitted Torah developed alongside the written Law. It was believed that God had given Moses not only a

Holiness and the Day of Atonement

One characteristic of the God of postexilic Judaism was his intrinsic holiness, a quality that his chosen people were to share. Those chapters in Leviticus that have been called "the Holiness Code" (Lev. 17–26) center on one key phrase: "You shall be holy, for I the Lord your God am holy" (Lev. 19.2). The nation's holiness was to be ensured by the faithful observance of the Law, and this brought about a new and far-reaching understanding of the purpose of worship. Sacrifice (see pp.122–3) came to be seen not so much as a rite that was effective in itself, but rather as a meritorious act of obedience that fulfilled the sacrificial regulations contained in the Law.

The whole system of sacrifice, and the one sanctuary where it was carried out (the Temple of Jerusalem), were seen as the means by which the nation could atone for its past failures to observe the Law. Sacrifice and the Temple were the focus of divine forgiveness. The Temple's center of holiness now became not the ark itself (see p.90), but its gold cover, the so-called "mercy seat." It was here that atonement was made, especially by the sprinkling of blood on the "Day of Atonement," Yom Kippur (Lev. 23.27–28; 25.9).

The Day of Atonement was probably an old observance, but it is only mentioned in writings from after the Exile, when it acquired the tremendous significance that it still possesses in Judaism. The rituals that marked this occasion had the effect of annulling all the sins committed by the nation during the previous year. As set out in Leviticus 16.6–22, the Temple and the priesthood were purified by the blood ritual. Afterward the high priest (see p.123) transferred the people's sins onto a goat (the scapegoat) in an ancient rite with many Near Eastern parallels. The animal was then released into the desert to carry its burden of sinfulness to a demonic spirit called Azazel.

written text but also a body of oral instructions that were handed down through a succession of teachers. In time, this body of teaching acquired almost the same status as the written Law. The oral Torah represented the work of the scribes, an important class of scholars that studied and interpreted the Law, and extended its scope to all areas of life.

The Pentateuch reflects the concerns of the Jewish community of the time. In the period after the Exile, Israel's Law came to be understood as an expression of God's will for the entire world, and obedience to the Law came to outweigh all other religious considerations and obligations. The Law tended to be viewed as something eternal, pre-existent, and independent, which had not come into being only when it was revealed to Moses on Mount Sinai (see pp.60–63). Some speak of the Law as an image of the cosmos, and the means through which the world was created.

After the Exile, the Jewish concept of God also became increasingly spiritual, transcendent, and universal. There was a growing tendency to avoid the use of God's personal name, Yahweh, which designated him specifically as the national deity of Israel, in favor of more abstract titles, such as "Lord" or "God of heaven." Alternatively, he might be referred to by circumlocutions such as "Name." The early Israelites pictured God essentially as a being that intervened directly in human affairs and spoke to men and women face to face. This picture became more and more unacceptable after the Exile, and the problem of how a transcendent being would communicate with his creation was resolved by assuming the existence of a number of intermediaries. For example, God's attributes, notably the divine Word, Wisdom, and Spirit, were personified as beings through whom he acted in the world (see also box, below).

Angels

Israel had always pictured Yahweh as surrounded by a court of heavenly spirits who acted as his agents and messengers. For example, when God says "Let *us* make humankind in *our* own image" in Genesis 1.26, he may be addressing these beings (see p.27; compare Pss. 29.1; 82.1). In postexilic Judaism there appeared a developed hierarchy of angels, with six or seven ranks. At the head of the angelic host were four senior angels (archangels) who had individual names containing the element *el*, meaning "god": Gabriel, Michael, Raphael, and Uriel. The functions of the angels were to praise God in heaven and to act as intermediaries between God and human beings (the word angel comes from the Greek *angelos*, "messenger"). They also protected Israel and other nations.

The influence of ancient Near Eastern religion can be seen in the way that angels are described. They are said to have wings, like the winged spirits that were a common feature of Mesopotamian belief. Their arrangement into six or seven groups recalls the seven Amesha Spentas, beings that personified aspects of the supreme Persian god Ahura Mazda (see p.154). This is reminiscent of the seven heavenly bodies that were worshiped by the Babylonians.

A winged deity, armed with a sword and wielding lightning bolts, attacks a monster; a reconstruction of a relief from the Assyrian royal palace at Nimrud, ca. 870 BCE.

THE SECOND TEMPLE

King David playing the bells, from a 13th-century Psalter produced at Norwich, England.

When the exiles returned to Judah from captivity in Babylon (see pp.114–15) they found that their homeland had been devastated during the Babylonian occupation and reduced in size to a small region around the former royal capital, Jerusalem. However, this meant that King Josiah's policy of eliminating all foreign elements in the national religion (see pp.106–7) could now be implemented without great difficulty. The Temple of Jerusalem, rebuilt by the former exiles (see pp.114–15), was at the heart of this national religious revival and took on a new significance and character. The growing belief in one universal God (see box on p.112) found a parallel in the doctrine of one Temple. The single universal deity manifested himself nowhere else on earth except in the Temple of Jerusalem, and the Temple was the only place where worshipers could offer sacrifice to their God.

The Temple also took on a more practical role in the postexilic period. With the disappearance of the monarchy, responsibility for the country's internal affairs was transferred to the authorities of the Temple (see p.123), and Judah took on the form of a "Temple state" or theocracy. Civil and religious matters, in so far as the two can be distinguished, were administered by the Temple personnel, and the Temple also functioned as the national treasury. The central role of the Temple in national life was reinforced by the ancient obligation on every male to attend the sanctuary on the occasion of the three great annual festivals, Passover, Weeks, and Tabernacles (Deut. 16.16). At such times, Jerusalem was crowded with pilgrims, and later, during the period of Roman rule (from the late first century BCE), the religious celebrations often led to riots against the occupying power.

The feasts themselves were now integrated into the nation's history. Originally, Passover and Unleavened Bread had been a celebration in the home, but by this time the sacrificial Passover meal could only be eaten in Jerusalem, where the Passover lambs were slaughtered in the Temple (cf. Deut. 16.6). The ceremony had come to be seen as a commemoration of the Exodus from Egypt (see sidebar on pp.58–9). The spring Israelite festival of Weeks (Deut. 16.9–12) marked the end of the wheat harvest, and although it always retained its agricultural character, it too came to be connected with events from Israel's past, such as the covenant with Noah after the great Flood (see p.33) or the giving of the Law to Moses at Sinai (see pp.60–61).

During the week of the great autumnal festival, the people were commanded to live in huts made out of branches, a custom which gave the celebration its name of Tabernacles or Booths. Agricultural laborers lived

A PSALM OF DAVID

The Lord is my shepherd, I shall not want.
He maketh me to lie down in green pastures: he leadeth me beside the still waters.
He restoreth my soul: he leadeth me in the paths of righteousness for his name's sake.
Yea, though I walk through the valley of the shadow of death, I will fear no evil: for thou art with me; thy rod and thy staff they comfort me.
Thou preparest a table before me in the presence of mine enemies: thou anointest my head with oil; my cup runneth over.
Surely goodness and mercy shall follow me all the days of my life: and I will dwell in the house of the Lord for ever.

Psalm 23 (King James Version, 1611)

A view of present-day Jerusalem from the Mount of Olives to the Temple Mount. The golden-domed Mosque of Omar (the Dome of the Rock), Islam's third holiest shrine, was built in ca. 690 CE on the site of the Temple's central sanctuary.

The Psalms and music in the Temple

The Psalter or book of Psalms has been called the hymn book of the postexilic Temple. However, it includes material spanning several centuries, and many of the psalms date from before the Exile, at least in their original form. The book contains a number of collections distinguished by their titles, such as the seventy-three psalms – almost half the entire Psalter – attributed to King David, the Songs of Ascent (Pss. 120–134), pilgrimage songs, and psalms which represent texts used by two groups of Temple choirs, those of Asaph (Pss. 73–83) and the sons of Korah (the Korahites; for example, Pss. 44–49). It is possible to distinguish different types of psalms, as well as the occasions on which they were used. Many psalms are laments of an individual in distress (for example, Pss. 42–43), or laments uttered by the community in the face of a national catastrophe (Pss. 44 and 74). There are psalms of thanksgiving (Pss. 116 and 118), and psalms celebrating God as sovereign (Pss. 96–99) and his choice of Zion, Jerusalem (Pss. 48, 87). Other psalms refer to the Davidic monarch himself.

Coins depicting instruments used in the Temple: trumpets (above) and a lyre (below). They were issued by the Jewish rebel leader Simon Bar Kochba during the anti-Roman revolt of 134–5 CE.

Most of the psalms are not the poems of particular individuals, but liturgical texts for use as part of the Temple cult. The Hebrew title of the Psalter is *Tehillim*, "Songs of Praise," which indicates the function of the Psalms in the Second Temple. During this period, the Psalms tended to be linked with the nation's past, and were given titles that related them to historical events, such as episodes in the life of King David (for example Pss. 51 and 63). The increasing predominance of the written Law (see pp.118–19) has also left its mark on the Psalter. The division of the Psalms into five books is probably a deliberate echo of the Pentateuch, the five books of Moses. Psalm 119, the Bible's longest chapter, is a glorification of the Law.

Music played a very important part in the worship of the Second Temple (the word psalm is from a Greek word meaning "song sung to the harp"). The rites of the Temple were not confined to the offering of sacrifice, but included prostrations, walking around the altar, processions in the sanctuary, and other rituals. These were generally accompanied by music. With the final destruction of the Temple in 70 CE, this liturgical music largely ceased, and we have very little idea of what it was actually like or how it sounded. The Bible provides a few clues, however. Psalm 68.24–27 speaks of a great ritual procession in the Temple, led by the ark, the dwelling place of Yahweh (alluded to in v. 24), and followed by singers, minstrels, and a group of girls beating tambourines. Other passages (such as Ps.150.3–5), refer to trumpets, stringed instruments, wind instruments, and percussion. Liturgical dancing was also a regular feature, and joyful cries and shouting (as in Ps.81.1), characterized most celebrations associated with the Temple.

Many Psalms apparently refer to the leader of a choir and to what are probably the names of tunes (Ps. 56), and suggest that the psalms were sung by guilds of professional singers. The author of 1 Chronicles 16 claims that King David organized these choirs under a director called Asaph, who led the singers in a hymn of praise (1 Chron. 16.36). The author is in fact referring to his own (postexilic) time. The membership of these choirs was hereditary and they were drawn from groups within the Temple's Levite personnel.

in such temporary shelters while the fruit was being harvested. In the Bible, however, they are no longer a mere agricultural necessity, but are constructed as a reminder of how the Israelites had once lived in booths in the wilderness during the period in which God brought his people out of slavery in Egypt (Deut. 16.13–15).

The postexilic period saw the beginnings of the Diaspora ("dispersal"), the establishment of permanent Jewish settlements outside Palestine. The Jerusalem Temple became the central authority for the regulation of the religious life of the scattered Jewish communities abroad. A tax of half a shekel was levied annually on every adult Jew throughout the world for the upkeep of the Temple.

An ancient Babylonian plaque from Nippur depicting a priest offering a libation to a seated god (above); worshipers leading animals to sacrifice (below).

SACRIFICE

Sacrifice, the offering up to a deity of an animal or other object, was a universal feature of early religion, and it played a prominent role in the cult of the Israelites. Many attempts have been made to explain its meaning, but there is probably no single theory that can take into account every aspect of sacrificial ritual. The Bible takes sacrifice for granted as a practice and provides no clear rationale for it. However, each type of sacrifice and its accompanying ritual is described in great detail in Leviticus 1–7, and these descriptions may provide some insight into the original purpose of the sacrifice.

One theory of sacrifice is that it is a gift to the deity. Certainly, in the societies of the ancient Near East it would have been unthinkable for an individual to approach a superior, whether human or divine, without bearing some sort of gift. Most types of animal sacrifice were no doubt originally intended as offerings of actual food. This idea especially fits the type of sacrifice known as the holocaust (from Greek *holos*, "whole" and *kaustos*, "burnt"), so called because the sacrificial victim was burnt on the altar (it is also known as the "whole-offering" or "burnt-offering"). Leviticus describes it as "an offering by fire of soothing odor to the Lord" (Lev. 3.16). This is reference to an ancient concept whereby the deity literally smelled the fragrant smoke of the offering and therefore looked favorably on those providing it. It occurs in Genesis, which relates that Noah made a burnt-offering after the Flood (see pp.32–3), and "when the Lord smelled the pleasing odor, the Lord said in his heart, 'I will never again curse the ground because of humankind' " (Gen. 8.20–21). This in turn recalls the Babylonian flood myth recounted in the epic of Gilgamesh. After the deluge, Utnapishtim emerged from his boat to offer a sacrifice, and the gods gathered "to smell the sweet savor." After the Exile, the holocaust became the predominant type of sacrifice. Elaborate daily offerings were made in the Second Temple that were viewed as gifts presented in homage to Yahweh.

The idea that through sacrifice the worshiper communes with God underlies another type of sacrifice, the "shared-offering." It is called this because part of the offering was given up to the deity, the worshipers eating the remainder. It was the most common form of sacrifice in early times, when every animal slaughtered was intended for a meal, in which the deity had his share. This was believed to create a bond between God and the human participants in the offering. In Leviticus much of the old communal character of this type of sacrifice has been lost: most shared-offerings are shared only by the male priests (compare Lev. 6.29; 7.6).

In the postexilic period, it came to be understood that the main aim of all types of sacrifice was expiation and atonement. Sacrifice neutralized the consequences of sin and uncleanness, and restored the people's relationship with God. When an animal was sacrificed, expiation was achieved by applying the blood of the victim to the altar of the Temple. The blood was "life," and its shedding represented not only death, but also the release of a vital force. This idea was applied primarily to the "sin-offering" or "purification-offering," in which the blood ritual was particularly prominent (Lev. 4.5–7), and to the "guilt-offering" or "reparation-offering." The compilers of Leviticus saw these types of offering, which are described in chapters 4–7, as central to the entire sacrificial system. For them, the overriding purpose of sacrifice was to cleanse the nation of sin and its punishment. All sacrifices were seen as expiatory, even those that did not involve the shedding of blood, as in Leviticus 5.11–13 and elsewhere (see also p.43 and pp.192–3).

The priesthood of the Second Temple

The structure of the Temple administration following the Exile was completed after a long process of evolution. Membership of the Temple personnel was hereditary, and the entire company of officiants was traditionally descended from Levi, a son of Leah and Jacob (see p.47) and the legendary ancestor of the Levites, the Israelite tribe which provided all members of the priesthood.

One group of Levites, the Temple servants, were not priests. The prophet Ezekiel states that they had been demoted from their priestly status because of their idolatry (Ezek. 44.10–13). It seems likely that they were descended from priests who had officiated at regional shrines, where foreign elements may have crept into the cult of Yahweh. Such shrines were abolished by King Josiah of Judah (see pp.106–7), who probably brought their redundant ex-priests to Jerusalem. The Temple servants carried out the more menial duties in the Temple, acting as doorkeepers and slaying the sacrificial animals (Ezek. 44.11). Even so, these Levites had an honored position: they provided the Temple music (see p.121) and retained the ancient priestly function of teaching the Law to the people. Some of the speeches in 1 and 2 Chronicles may reflect Levitical sermons of this postexilic period.

Above: *Aaron turning the Nile into blood (Exod. 7.21), from a Hebrew manuscript produced at Barcelona, Spain, in 1320.*

Below: *Aaron portrayed as high priest of the Israelites; a stained-glass window in Chartres cathedral, France.*

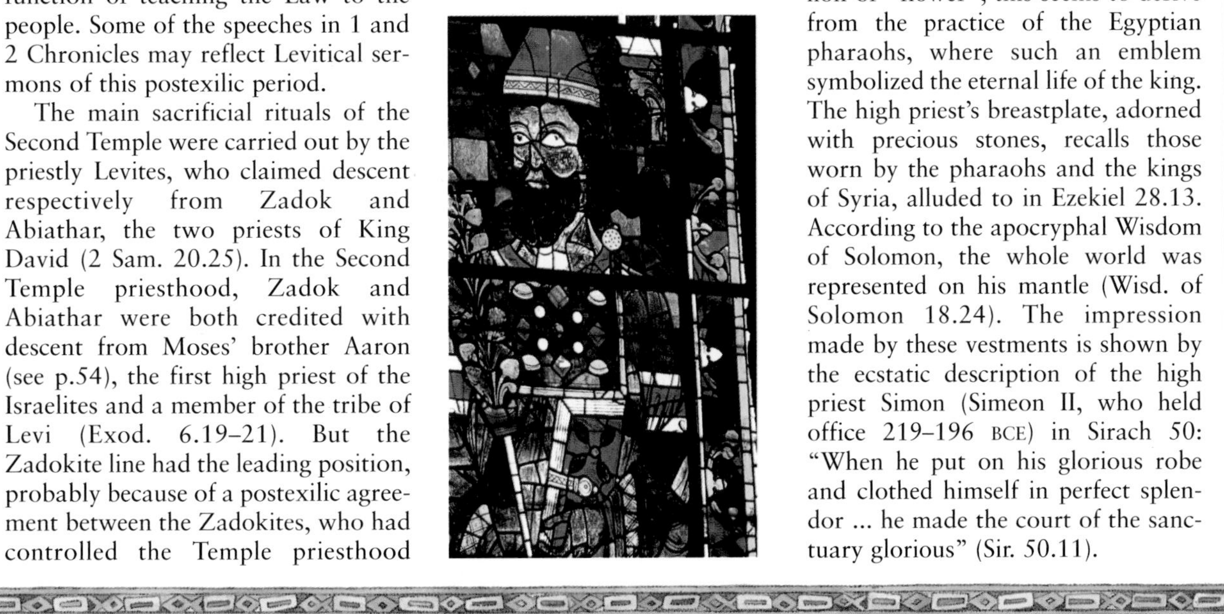

The main sacrificial rituals of the Second Temple were carried out by the priestly Levites, who claimed descent respectively from Zadok and Abiathar, the two priests of King David (2 Sam. 20.25). In the Second Temple priesthood, Zadok and Abiathar were both credited with descent from Moses' brother Aaron (see p.54), the first high priest of the Israelites and a member of the tribe of Levi (Exod. 6.19–21). But the Zadokite line had the leading position, probably because of a postexilic agreement between the Zadokites, who had controlled the Temple priesthood before the Babylonian captivity, and the priesthood of the older Israelite shrines, such as Shiloh. During the Exile, the regional clergy had taken over the Jerusalem Temple priesthood from the Zadokites. When the Zadokites returned from Babylon, the two groups came to an arrangement that restored the leading role of the Zadokites, but also enabled the regional clergy to share in the running of the Temple.

Before the Exile, the Israelite cult was headed by the king, although there was also a leader of the Jerusalem clergy, who was usually called simply "the priest." With the disappearance of the monarchy, he came to be referred to by a new term, "high priest." The high priest took over from the king as the head of the nation and inherited much of the royal ideology. Originally, only the king was anointed: in the postexilic period the high priest was anointed at his installation. (Some passages in the Bible suggest that ordinary priests were also anointed, but this does not appear to have happened in practice.) The high priest dressed in splendid vestments like those once worn by the king (see, for example, Sir. 45.8–13). He wore a turban, a royal headdress, adorned with a gold medallion or "flower"; this seems to derive from the practice of the Egyptian pharaohs, where such an emblem symbolized the eternal life of the king. The high priest's breastplate, adorned with precious stones, recalls those worn by the pharaohs and the kings of Syria, alluded to in Ezekiel 28.13. According to the apocryphal Wisdom of Solomon, the whole world was represented on his mantle (Wisd. of Solomon 18.24). The impression made by these vestments is shown by the ecstatic description of the high priest Simon (Simeon II, who held office 219–196 BCE) in Sirach 50: "When he put on his glorious robe and clothed himself in perfect splendor ... he made the court of the sanctuary glorious" (Sir. 50.11).

THE GOD OF ALL NATIONS

Jonah and the Great Fish *(top), an illustration from* The Fine Flower of Histories, *a Turkish manuscript of 1583.*

JONAH

The author of the book of Jonah presents a novel and remarkable picture of God: when sinful human beings show contrition, God himself can "repent," because his nature is always to be merciful to Jews and gentiles alike.

The author weaves a tale around the prophet Jonah, who is mentioned in 2 Kings 14.25. God sends Jonah to the sinful city of Nineveh to announce its destruction. At first, the prophet tries to escape from this obligation by fleeing on a ship, but God sends a storm. Jonah admits that Yahweh is angry with him and asks to be thrown overboard. He is swallowed by a "great fish" (Jon. 1.17), but prays for deliverance (Jon. 2) and is released on dry land. He goes to Nineveh, but the inhabitants repent of their sins and God spares the city. By a sign, God shows Jonah that he cares for and pities even non-Israelites (Jon. 4.6–11).

The strict monotheism (see box on p.112) that became a distinctive feature of Judaism after the Exile involved a missionary task: to bring about the recognition of Yahweh by all peoples. In the Bible, this universalistic vision is expressed in some passages of the postexilic prophets. Zechariah (see p.114), for example, looked to a time when all nations would go on pilgrimage to Jerusalem to pay homage to Yahweh (Zech. 8.22), while Third Isaiah (see p.117) speaks of the Temple as "a house of prayer for all peoples" (Isa. 56.7). Malachi (see p.117) even claimed that Yahweh was actually worshiped by the gentile nations (Mal. 1.11).

It has sometimes been suggested that Ruth, the protagonist of the book of that name, is depicted as the ideal proselyte, a symbol of Israel's mission to convert the gentiles. The charming story of Ruth is set "in the days when the judges ruled" (Ruth 1.1) and deals with the ancestors of King David. It is placed in the Christian Bible between the books of Judges and Samuel, although in the Jewish canon it comes between Song of Solomon and Lamentations in the books called the Writings (Psalms, Proverbs, Job, Song of Solomon, Ruth, Lamentations, Ecclesiastes, Esther, Daniel, Ezra-Nehemiah, and Chronicles).

At the beginning of the story, the Bethlehemite Naomi and her husband Elimelech leave Judah during a famine to go to the land of Moab. While they are there, their two sons marry local women, Orpah and Ruth. Elimelech and the two sons die, and when the famine in Judah is over, Naomi prepares to return to Bethlehem alone, instructing her Moabite daughters-in-law to return to their families (Ruth 1.8). But Ruth insists on accompanying Naomi, and when they arrive in Bethlehem the story of her devotion to her mother-in-law soon spreads. It is harvest time, and Ruth finds work gleaning in the fields of Boaz, a rich farmer and kinsman of Naomi's late husband. Boaz comes across Ruth and extols her virtues in caring for Naomi and seeking refuge under the wings of the God of Israel (Ruth 2.11–12). Ultimately they are married and, to Naomi's joy,

Naomi Entreating Ruth and Orpah to Return to the Land of Moab, *a print (1795) by William Blake.*

The Samaritans

As God's unique gift to Israel, the Law set his people apart from all other nations, and defined what it meant to be a true Israelite, a question which was central to the spiritual life of the nation after the return from Babylon. But the Law required constant study and interpretation, and various groups within the community differed in their point of view. Postexilic Judaism was not a monolithic faith, but contained a number of sects and parties with different understandings of the Law. This gave rise to internal tensions that are perhaps most clearly illustrated in the case of the Samaritans.

The remains of the Samaritan temple on Mount Gerizim. The Samaritan community, which today numbers in the hundreds, still carries out its traditional Passover sacrifices on the mountain.

In the New Testament, the Samaritans appear as an ethno-religious community living in the area of the former northern kingdom of Israel (from the capital of which, Samaria, they took their name). They claimed to be the descendants of the northern Israelite tribes of Ephraim and Manasseh, but the Jews by this time had long regarded them as a foreign people (see, for example, Sirach 50.26). Nevertheless, there was an awareness of a shared heritage – the gospel story of Jesus and the woman of Samaria (see p.177) refers to their common ancestor, Jacob – and it is best to view the Samaritans as a sect that emerged from the religious situation after the Exile.

The written Pentateuch was the chief authority for both Samaritans (who regard nothing else as Scripture) and Jews, although the Samaritan Pentateuch differs in a few respects from the Jewish one. Both groups held that there was only one legitimate sanctuary where sacrifice could be offered: for the Samaritans this was not Jerusalem but Mount Gerizim, near the city of Shechem (modern Nablus).

When and how the final break between the two groups took place is difficult to determine. The Bible regards the Samaritans as an ethnically-mixed population originally introduced by the Assyrians after the fall of Israel in 721 BCE (see p.101). They were thus tainted by foreign religious influences. But Samaritan tradition dates the split to the much earlier time of the Judges (see p.72), when the priest Eli, the predecessor of the prophet Samuel, established the sanctuary of Shiloh as a rival to Gerizim. Neither of these claims contains the whole truth, and two factors probably contributed to a separation which in fact occurred gradually over some centuries. Political considerations must have played a part. The opposition to the reconstruction of Jerusalem and its Temple was led by the authorities of the old northern kingdom (see p.114). They doubtless adhered to their ancient sanctuary of Shechem, which they claimed had been founded by Joshua. However, the first-century CE Jewish historian Josephus claimed that the split was begun by dissident priests from Jerusalem. This may explain how the Samaritans came to possess the Pentateuch, which had been codified in Jerusalem (see p.118). It may also account for the erection on Mount Gerizim of a Samaritan temple (destroyed by the Jewish ruler John Hyrcanus in 128 BCE).

The truth may involve a combination of both factors. Samaritanism possesses features of ancient Israelite tradition, such as the slaughter of Passover lambs (which occurs to this day on Mount Gerizim), alongside aspects of postexilic Temple-based Judaism, such as the Pentateuch.

Ruth bears a son to continue the line of Elimelech. That son is Obed, who was to become the father of Jesse and grandfather of King David.

The date of Ruth is not certain, but it seems likely to have been a postexilic work. It has often been claimed that the story is a conscious protest against the policies of Ezra, in particular his ban on marriages between Israelites and foreigners (see p.116), and is therefore to be dated around the period of his activity. The author's purpose was to present the Moabite heroine as one who is not only welcomed into the chosen people as a devout adherent of Yahweh, but who also becomes the ancestor of Israel's most famous king. However, there is no clear evidence of polemic in Ruth, which has a distinctly folktale flavor. The story is probably best seen as the product of a group of writers who took folktales as their models.

WISDOM

The Egyptian goddess Isis suckling the young god Horus, a bronze and gold statuette of the period 664–332 BCE.

The book of Proverbs represents the Israelite fund of popular wisdom. As its title implies, most of the work consists of strings of proverbial sayings, such as "A soft answer turns away wrath, but a harsh word stirs up anger" (Prov. 15.1) and "Pride goes before destruction, and a haughty spirit before a fall" (Prov. 16.18). The Bible attributes the book to King Solomon, the archetypal wise man (see pp.86–7), but in fact it consists of several collections dating from different periods.

It has sometimes been claimed that the sayings in Proverbs originated among pre-exilic royal courtiers and reflect their concerns. This is unlikely, since the great majority of the proverbs deal with agricultural, family, and personal matters. Others have suggested that those referred to in the Bible as "the wise," as in Proverbs 1.5, were a professional group of scribes. According to this theory, the proverbs formed the teaching material of scribal schools, which were probably attached to the pre-exilic Judean court. There were such schools in ancient Egypt and Mesopotamia, but there is no evidence for their existence in Israel. The

Ecclesiastes

Israel's sages were optimistic: they perceived an order in the universe, which human beings could also discover. If people aligned themselves with it, they would find happiness and prosperity; if not, misery and disaster would be their lot. This teaching is challenged in the remarkable book of Ecclesiastes. The author is described by the Hebrew term *Koheleth*, which was translated into Greek as *Ekklesiastes* ("a person who addresses a gathering"), hence the title of the book. This is traditionally rendered as "Preacher," (NRSV: "Teacher"). *Koheleth* in fact probably means something like "Speaker," and the work perhaps represents notes made for public lectures, although the writer identifies himself as King Solomon, the ideal wise man. He presents a view of Solomon that is very different than the traditional image of the king. The book opens with a famous declaration: "Vanity of vanities! All is vanity" (Eccles. 1.2), a phrase which recurs near the end (Eccles. 12.8). Solomon's wealth and wisdom are said to be total futility, "vanity and a chasing after wind" (Eccles. 1.16 and elsewhere). A world order exists, but the human mind cannot discover it (see sidebar, left), and what happens in life is very much a matter of chance. Above all, every human achievement is frustrated by the inevitability of death, a theme which haunts the writer and finds expression in a moving poem on old age and death that rounds off the book (Eccles. 11.7–12.8).

There are parallels with Ecclesiastes in ancient Near Eastern culture. The inexorability of death is the central concern of the Mesopotamian epic of Gilgamesh, and several Egyptian and Mesopotamian writings express utter skepticism about the possibility of achieving the good life. But Ecclesiastes is not wholly pessimistic. Human beings are not able to discover the ultimate secret of existence, but because God sends both good and evil, they can still prepare themselves for whatever benefits he may grant. Israel's old body of practical proverbial wisdom (see main text) can help them to do this (Eccles. 12.9–14).

A TIME FOR ALL THINGS

"For everything there is a season, and a time for every matter under heaven:
a time to be born, and a time to die;
a time to plant, and a time to pluck up what is planted;
a time to kill, and a time to heal;
a time to break down, and a time to build up;
a time to weep, and a time to laugh;
a time to mourn, and a time to dance;
a time to throw away stones, and a time to gather stones together;
a time to embrace, and a time to refrain from embracing;
a time to seek, and a time to lose;
a time to keep, and a time to throw away;
a time to tear, and a time to sew;
a time to keep silence, and a time to speak;
a time to love, and a time to hate;
a time for war, and a time for peace.
What gain have the workers from their toil? I have seen the business that God has given to everyone to be busy with. He has made everything suitable for its time; moreover he has put a sense of past and future into their minds, yet they cannot find out what God has done from beginning to end."

Ecclesiastes 3.1–8

book of Proverbs itself makes clear (Prov. 4.3–4) that proverbial wisdom was learned in Israel in the context of the family. "The wise" should not, therefore, be seen as a distinct professional class.

Nevertheless, the sophisticated poetical form of the sayings in Proverbs suggests that they were transmitted among intellectuals, such as "the officials of King Hezekiah of Judah," to whom one of the collections in Proverbs is ascribed (Prov. 25.1). Such intellectuals would have been influenced by the broader Near Eastern wisdom tradition, and Proverbs contains two collections attributed to non-Israelite authors, who are possibly Edomite sages (Prov. 30 and 31). In addition, the thirty sayings in Proverbs 22.17–24.22 ("The words of the wise") are adapted from the thirty chapters of moral instruction by the Egyptian scribe Amenemope.

The influence of a wider wisdom tradition is apparent in Proverbs 1–9, which are very different in character from the rest of the book. They consist of extended discourses that are typical of the wisdom literature of the ancient Near East, and "wisdom" represents a concept of order that is to be found both in nature and human society. A prominent word in Egyptian wisdom texts is *maat*, "world order," which was also the name of a goddess embodying divine order, harmony, truth, and justice. There are some parallels between the goddess Maat and the female personification of Wisdom (the Hebrew word for "wisdom" is feminine) who appears in the first nine chapters of Proverbs (from Prov. 1.20). She has also been compared with the Canaanite goddess Asherah, whose symbol was the tree of life: in Proverbs 3.18 Wisdom is called "a tree of life to those who lay hold of her." Wisdom is contrasted with another female figure, the "foolish woman" (Prov. 9.13–18) or "loose woman" (Prov. 5.3 and elsewhere). She personifies stupidity and also has features that recall a type of Near Eastern goddess whose dangerous charms often lured young lovers.

The oldest wisdom tradition in Israel has often been described as human-centered rather than God-centered. But later texts developed the concept of wisdom as an inseparable attribute of Yahweh, functioning as God's agent in creation and in the government of the world. The personified figure of Wisdom is described in Proverbs 8.23 as existing before the universe was made (a concept commonly found in Egyptian and Mesopotamian texts), and in Proverbs 8.27 as being with Yahweh in his work of creation. In 8.30, according to one reading of the Hebrew text, Wisdom is also depicted as a "little child" playing in the divine presence, which recalls the goddess Maat's position as daughter of Ra, the great sun-god of ancient Egypt. Even more mythological elements surround another Wisdom figure, who appears in chapter 24 of the apocryphal book of Sirach (Ecclesiasticus). The chapter is headed "The praise of Wisdom." Like the great Egyptian mother goddess Isis, Wisdom praises herself in language which links her closely with nature and fertility. For example: "I will water my garden and drench my flower-beds" (Sir. 24.31). But in spite of these foreign mythological resonances, Wisdom is linked with the central elements in Israel's national life. Like Yahweh, her dwelling is in Israel (Sir. 24.8), and she is identified with the written Law (Sir. 24.23).

A distinctive picture of divine wisdom can be found in another apocryphal book, the Wisdom of Solomon, a product of the Greek-speaking Jewish community of Alexandria in Egypt (see sidebar on p.132). In this book, Wisdom is described in terms that are drawn from Greek science and philosophy. She is the creative agent, the expression of the divine presence in everything (Wisd. of Sol. 7.22–24), and the director of Israel's history, especially at the Exodus (Wisd. of Sol. 10). At the same time she is an emanation of God, the outpouring of the divine glory, the radiance of eternal light, the mirror, and the image of God himself (Wisd. of Sol. 7.25–26).

A pharaoh making an offering before the goddess Maat, a silver gilt statuette of the period 1303–1224 BCE. The large feather in the goddess's hair represents justice and truth. In Egyptian belief, the heart of every dead person was weighed against this feather on scales in the underworld. If the heart was heavier, it was burdened by sin, and the soul of the deceased was handed over to a monster to be devoured.

THE SUFFERINGS OF JOB

The book of Job tells the story of a righteous and wealthy man, Job, who never relinquishes his faith in God, in spite of suffering a series of terrible personal misfortunes. The work can be seen as a theodicy, a vindication of the goodness of God in a world where he allows the existence of evil (see also p.40). Its date of composition is uncertain, but it was probably inspired by the postexilic doctrines of a transcendent, wholly righteous deity and of the central role of the Law. At this time, society came to be seen as fundamentally divided between the pious, who were the faithful observers of the Law, and the wicked, who were not. Divine justice demanded that the pious should be rewarded with God's favor, but clearly this did not always happen. The question arose as to how the sufferings of the righteous could be squared with the justice that was to be expected from God. It is this particular issue that the book of Job addresses. The book is not concerned with the problem of suffering in general (in other words, why there is pain, disease and death in the world): the whole work takes for granted that Job's suffering is the result of divine action. The figure of Job himself appears in the book of Ezekiel (Ezek. 14.14, 20), together with Noah and Daniel, as the epitome of a righteous man. He is the symbol of those who suffer, wherever and whenever they live.

Most of the work is poetry, but the verse narrative is contained within a prose prologue and epilogue which, taken together, have often been seen as constituting an ancient folktale. Following a common strand of Israelite

Satan

In the central part of the book of Job it is taken for granted that the source of Job's afflictions is God himself. This is in keeping with the Israelite belief that everything, both good and evil, comes directly from Yahweh. However, according to the opening narrative, it is Satan who causes the disasters that befall Job (Job 1.12; 2.6). He is not a figure of evil, but one of the heavenly court (Job 1.6), who acts only with God's permission. "Satan" is in fact a title, meaning the "Adversary" or "Accuser" (Hebrew ha-satan), a legal term roughly equivalent to "prosecuting attorney."

The development of Satan into a negative character can be seen in the Bible itself. In 2 Samuel, God incites David to take a census (2. Sam. 24.1), a sin which results in a devastating plague on Israel. But in the parallel passage of 1 Chronicles, it is Satan – here the term has become a proper name – who prompts the census and the plague (1 Chron. 21.1). This reflects the view of postexilic Judaism that God is the only source of goodness, and that the source of evil must therefore be sought elsewhere. As the being who tempted humanity to act against the divine will, Satan became God's enemy. He subsequently appeared in Jewish writings as the chief of the fallen angels (see sidebar on p.32), who were expelled from heaven for rebellion against God, and are the cause of evil in the world.

In the New Testament, Satan is the head of a realm of evil that stands in opposition to the kingdom of God. The satanic kingdom is manifested in the activity of the demons who afflict humankind.

Satan and the Rebel Angels *(1808), by William Blake (1757–1827).*

Wisdom teaching (see pp.126–7), this folktale is said to present an individual who patiently accepts his sufferings and, as a result, is finally rewarded by God. According to this theory, the author of Job replaced the central part of the old tale with a long poetic dialogue (Job 38.1–42.6). However, it is probably better to consider the book as a whole. It has often been asked what solution it offers to the problem of suffering, but it should perhaps rather be seen as a work of poetical imagination that offers a variety of insights, all of which illuminate the problem. When Job is overwhelmed by a manifestation of God (Job 38.1–42.6), the author points to the ultimate mystery and freedom of God's actions, but also stresses the validity of the human quest to grasp that mystery.

Job's friends, Eliphaz, Bildad, and Zophar, go to "console and comfort" him (Job 2.11) but are so shocked by his state that at first they cannot speak to him. When they do, they try, unsuccessfully, to persuade Job that he must be in a state of sinfulness to have suffered so much. The friends are often viewed as straw men, who reproduce stock arguments for Job (and the author) to refute. But what they say, although certainly not the last word, has its own worth. They stress the positive value of acceptance and of trust in the ultimate divine benevolence. As Eliphaz says to Job: "Agree with God, and be at peace; in this way good will come to you" (Job 22.21).

NEAR EASTERN TRADITION AND THE BOOK OF JOB

The kind of problem which concerns Job was also present to the minds of thinkers elsewhere in the ancient Near East, and a series of writings from Egypt and Mesopotamia suggests that the book of Job was partly influenced by this wider tradition. The texts in question differ considerably among themselves, both in form and content, but they contain elements which resemble features in the biblical work. The dialogue form occurs frequently to present a debate or dispute between a sufferer and another participant. The Babylonian poem *A Dialogue about Human Misery* recalls the pattern of dialogue between Job and his three friends: the sufferer and his friend speak in alternate stanzas, the sufferer uttering bitter complaints against apparent divine injustice and the friend responding with moralistic admonitions and general observations. Like Job, the sufferer often protests his innocence.

Both in Egypt and Mesopotamia, such dialogues are sometimes framed by a narrative describing the nature of the sufferer's distress. This is not to say that the book of Job depends directly on all or any of these texts. Its extraordinary poetical power and beauty set it apart from any other composition, either in the culture of the ancient Near East or in the rest of the Bible.

Left: Job on the Dungheap, *from the* Heures d'Etienne Chevalier, *ca. 1445, by Jean Fouquet (ca. 1420–ca. 1480). Job sitting on his dungheap (a reference to Job 2.8) in the company of his friends was very commonly depicted in this period.*

Portrait of Queen Esther, *by Andrea del Castagno (ca. 1421–1457).*

POPULAR STORIES AND LOVE SONGS

The Diaspora – the dispersal of many Jews from Palestine to various areas of the Near East (see glossary) – brought the Jews into contact with a wider world. One result of this was the production of a number of short stories, intended as popular entertainment, based on supposedly historical events involving the Jews and other nations. Such stories recount, with humor and irony, dramatic and bloodthirsty episodes, court intrigues, and the perils undergone by beautiful and virtuous heroines. They showed the (Jewish) underdog getting the better of mighty oppressors, a common folklore theme. As well as providing entertainment, therefore, the stories served as encouragement to Jewish communities living abroad in a sometimes hostile environment.

The tale of Esther, told in the book of that name, is set at the splendid court of Ahasuerus, the historical King Xerxes of Persia (486–465 BCE). It probably originates from Jews living in Persia, and it may well be a Jewish adaptation of a Persian composition: the names of the Jewish heroine, Esther, and her foster-father, Mordecai, seem to be derived from the Babylonian deities Ishtar and Marduk. Also, the book never mentions God, a lack that was filled by additions to the Greek translation of the work, which are now found in the Apocrypha. Esther lives with her adoptive father in the Persian capital, Susa. Following the disgrace of the Persian queen, Vashti (Esther 1.19), Esther's beauty and charm captivate the king and he makes her his new queen (Esther 2.17). Esther is pictured as a national savior, who successfully defends Mordecai and her fellow Jews against the plots of the anti-Semitic royal counselor, Haman. She dis-

Tobit and Tobias

The apocryphal book of Tobit consists of two interwoven narratives about the elderly Tobit and his son Tobias. It combines two common folklore motives, that of "the grateful dead" and "the unfortunate bride," and it also draws on the non-biblical story of Ahiqar, a wise man at the court of the Assyrian kings Sennacherib and Esarhaddon (see pp.100–101). It is a product of the Jews of the Diaspora, and a version of the story was found among the documents of the Jewish colony at Elephantine in Egypt (see p.117).

A limestone amulet, dating from ca. 650 BCE, which depicts the Mesopotamian demon Pazuzu.

The tale is set in Nineveh and Mesopotamia. Tobit, a righteous man, goes blind (Tob. 2.10) and is reduced to poverty as a consequence of fulfilling the pious duty of burying the dead. But in spite of everything, he preserves his integrity and his piety is rewarded. His sight and prosperity are finally restored and he meets with a happy and peaceful end.

The story of the young Tobias is a typical folktale. It takes the form of the hero's quest for a bride, Sarah, whose seven husbands have been killed by a demon, Asmodeus, on their wedding night (Tob. 3.8). (The idea that the wedding night is a dangerous time, when evil forces may attack the partners, is common in folklore.) Tobias's quest involves a long journey and he encounters various dangers. He is accompanied and aided on his mission to exorcise the demon (and also to restore his father's sight) by a young man called Azariah, who is Raphael, one of the archangels (see p.119), in disguise. The figure of Asmodeus is typical of a postexilic Jewish concern with demons, and reflects a belief that a realm of evil powers exists which acts against the agents of the true God.

plays both skill and courage in carrying out the advice of Mordecai and in confronting Xerxes when her people are threatened. Haman is a stock villain, vain and scheming, who in the end is hanged on the gallows he had originally erected for Mordecai (Esther 6.10). The author displays a sound knowledge of Persian customs, but is not recording actual history. Rather, the narrative explains the origin of the Jewish festival of Purim ("Lots"; see Esther 9.24–28), a popular celebration with the character of a lively carnival. Purim is not mentioned in the Law (see pp.118–19), and probably represents the Jewish version of a festival that occurs in many cultures at around the time of the spring equinox.

The apocryphal book of Judith has much in common with Esther, but its heroine is a considerably more vigorous character, who recalls such figures as Deborah and Jael in Judges 4–5. The book is set against the background of a massive invasion of Judah under the tyrant Holofernes, the commander-in-chief of "King Nebuchadnezzar of Assyria," presumably a reference to King Nebuchadrezzar II of Babylon (605–562 BCE), the ruler who exiled the Jews (see pp.110–11). However, the author probably has in mind the events of the time of Antiochus IV Epiphanes (175–164 BCE), the ruler of the Hellenistic Seleucid empire, who prompted the revolt of the Maccabees (see pp.138–41).

Judith, a wealthy widow, singlehandedly averted the threat to her people by offering herself to Holofernes. When he fell asleep, dead drunk, she beheaded him with a sword (Jth. 13.8), and returned with his head in a bag to her home town of Bethulia. The assassination put the Assyrians into disarray and they were routed by the Israelites, who celebrated with a great song in praise of God and his agent, Judith (Jth. 16.1–17). The story portrays her as a model of Jewish womanhood, beautiful, incorruptible, strict in all religious observance, and ready to "put away her widow's clothing" (Jth. 16.7) and risk danger for the sake of her people and religion.

FROM THE SONG OF SOLOMON

"My beloved speaks and says to me:
'Arise, my love, my fair one,
and come away;
for now the winter is past,
the rain is over and gone.
The flowers appear on the earth;
the time of singing has come,
and the voice of the turtledove
is heard in our land.
The fig tree puts forth its figs,
and the vines are in blossom;
they give forth fragrance.
Arise, my love, my fair one,
and come away.
O my dove, in the clefts of the rock,
in the covert of the cliff,
let me see your face,
let me hear your voice;
for your voice is sweet,
and your face is lovely.' "

Song of Solomon 2.10–14

The Song of Solomon

The Song of Solomon (or Song of Songs) is unique in the Bible. It is a series of poems celebrating, in remarkably exotic and erotic language, the physical love between a man and a woman. Not surprisingly, it was long debated whether or not it should form part of Holy Scripture, and it probably owed its place in the canon to its attribution to Solomon, who is mentioned several times in the text. The traditional interpretation, both in Judaism and Christianity, is that it is an allegory of Yahweh's love for Israel. As far as Christians are concerned, it also expresses the love of Christ for his Church.

The work resembles some Egyptian love poems and Arabic wedding songs praising the physical charms of the bride. There are also parallels with sacred marriage texts concerning the union of the Near Eastern goddess Ishtar and her consort Tammuz, a shepherd (or the union of a king and queen as their representatives). Some have seen the book as an actual liturgy for such a divine marriage, but this is unlikely. It is more widely accepted that the work simply celebrates human love.

A relief of ca. 1750 BCE from Ashur depicting a god, possibly Tammuz (who is also known as Dumuzi).

Nevertheless, if the Song of Solomon is not a royal marriage liturgy, it may well be derived from one. The poems form a kind of antiphonal exchange between the man, the woman, and a chorus of her companions, the daughters of Jerusalem. The man is sometimes identified as King Solomon and appears as a shepherd, a regular royal title that recalls Tammuz. The poems themselves may go back to the period of the monarchy, while their final version probably belong to the postexilic age.

HELLENISM

THE JEWS OF ALEXANDRIA

Jewish communities grew up in many Hellenistic cities (see right), but the most important of them, in terms of both its size and influence, was Alexandria in Egypt. The Jews there did not abandon their ancestral faith and distinctive ways, but at the same time they sought to learn and interpret them through the medium of Greek learning. One significant product of this process was the translation of the Scriptures into Greek. This translation, the Septuagint, was so called because it was said to have been carried out by seventy (Latin *septuaginta*) scholars at the request of king Ptolemy II of Egypt. It aimed primarily to serve the needs of the Greek-speaking Jewish community, and it reveals how they understood their ancient Scriptures. But it also served an apologetic purpose, demonstrating that Jewish monotheism was not inconsistent with Greek philosophy.

The Alexandrian Jews also produced new religious writings in Greek, including the apocryphal Wisdom of Solomon (see p.127). The work is steeped in biblical history and Jewish tradition, but it also draws upon Hellenistic universalist ideas and philosophical forms. Notably, it treats the concept of Wisdom in terms of the doctrine of the *Logos*, the controlling rational principle of the universe, propounded by the Stoic philosophers.

Alexander the Great (see box on opposite page) was more than simply a remarkable military conqueror. He wished to unite the East and West on the basis of the Greek civilization of which he was the champion. The result was that a new culture, broadly called "Hellenism," arose. It combined Greek and Oriental elements and was dominant in the Near East from ca. 330 BCE until the final establishment of Roman supremacy in 30 BCE. Greek language and culture remained significant factors in the civilization of the Near East for many centuries thereafter.

Hellenism created an intellectual climate dominated by Greek literature, science, and, especially, philosophy, through which the idea grew up of one world that was the common possession of "civilized" human beings. The leading Greek philosophical schools of the time, the Cynics, Stoics, and Epicureans, emphasized cosmopolitanism and individualism. Barriers between peoples were breaking down, not least through the widespread adoption of *koine*, the everyday Greek language, which became the lingua franca of the civilized world.

The main impetus for the spread of Hellenism was the building or rebuilding of Near Eastern cities on the model of the cities of Greece. Alexander established several new cities (see map) where he settled his mercenary soldiers, and this policy was continued by his successors. Hellenistic towns sprang up throughout the region; they were attractive, open places, typically with a marketplace (*agora*) surrounded by law courts, temples, a theater, and a gymnasium. They provided a humane and elegant environment, in contrast to the cramped squalor typical of earlier New Eastern towns. Chief among the new cities was Alexandria in Egypt, founded by Alexander in 331 BCE. Alexandria, with its famous Library and Academy built by the Greek king of Egypt, Ptolemy II, came to epitomize the best of Hellenistic culture. However, Hellenistic civilization was neither universally welcomed nor open to everyone. Its economy rested on slavery, which in turn disenfranchised many. Administrative power was confined to Greeks and to those who adopted Greek ways. This factor played a central part in the reaction to Hellenism among the Jews of Palestine.

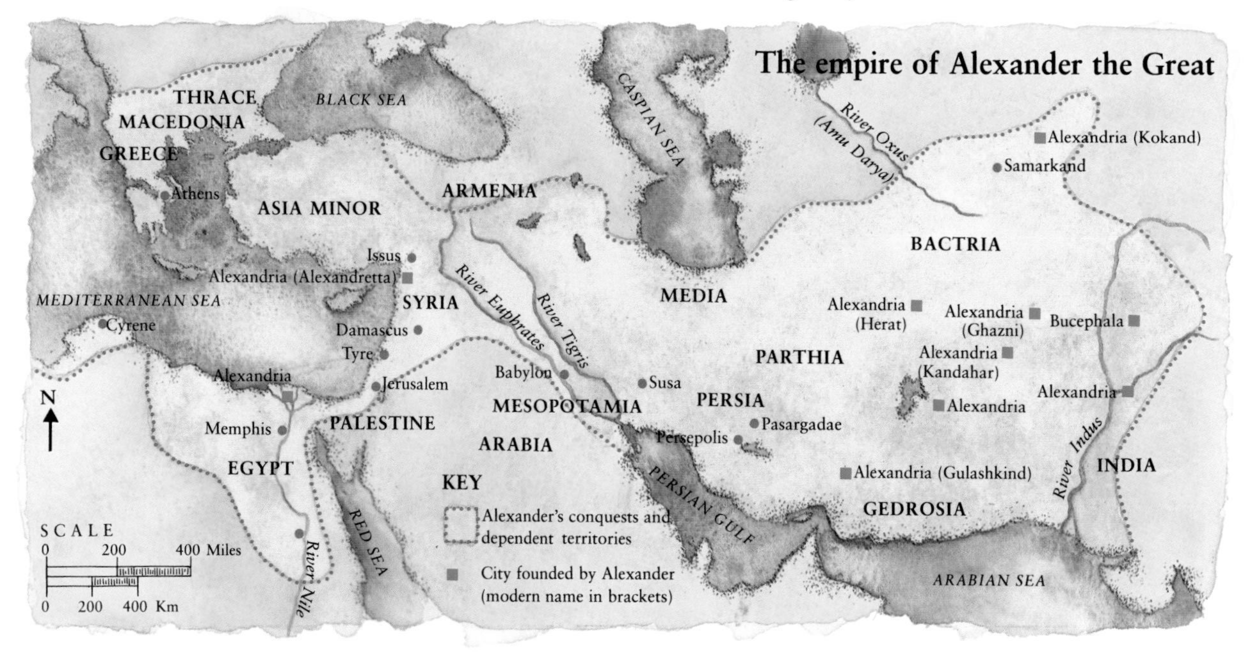

Alexander the Great and his successors

In 333 BCE, King Alexander III of Macedonia (Alexander the Great), defeated the Persian king Darius III at Issus in Syria (see map). In 331 BCE the whole Persian empire was in Alexander's hands, and, with visions of a world empire before him, he campaigned as far as India. Alexander's empire was shortlived: in 323 BCE he died in Babylon aged just thirty-three. His conquests were divided up among five of his generals, who embarked on several years of struggle. But by 312 BCE two of them, had emerged as the leading powers in the Near East. Ptolemy was established in Egypt, while Seleucus controlled roughly the territory of the old Babylonian empire (see p.108). Thus the old political pattern of the Near East was restored, with its two great imperial centers of Egypt and Mesopotamia. As in the past, Palestine lay between the two powers, and changed hands several times before the Seleucid ruler Antiochus III Megas ("the Great") won control of the area by his victory in 199 BCE at Paneas (later Caesarea Philippi; see map on p.165).

THE SELEUCID DYNASTY
All dates are approximate and BCE.

Ruler	Dates	Ruler	Dates
Seleucus I Nicator	312–281	Demetrius II Nicator (again)	129–125
Antiochus I Soter	281–261	Cleopatra Thea	125–123
Antiochus II Theos	261–246	Antiochus VIII Grypus	125–96
Seleucus II Callinicus	246–225	Seleucus V	125
Seleucus III Soter	225–223	Antiochus IX Philopator	114–95
Antiochus III the Great	223–187	Seleucus VI	95
Seleucus IV Philopator	187–175	Antiochus X Eusebes	95
Antiochus IV Epiphanes	175–164	Antiochus XI Epiphanes Philadelphus	95
Antiochus V Eupator	164–162		
Demetrius I Soter	162–150	Demetrius III Philopator Soter	95–88
Alexander Balas (Epiphanes)	150–145	Philip I	95–84
Demetrius II Nicator	145–139	Antiochus XII Dionysios	87
Antiochus VI Epiphanes	139–138	Philip II	84
Antiochus VII Sidetes	138–129		

Ruins of the Hellenistic city of Gerasa (modern Jerash, Jordan), one of a league of cities east of the Jordan known as the Decapolis (see map on p.143).

A mosaic from the Roman city of Pompeii depicting Alexander the Great at the battle of Issus. Alexander is on the left, his adversary, King Darius, in the center.

DANIEL

Daniel, the hero of the book of that name, begins as a legendary figure from the remote past. In the book of Ezekiel, he appears as a national hero, the embodiment of wisdom, perhaps from the city of Tyre. He is therefore of Canaanite origin: a figure called Daniel is the subject of one of the Canaanite texts from Ugarit (see p.72). In the Bible, he becomes one of the Jewish exiles in Babylon and the tales about him are of the same general type that became popular in the postexilic period (see, for example, pp.130–31). They illustrate how Jews ought to behave in a foreign and idolatrous environment, and how their trust in their faith and Law will be vindicated. In spite of severe crises from time to time, the Jews live mainly in peace, and some even attain high positions at the court of the Babylonians or their successors, the Medes and Persians.

The friendship and respect between Daniel and these foreign kings contrasts with the hostility between the Jews and the second-century BCE Hellenistic king Antiochus IV Epiphanes (see pp.138–9). Probably between 150 BCE and 100 BCE, an editor combined the popular cycle of Daniel tales with the remaining chapters of the present book, so that the stories became relevant to the persecutions of Antiochus IV. Nebuchadnezzar's golden image, which everyone was ordered to worship (Dan. 3.5), recalls Antiochus's decree that the Jews must "build altars and sacred precincts and shrines for idols" (1 Macc. 1.46), in particular his erection of "a desolating sacrilege" upon the sacrificial altar in the Temple of Jerusalem (1 Macc. 1.54). Similarly, Darius's claim of worship for himself (Dan. 6.7) may refer to Antiochus's assumption of divinity.

The stories reflect above all Daniel's traditional character as the supremely wise man. It is by his wisdom that he is able to interpret dreams, show his superiority to Babylon's magicians and soothsayers, and win the royal confidence, all of which is probably a conscious echo of the biblical story of Joseph (see pp.48–51). There is a powerful theological purpose underlying all the episodes: no adversary, however powerful, can successfully withstand the one true God, and even the greatest of kings

BEL, SUSANNA, AND THE SONG OF THE THREE JEWS

As well as the work bearing his name, other stories about Daniel have survived. The apocryphal tales of Bel and the Dragon and of Susanna and the Elders, which are attached to the book of Daniel in the Septuagint, the Greek translation of the Hebrew Scriptures (see sidebar on p.132), display Daniel's wisdom not as the ability to interpret dreams, but rather as the practical astuteness reminiscent of King Solomon (see p.86). They have been described as the world's earliest detective stories.

The story of Bel and the Dragon is an example of Jewish propaganda against idolatry, which it mocks with a broad humor that would have delighted a popular audience. Similarly, the story of Susanna has elements that would appeal to popular taste: the attempted seduction of a beautiful young woman by two elders supposedly responsible for maintaining high standards in the community, and their clever unmasking by Daniel. The story ends with the comforting assurance that God will always protect and preserve virtue.

In order to lend the story of the three Jews in the fiery furnace a more overtly religious tone, the Septuagint inserts a prayer and a hymn (*The Prayer of Azariah* and the *Song of the Three Jews*) between Daniel 3.23 and 3.24. The inserted text, which is generally placed in the Apocrypha, is put into the mouths of Daniel's three Jewish companions (Azariah is the Hebrew name of Abednego) when they are thrown into the fiery furnace by Nebuchadnezzar. The three are made into models of Jewish piety, and their escape from the furnace is explained by the intervention of an angel who dampens the flames. *The Song of the Three Jews* (also translated as *The Song of the Three Children* and *The Song of the Three Young Men)* was adopted into the liturgy of the western Christian Church as two hymns known from their opening words in Latin as the *Benedictus es* (verses 29–34) and *Benedicite* (verses 35–68).

Susanna and the Elders *(ca. 1560), by* Iacopo Tintoretto *(1518–1594). Before the Renaissance, artists treated the scene of Susanna bathing with great modesty. However, from ca. 1500 the erotic aspect of the story tended to be emphasized.*

Belshazzar's Feast *(1820), by John Martin (1789–1854).*

are compelled to recognize that their authority derives from God alone.

A recent discovery has thrown light on two of the most striking narratives in the book of Daniel. The first narrative relates how King Nebuchadnezzar of Babylon was afflicted with madness and banished from human society for seven years, until he turned to the God of Israel and was restored (Dan. 5.31–37). Nothing corresponding to this episode is known of the historical King Nebuchadrezzar II (605–562 BCE; see p.108). But something similar may have happened to Nabunaid or Nabonidus, the last Neo-Babylonian king (555–539 BCE), who left Babylon for ten years to live in the remote desert town of Teman in Arabia. Among the Dead Sea Scrolls (see p.145) there is a prayer attributed to Nabonidus, in which he relates that in Teman he suffered from an ulcer for seven years. After praying in vain to gods of gold, silver, bronze, iron, wood, and stone, he was cured by a Jewish exorcist, who ordered him to write his prayer to the glory of the true God.

There is another link with Nabonidus in the story of Belshazzar's Feast (Dan. 5). Nabonidus, not Nebuchadnezzar as in Daniel, was the father of Belshazzar (Bel-sharra-usur), who ruled as regent 552–542 BCE during his father's absence. In Daniel, Belshazzar holds a great feast at which the guests drink from sacred vessels looted from the Temple of Jerusalem and praise "the gods of gold and silver, bronze, iron, wood, and stone" (Dan. 5.3–4). Daniel condemns Belshazzar for exalting himself "over the Lord of heaven" (Dan. 5.23) and foretells the fall of Babylon. Belshazzar is killed "that very night" (Dan. 5.30). The gods praised at the feast are the same as those in Nabonidus' prayer, and the story of Belshazzar's idolatry and downfall may be adapted from a piece of propaganda by the Babylonian clergy originally aimed at the religious policies of Nabonidus' regime. The priesthood of Babylon objected to Nabonidus' preference for the moon god Sin over Marduk, Babylon's national god.

The Four Beasts and God Enthroned, *an illustration to Daniel's vision from the Beatus Apocalypse, produced at Santo Domingo de Silos, Spain, ca. 1100.*

APOCALYPTIC

In the later chapters of the book of Daniel, the figure of Daniel changes from an interpreter of other people's dreams to someone who himself has dreams and visions. The four visions which he sees represent a distinct literary genre known as "Apocalyptic." Daniel is probably the first fully apocalyptic work, although elements and anticipations of the genre are found in some of the prophets. The Greek term *apokalypsis* means the "revelation" or "unveiling" of some secret message at the point where it is about to be fulfilled. So Daniel is commanded to write down his words and keep them "secret and sealed until the time of the end" (Dan. 12.9). Daniel is associated with the Jewish Exile in Babylon in the sixth century BCE (see pp.112–13), but the apocalyptic message relates to the time of the persecution of the Jews by the Seleucid king Antiochus IV Epiphanes in the mid-second century BCE (see pp.138–9), when the book of Daniel was actually produced in its present form. Apocalyptic works often contain surveys of past world history in the form of predictions. Since these past events are known to have come true, they guarantee the seer's trustworthiness when he foretells what is yet to happen. All four of Daniel's visions are concerned with the empires of the ancient Near East, culminating in Daniel 11 with a detailed history of Palestine under the Ptolemies and the Seleucids and an even more detailed account of the career of King Antiochus IV Epiphanes.

It is widely accepted that Daniel represents the outlook of the Hasidean

The second book of Esdras

Daniel is the only apocalyptic writing in the biblical canon, but the apocryphal second book of Esdras is a work with many resemblances and, indeed, references to it. Originally composed in Hebrew, it has extensive later Christian additions. It is attributed to the prophet Ezra (see pp.116–17), who is called Esdras in Greek. It is set in the time of the Babylonian Exile, but it was probably written ca. 100 CE and is actually concerned with the persecutions of the Roman empire and the fall of Jerusalem in 70 CE.

As with Daniel, the book mainly consists of the seer's visions. The most remarkable are the first four, in which Ezra engages in a dialogue with an angel, not simply about the sufferings of the Jews, but also about the general problem of the existence of evil in the world. The conclusion is pessimistic: the wrongs of the present can only be rectified in a new age still to come, which only a few will enjoy (2 Esd. 8.1). In other visions, there is animal symbolism. Rome is represented as an eagle, which is confronted by a lion, a nationalistic Messiah of the house of King David (2 Esd. 11.37–12.3), while the author takes up the human-like figure from Daniel and makes him into a pre-existent Messiah (2 Esd. 12.31–34).

Ezra becomes a second Moses, set to rewrite the Scriptures, which have been destroyed (2 Esd. 14.21). Five helpers produce ninety-four books in forty days. Twenty-four of these, representing the present Hebrew canon, are to be available for everyone, but the remaining seventy, representing apocalyptic and apocryphal writings, are only to be given to the "wise," for in them is "the spring of understanding, the fountain of wisdom, and the river of knowledge" (2 Esd. 14.47).

party, represented by the "wise" who alone will understand the significance of his message (Dan. 12.10). They expect deliverance through divine intervention rather than through the military prowess of the Maccabees (see pp.140–41), which the book dismisses as affording only "a little help." The work is written to encourage the faithful followers of the Law, by promising them that Antiochus's persecution will be short-lived and that the true religion will soon triumph.

Apocalyptic is the heir to the prophetic assurances that God will overthrow Israel's enemies and restore the nation, but its message of hope goes much further. It envisages a renewal of the whole cosmos, when the world will be restored to the ideal state in which it left the divine hand, and God's intervention will signal the final end of history. The powers of evil will be eradicated forever and the righteous dead will rise to enjoy eternal bliss. A belief in bodily resurrection appears for the first time in Daniel (Dan. 12.2), and the increasing acceptance of the doctrine probably owed much to the feeling that a just God would reward those, such as the Maccabee martyrs, who had died for their faith (see pp.140–41).

One immediately striking feature of Apocalyptic is its use of symbolic language and bizarre imagery. In part, this indicates the esoteric character of the Apocalyptic genre: its language is a code which only those who understand its secrets can unravel. However, the sources of this imagery lie in the myths and rituals of the ancient Near East. One example is the prevalence of animal symbolism, as in Daniel's opening vision, where the world empires are depicted as four terrible beasts: a winged lion; a tusked bear; a four-headed, four-winged leopard; and a "terrifying and dreadful" creature with iron teeth and horns (Dan. 7.3–7). They recall the fantastic monsters that appear frequently in Mesopotamian art, representing evil and demonic powers. In Daniel's vision they emerge from "the great sea," an image related to "the deep" of Genesis 1.2 and corresponding to the hostile watery chaos that appears in both the Babylonian Creation Epic and the Canaanite myth of Baal (see also pp.20–23 and pp.26–27).

In such texts, the deity defeats the monster of chaos, and this motif is also found in Daniel. The last beast is destroyed, and then "one like a human being" is presented to the high god, "the Ancient One" (or "Ancient of Days") and given a universal kingship (Dan. 7.13–14). One epithet of the Canaanite supreme god El is "Father of Years," and the fact that the "one like a human being" comes with the clouds of heaven shows that he, too, is a divine being. The scene is based on El's grant of kingship to the young storm god Baal after Baal has defeated Yam, the Sea in a battle for sovereignty. An angel explains the vision to Daniel and identifies the "one like a human being" with "the holy ones of the Most High" – the pious of Israel – who "shall receive the kingdom and possess the kingdom forever – forever and ever" (Dan. 7.18).

The monstrous lion-headed bird Imdugud, whose beating wings were said to cause storms; a relief from the temple of the goddess Ninhursag at Tell-al-Ubaid (ca. 3000 BCE). Such fabulous beasts were common in Mesopotamian art and inspired the fantastic creatures in Daniel and other apocalyptic writings.

A Greek youth riding bareback and wearing a cloak, short skirt, and a wide-brimmed "cap of Hermes," from a bowl. This style of dress was widely adopted by young Jewish men in the hellenized cities of Palestine.

ANTIOCHUS IV EPIPHANES

Little is known of conditions in Judah after Ezra and Nehemiah (see pp.116–17), for the remainder of the Persian age, and the period of Ptolemaic Egyptian rule. It was probably a time of some stability for the Jewish population. Internally, the country was governed by a group of Jewish aristocratic and priestly families who formed the Sanhedrin or senate (Greek *synhedrion*, "council") under the high priest of the Temple.

With the coming of Seleucid rule (see p.133), dissension broke out among these families, some supporting the Ptolemies in Egypt and others the Seleucids in Syria. Rivalry was particularly intense between the Tobiads and the Oniads. The Ptolemies had appointed the Tobiads tax collectors for the whole of Palestine, while the Oniads held the high-priesthood and were usually pro-Syrian. However, both families switched allegiances whenever it seemed to their advantage. Their rivalry for the high-priesthood came to be a regular source of trouble, and was one cause of the repressive religious measures imposed on the Jews by Antiochus IV Epiphanes (175–164 BCE).

One dispute between the Tobiads and Oniads prompted a raid on the Temple treasures by Heliodorus, the minister of the Seleucid king Seleucus IV Philopator (187–175 BCE). The high priest Onias III objected to the raid on the Temple and was replaced by Jason, his brother. Jason paid higher tribute to Seleucus IV's successor, Antiochus IV, and promoted a policy of hellenization. He built a Greek-style gymnasium in Jerusalem, where even young priests took part in sports, running naked and sporting the Greek cap of Hermes. The Jewish enthusiasts for Hellenism (see pp.132–3), who were mainly among the Jerusalem upper classes, exchanged Hebrew names for Greek ones, abandoned circumcision, and regarded themselves in every respect as residents of a Greek city.

In 171 BCE, Jason was deposed in favor of Menelaus, who promised Antiochus even more tribute, which was paid for with some of the Temple treasures. When Jason attempted a return to power, Antiochus responded

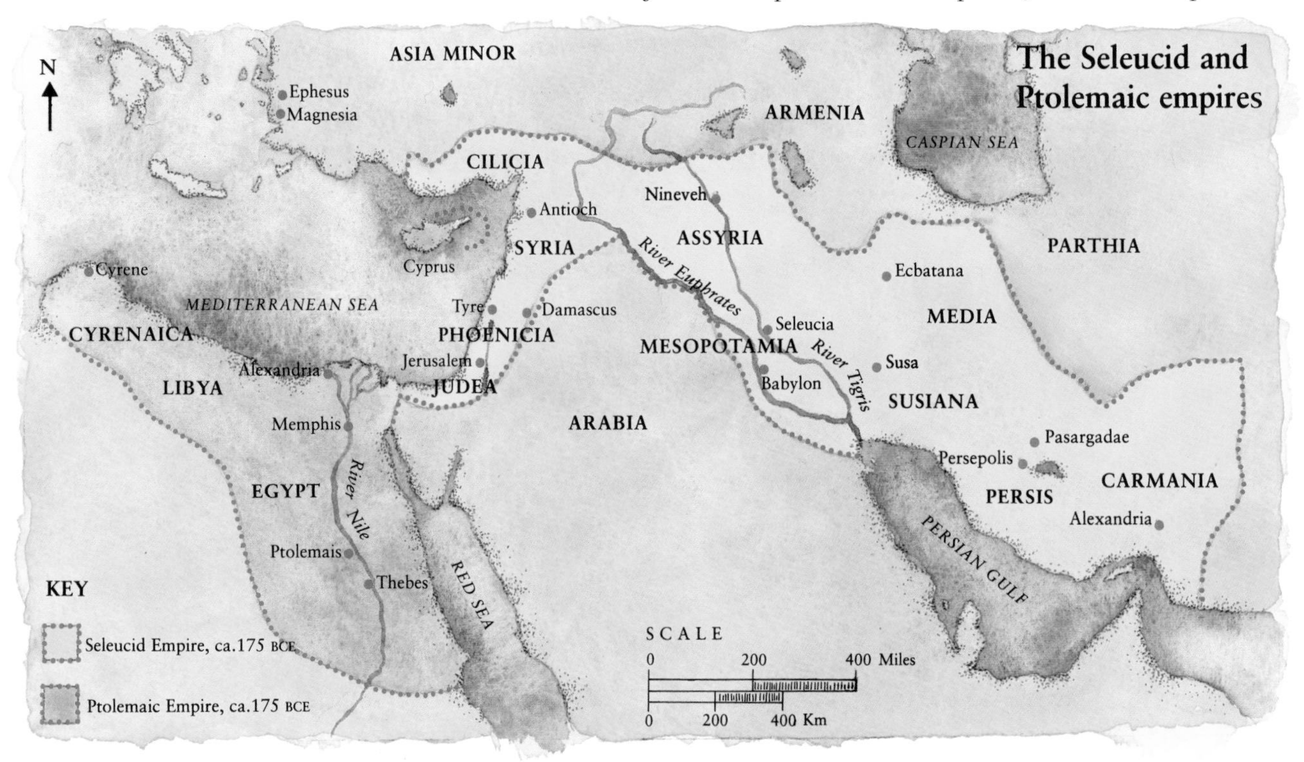

in 169 or 168 BCE by sacking the Temple. Then, in 167 BCE, he followed up this attack with a series of sweeping decrees that aimed to assimilate the Jews into his empire. Antiochus banned all the distinctive features of Judaism: circumcision, Sabbath observance, the keeping of feasts, and the purity laws. At the same time, sacred places were defiled, books of the Law were burned, and Jews were forced to attend sacrificial meals of pork (which was unclean to Jews) and ceremonies honoring pagan deities. In imposing these drastic measures, Antiochus would only have been in contact with the pro-hellenizing party in Jerusalem, who shared his objectives, and would not have anticipated any serious resistance. But in this he proved to be very much mistaken (see pp.140–41).

A silver four-drachma coin of Antiochus IV Epiphanes. The portrait of the king on the obverse (left) depicts him as the god Zeus; the reverse shows Zeus enthroned carrying the goddess Nike (Victory). The legend reads: "King Antiochus, God Manifest, Bearing Victory."

Unsurprisingly, Antiochus appears in the blackest light in Jewish tradition, but he was in fact an able and far-sighted ruler, who endeavored to counter the threat to his empire posed by the growing power of Rome. His father, Antiochus III, had been defeated by the Roman army at Magnesia in 190 BCE. Under the terms of the peace treaty, he had lost Asia Minor, been forced to pay a crippling indemnity, and had to send the young Antiochus as a hostage to Rome, where he remained for twelve years. The lesson was driven home for Antiochus himself when a Roman general ignominiously thwarted his hopes of conquering Egypt in 168 BCE.

To meet the Roman menace, Antiochus pursued a vigorous policy of hellenization in order to unite the various peoples of his heterogeneous kingdom. His aim was not to suppress the worship of local deities, although this was how it appeared to many Jews, but to assimilate them all to "Zeus," a name which had come to mean little more than "God." Antiochus himself took the title *Epiphanes* ("[God] Manifest"): he was the manifestation of God, by whatever name God might be worshiped.

The altar erected by Antiochus in the Temple of Jerusalem was called by the Jews "a desolating sacrilege" (1 Macc. 1.54). This is in fact a malformation of the name Baalshamaim, the Lord of heaven, a widely venerated Semitic god. Under Antiochus, the Temple was dedicated to Olympian Zeus, Zeus Olympios. Hence Antiochus sought to claim that Yahweh, Baal, and Zeus were the same deity under different names, depending on whether the worshiper was Semitic or Greek.

The Sack of Jerusalem *by Antiochus IV Epiphanes, a 15th-century French manuscript illumination.*

The Triumph of Judas Maccabeus *(1635), by Peter Paul Rubens (1577–1640).*

THE MACCABEES

Seleucid rule in Palestine came to an end as a result of the Maccabees, so-called from the rebel leader Judas Maccabeus. The Maccabee (or Maccabean) movement restored Jewish religious freedom after the repression of Antiochus IV (see pp.138–9), achieved political independence, and re-established a Jewish monarchy, albeit of a typically Hellenistic kind.

In one respect, the Maccabee rebellion was an expression of the long-standing tension between the more conservative Jews of the countryside and the cosmopolitan, hellenized capital. It was sparked off by a country priest, Mattathias, of a family called the Hasmoneans after his great-grandfather Hasmoneus. Initially, the struggle took the form of a guerrilla war to rescue the Law "out of the hands of the Gentiles and kings" (1 Macc. 2.48). The group around Mattathias and his family were joined by the Hasideans, the "pious" or "faithful," a group mentioned for the first time in 1 Maccabees 2.42. The Hasideans were specially devoted to the Law, and were prepared to die for it. They seem to have been the ancestors of similar groups within Judaism in the Roman period, such as the Pharisees and Essenes, and they always retained a separate identity from the Maccabees, from whom they eventually withdrew. The Maccabees were more concerned with the survival of the national religion than with the strict observance of the Law. When they heard of the massacre of a thousand people who had refused to take up arms on the sabbath, they resolved to retaliate if they were attacked on the day of rest (1 Macc. 2.41).

After the death of Mattathias, leadership passed to his third son Judas, nicknamed *Maccabeus*, "The Hammer," because under him the rebellion moved to full-scale war with the government forces, which Judas defeated in a series of battles. He eventually occupied Mount Zion (1 Macc. 4.37) and purified the Temple, which was re-dedicated, with great and joyful ceremony, on the twenty-fifth of Kislev (November–December) 164 BCE, the year Antiochus IV died. The day was ordered to be observed annually (1 Macc. 4.59), and remains as the Jewish feast of Hanukkah, "Dedication."

Recognizing Judas's success, Lysias, the regent of the boy king Antiochus V, rescinded Antiochus IV's religious decrees and deposed and executed the pro-hellenistic high priest Menelaus. Jewish religious liberty was at last restored, but it did not mean political freedom. The Jews were still subject to the Seleucid empire and obliged to pay taxes to it, and the Akra, the Greek garrison that Antiochus IV had established in Jerusalem, remained. The office of high priest continued to be in the Seleucid king's hands. On the appointment of the high priest Alcimus, who was of correct priestly lineage (unlike earlier royal choices), the Hasideans broke with Judas and backed the royal appointee. However, when they discovered that Alcimus was an ardent hellenizer, they soon returned to the Maccabee cause.

Independence now became the Maccabees' goal. The Seleucids still had enough force to defeat Judas, who fell in battle in 160 BCE (1 Macc. 9.18), but the kingdom then became increasingly weakened by dynastic rivalries. Jonathan, Judas's brother and successor, played one claimant to the throne off against another and, in return for his support, gained the office of high priest for himself and won more privileges for what was rapidly becoming a Jewish state. Full independence was achieved under the last Maccabee brother, Simon Hasmoneus. The Seleucid king Demetrius II (145–139 BCE) recognized Simon as high priest and granted him exemption from tax, effectively acknowledging the sovereignty of Judah. Simon secured his rule by taking the Akra, the last stronghold of foreign domination. The Jews began to date contracts and agreements from 142 BCE, "the first year of Simon the great high priest and commander and leader of the Jews" (1 Macc. 13.42).

A tomb built in Jewish-Hellenistic style in the Kidron valley, Jerusalem (see map on p.190). It is traditionally known as the "Tomb of Absalom," the son of King Saul, but it was probably built in the 1st century BCE.

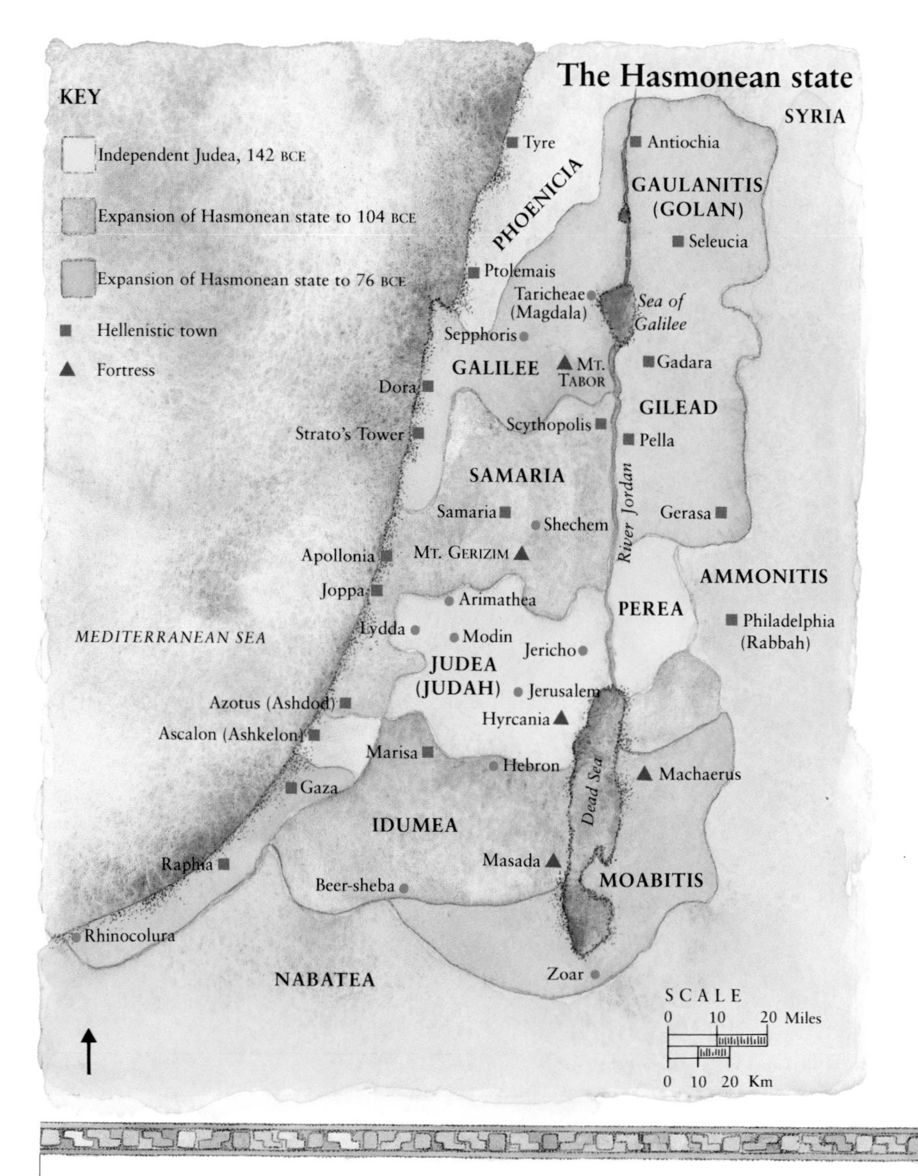

The first two books of Maccabees

The apocryphal first book of Maccabees begins with the career of Alexander the Great and ends with the death of the Hasmonean leader Simon, under whom the Jews obtained effective independence (see main text), and whose reign is depicted in idealistic language. The book exists only in Greek, but it was almost certainly written in Hebrew by a supporter of the Hasmonean dynasty founded by the Maccabees (see main text). It is recognized as a basically reliable and sober narrative, but it has its own bias. Its aim is to present the Hasmoneans as the divinely chosen saviors of the nation, and the author models himself on earlier Hebrew historical writings, such as the books of Samuel and Kings. Not surprisingly, he tends to exaggerate the successes of his heroes and to play down their defeats.

The second book of Maccabees is a Greek writing, both in language and style, an abridgement of a five-volume work by Jason of Cyrene in North Africa. This kind of Hellenistic historiography, as the author himself tells us in a preface, aimed to give pleasure to readers and to incite their emotions by concentrating on incidents of drama and pathos (2 Macc. 2.29–32). The book is marked by a number of highly rhetorical set pieces, such as Heliodorus's attempt to despoil the Temple and, especially, the account of the martyrdom of a mother and her seven sons.

The author sees the Maccabean revolt as a religious conflict between Hellenism and Judaism, and he is the first writer to use these terms in contrast. He carries the story no further than what he presents as the final decisive victory of Judas Maccabeus.

THE COMING OF ROME

There is no biblical record for the period between the accession of the Hasmonean ruler John, son of Simon, in 134 BCE (1 Macc. 16) and the reign of Herod the Great. Like the Seleucids, whose rule they had thrown off, the Hasmoneans ultimately fell victim to dynastic rivalry and infighting that was exploited by the growing imperial power of the Mediterranean region, Rome. By the period of the New Testament, Judea had become part of the Roman empire, governed by a new dynasty of client kings: the Herods.

HYRCANUS TO HEROD

In the final years of Hasmonean rule, those who had begun as the opponents of Hellenism ended as typical Hellenistic princes, although always with a clear Jewish identity. John, the son of Simon (see p.140), succeeded in 134 BCE, took the Greek name of Hyrcanus, and embarked on a policy of aggression. He extended his borders and destroyed the Samaritan temple on Mount Gerizim (see p.125). Like other Hellenistic rulers, he minted coins bearing his name and Greek fertility symbols. He was opposed by the Pharisees (see p.185), the spiritual descendants of the Hasideans (see pp.140–41), who plagued the dynasty from now until its fall. The Pharisees accused the Hasmoneans, who were not of priestly lineage, of profaning the office of high priest out of a desire for military and secular power.

John died in 104 BCE, but his son, Alexander Jannaeus (103–76 BCE), continued his militaristic tendencies. He acquired a reputation for his brutal suppression of internal opposition. By conquest and forced proselytization, he restored the Jewish state to its extent under David and Solomon. As well as being high priest, he assumed the title of king, which appeared in Greek (*basileos*) on his coins. But after Jannaeus died the

Herod the Great

Whatever his personal faults, and they were many, Herod was an outstanding figure among Rome's client kings and not undeserving of his epithet "the Great." He strengthened his realm economically and extended its borders (see map). He understood that his authority rested on Rome's favor and preserved his position by skillful diplomacy in the face of the power struggles that took place in Rome during his reign.

The remains of the fortress and palace of Herodium, built on an artificial hill by Herod the Great who was also buried here.

Herod's state had a very mixed population, which he sought to unify by a thoroughgoing and largely successful policy of hellenization. He founded a number of splendid cities as standard bearers of Hellenistic culture, with theaters, baths, marketplaces, gymnasia, and temples. However, unlike the Seleucids (see pp.138–9), he respected Jewish religious sensitivities: he rebuilt the Temple on a magnificent scale (see p.91 and map on p.190), and recognized the respect in which the Pharisees were held by the people. He tried to win the Pharisees' favor by observing their ritual laws, and exempted them from taking an oath of allegiance to him.

In spite of all his efforts, Herod never really won the hearts of the Jews, who regarded him as a foreigner (he was half-Idumean), and resented his dictatorial rule, control of the high-priesthood, and hellenizing ways. His ruthless suppression of potential opposition within his own family meant that he was unable to organize an orderly succession. In the following century, this failure ultimately spelled disaster for the nation.

Hasmonean state lapsed into prolonged dynastic rivalry. Increasing popular unrest was spearheaded by the Pharisees, who campaigned for the abolition of the kingdom and the establishment of a theocracy under a high priest of the legitimate line. The collapsing Seleucid state was powerless to intervene in Judah, so the Hasmoneans survived for a time. In 64 BCE, a brilliant young Roman consul, Pompey, captured Antioch in Syria and finished off the Seleucid empire. The arrival of the expanding Roman empire sealed the fate of the Hasmoneans. Pompey turned to Judah, where Jannaeus's sons, Hyrcanus II and Aristobulus, sought to outbid each other to win Roman support. Pompey sent the legate of Syria, Scaurus, who supported Aristobulus, but then Pompey himself took Jerusalem after a siege. He entered the Holy of Holies in the Temple – a sacrilege which pious Jews never forgot – but also restored the regular sacrifices and confirmed Hyrcanus II as high priest, exercising authority under the Syrian legate.

The real power behind Hyrcanus II emerged as Antipater from Idumea, whose people, the descendants of the Edomites, had been forcibly converted to Judaism by John Hyrcanus I. When Julius Caesar supplanted Pompey as Roman leader, Antipater was appointed Roman procurator of Judea (the Roman name for Judah) and his son, Herod, became military governor of Galilee. Herod's success in this post led the Roman Senate to appoint him king of Judea in 40 BCE. Although a client of Rome, Judea was once again a kingdom (see box on opposite page).

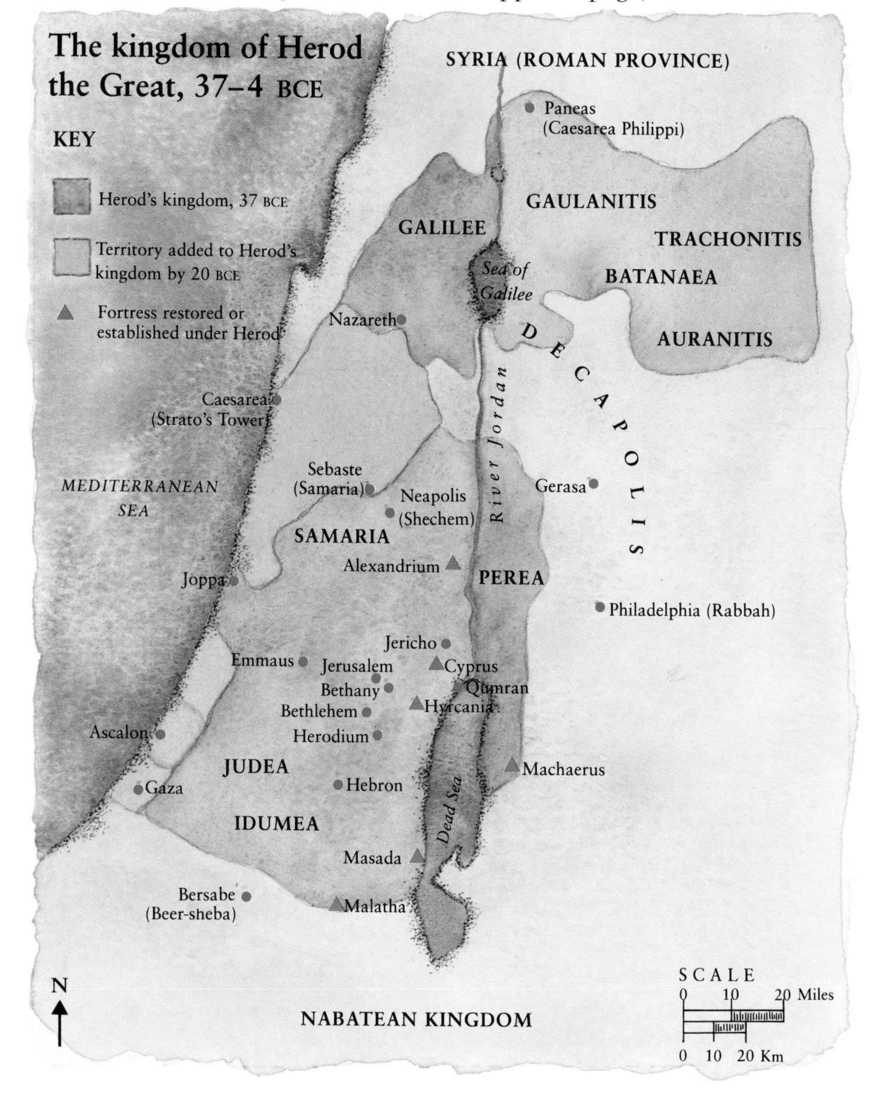

THE SYNAGOGUE

One of the most significant developments in Judaism by the beginning of the Roman period was the rise of the synagogue. Its origins are obscure: it is first mentioned in texts from the first century CE, and the earliest remains of synagogues in Palestine date from this time. But by then it was clearly a well-established institution, and had probably come into being during the two preceding centuries. In any case, synagogues were widespread both in Palestine and in the Diaspora.

The synagogue reflected the central place held by the Law in Judaism of the period, for its focal point was the reading of Scripture and the exposition of the Law. It was thus essentially a place of instruction, not of worship. Synagogue worship developed fully only after the destruction of the Temple in 70 CE, when aspects of the Temple cult, such as psalmody and prayer, were transferred to it. Synagogue and Temple should not be seen as in opposition to one another, but rather as complementary, not least because the Law taught in the synagogue authorized the sacrifices and priesthood of the Temple. The synagogue was one sphere of activity of the Pharisees (see pp.184–5), who employed it to teach the Law, and their interpretation of it, to the mass of the people. After the fall of the Temple, Pharisaic doctrine and the synagogue were key elements in the development of Judaism.

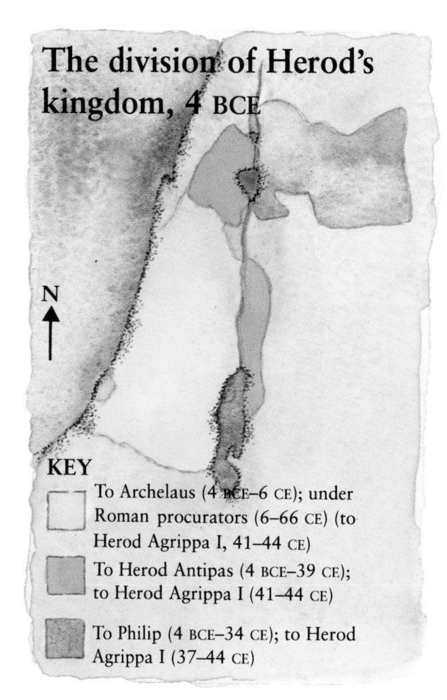

THE HERODIAN DYNASTY
(See also maps on p.143.) All dates are approximate and CE unless stated.

Herod the Great, *king of the Jews*	37–4BCE
Archelaus, *ruler of Judea*	4BCE–6
Herod Antipas, *ruler of Galilee, Perea*	4BCE–39
Herod Philip (Philip), *ruler of lands east of Galilee*	4BCE–34
Herod Agrippa I, *king of lands east of Galilee*	37–44
king of Judea, Galilee, Perea	41–44
Herod Agrippa II, *king of lands east of Galilee*	53–100
king of part of Galilee, Perea	56/61–100

Herod the Great
- Aristobulus (died 7 BCE)
  - Herod Agrippa I
    - Herod Agrippa II
- Archelaus
- Herod Antipas
- Philip

Two "suicide lots" (above) found at Masada (right), the last outpost of Jewish resistance to the Romans during the war of 66–73 CE. When faced with defeat, the Zealots defending Masada wrote their names on shards of pottery and drew lots to determine the order in which they would commit suicide. On the top right of the aerial view can be seen the ramp built by the Romans to break the siege.

JUDEA UNDER THE ROMANS

In his will, Herod the Great specified that his kingdom was to be divided among his three surviving sons (see map on p.143). This arrangement was ratified by the Roman emperor Augustus, but he withheld the title of king from Herod's successors: there was to be no danger of their claiming independence from Rome. Archelaus was given Judea, Samaria, and Idumea; Antipas received Galilee and Perea; and Philip gained the mainly non-Jewish region to the north and east of Lake Galilee.

All three behaved as typical Hellenistic rulers, and while this worked well for areas with mixed populations, Archelaus was no more acceptable to the strongly Jewish populace of Judea than his father had been. After a troubled reign, complaints from his subjects led to his deposition by Augustus in 6 CE, and Judea became an imperial province under a Roman procurator, who resided in Caesarea. It remained a procuratorial province until 66 CE, apart from the period 37–44 BCE, when the emperor Caligula made Herod's grandson, Herod Agrippa I, king of the region.

Procuratorial rule proved a disastrous failure. Most of the procurators proved deeply insensitive to Jewish religious feelings, and their actions were often tactless or even deliberately provocative. The most famous procurator, Pontius Pilate (in office 26–36 CE), took Roman military standards bearing pagan images into Jerusalem, and plundered the Temple treasury to build an aqueduct. Such acts inevitably provoked riots, and the procurators tended to overreact and put them down with great brutality. There was often disorder during the great festivals, when Jerusalem was crowded with pilgrims. Things might have gone better if the Romans

had been able to govern, as they did elsewhere, through a ruling class that was acceptable to the people. But the Judean aristocracy enjoyed little popular support, and the high priests and the Sanhedrin (see p.138) were regarded as Roman puppets. Economic troubles added to the general misery, and Roman taxation was a heavy burden on small farmers and merchants.

As a consequence, there were constant attempts at rebellion, led by individuals with messianic claims, who gathered large followings by promising that God was about to overthrow the hated Romans. Law and order became precarious: sources speak of outright banditry and of the *Sicarii*, "Dagger Men," who assassinated those associated with the Romans, including, on one occasion, the high priest. The procurators countered the unrest with harsh measures that only fueled Jewish resentment, setting the scene for the final catastrophe. The great Jewish War (66–73 CE) broke out when the last procurator, Gessius Florus, requisitioned a heavy payment of gold from the Temple treasury and provoked a riot which proved uncontrollable. Florus was forced to abandon the city to the rebel. The revolt was in part a civil war, a popular uprising against the Jewish upper classes, whose authority in Jerusalem was destroyed.

Rebellion spread to the whole country and at first the Jews were successful, defeating an army sent by Cestius Gallus, the military governor of Syria. In Rome, the seriousness of the situation was quickly realized, and the emperor Nero (54–68 CE) dispatched a general, Vespasian, to quell the revolt. He first subdued Galilee, and by 68 CE prepared to advance on Jerusalem. There was a respite for the rebels when, after Nero's death, Vespasian left for Rome to assume the throne, leaving his son Titus (later emperor) to continue operations. In 70 CE, after a long siege, Titus captured Jerusalem, destroyed the Temple, and took its treasures as booty. Resistance to Rome was now led by the Zealots, a fanatical nationalist party. They held out until 73 CE, when their last stronghold, Masada, the former mountain fortress of King Herod, finally fell. Almost the whole garrison – a thousand men, women, and children – committed suicide rather than surrender. Jewish hopes for an independent state in Palestine were to remain unfulfilled for another nineteen hundred years.

Part of the arch erected in Rome in honor of the emperor Titus (79–81 CE). It shows Temple treasures paraded through Rome after Titus's sack of Jerusalem in 70 CE.

ROMAN PROCURATORS OF JUDEA
All dates are approximate and CE.

Coponius	6–8
Marcus Ambivius	9–12
Annius Rufus	12–15
Valerius Gratus	15–26
Pontius Pilatus (Pilate)	26–36
Marullus	37
Herennius Capito	37–41
[King Herod Agrippa I ruled]	37–44
Cuspius Fadus	44–46
Tiberius Alexander	46–48
Ventidius Cumanus	48–52
Marcus Antonius Felix	52–60
Porcius Festus	60–62
Clodius Albinus	62–64
Gessius Florus	64–66

Qumran and the Dead Sea Scrolls

Not everyone in Judea responded to Roman rule with violence, and one group in particular simply withdrew into its own idealistic community. They were the Essenes, whose writings – the so-called "Dead Sea Scrolls" – and ruined buildings have been recovered since 1947 at the remote desert site of Qumran by the Dead Sea.

The Essenes probably began as a protest movement by priests of true, that is Zadokite, priestly descent (see p.123), against the non-Zadokite priesthood of the Hasmoneans (see p.140). They were a strongly ascetic group, with strict disciplinary rules, holding possessions in common and with their own distinctive worship, in which communal meals figured prominently. Their scrolls show that their main occupation was the study of the Scriptures, which in their own writings they interpreted according to the teachings of their founder, who was called the Teacher of Righteousness. They regarded themselves alone as the true Israel, and their separation from the rest of Jewish society was absolute. Essentially, they were a quietist apocalyptic sect, preparing themselves for the coming judgment, when they would inherit God's promises for Israel. A new Temple, they believed, would then take the place of the unclean Temple of Jerusalem.

Not all Essenes lived at Qumran and there were many Essene communities all over the country. It has been suggested that both John the Baptist and Jesus may have been linked with the Essenes, at least for part of their careers. There are a number of similarities between the outlook of the Essenes and the teachings of John (see pp.160–61), Jesus, and the early Church. However, there are also many profound differences, and any close connection is unlikely.

THE NEW TESTAMENT

Left: The Tree of Jesse, *from the Ingeborg Psalter (French, ca. 1210). The descent of Jesus from Jesse, the father of King David (Matt. 1.6–16; compare Luke 3.23–38) was commonly represented in European illuminated manuscripts of the 12th–14th centuries. The genealogy was shown as a tree with Jesse sleeping at its roots and Jesus at the top. Between them are (from below) David, Solomon, and Mary.*

I THE "GOOD NEWS"

The New Testament opens with the gospels of Matthew, Mark, Luke, and John. "Gospel" is an old English word for "good tidings," and directly translates the Greek euangelion, *meaning good news in general. In the New Testament it takes on the specific sense of the message brought by Jesus and preached by his followers. To Christians,* euangelion *came to mean a written account of the life, death, and resurrection of Jesus. The gospels tell similar stories, but often exhibit differences in detail and outlook. All four begin with the preaching of John the Baptist.*

St. Mark, *from the Ebbo gospels, produced at Rheims, France, 816–835* CE.

WRITING THE LIFE OF JESUS

For most of Christian history, the gospels have been interpreted as biographies, a view that goes back to the writer Justin Martyr (ca. 150 CE), who described them as "memoirs of the apostles." However, the common critical approach to the four New Testament gospels – Matthew, Mark, Luke, and John – has been to see them as works with no real parallels in the literature of their own time. According to this view, their authors, the evangelists, were primarily transmitting popular oral traditions. More recently, they have again come to be understood as some form of biography, although a number of resemblances to other writings of the period have also been addressed. In addition, the gospel writers are now seen not merely as transmitters of existing traditions, but as creative authors, each of whom contributed his own distinct understanding to the material he had available.

For the first Christians, the only real "Scripture" was the Jewish canon, and it was probably not until the end of the second century CE that what we know as the four New Testament gospels were generally accepted as an authoritative group. At this time, the theologian Irenaeus of Lyons (ca. 180 CE) identified their authors with the four beasts of Ezekiel 1, which have since become the symbolic figures for the evangelists in Christian iconography: "The four had the face of a human being, the face of a lion on the right side, the face of an ox on the left side, and the face of an eagle; ... each creature had two wings" (Ezek. 1.10–11).

Various factors probably contributed to the four gospels' achievement of canonical status. The books were recognized as having been composed at an early date: all of them must have existed by 100 CE, and scholars now tend to date them considerably before this. Despite their differences, especially between John and the three "Synoptic" Gospels (see below), they share a common form and content, recounting the career of Jesus in a connected narrative that begins with the activities of John the Baptist and culminates in the arrest, death, and resurrection of Jesus. The gospels present essentially the same story in a manner that is more sober and less legendary than other early Christian writings that claim gospel status (see box). The authenticity, as perceived by the early Church, of the four gospels is seen in the way in which each book came to be titled ("*The* Gospel according to Matthew," and so on). Second-century Christians recognized that these gospels could be brought together as a single narrative: one such harmonization, the *Diatessaron* produced by the writer Tatian in the mid-second century CE, was widely popular.

It has always been recognized that the gospels of Matthew, Mark, and Luke are so similar in form and substance that they might be laid side by side and read in parallel. For this reason they are called the "Synoptic" Gospels, from a Greek word meaning "seen together." But while it is generally agreed that there is a relationship between the three texts, there is no consensus on the exact nature of the relationship. The question at the heart of what is called the "Synoptic Problem" is the origin of the

Apocryphal gospels

From the second century CE onward, if not earlier, Christians produced a vast body of literature to supplement the New Testament writings, including a number of gospels written in imitation of the canonical texts. Some of these apocryphal gospels, such as the "Gospel according to the Hebrews," are known only from brief quotations in other writings, but essentially they fall into two main categories: "Infancy gospels" and "Passion gospels." Their purpose was to satisfy popular curiosity about periods in the life of Jesus of which the New Testament records comparatively little, such as the "hidden years" between his birth and the beginning of his ministry, or the time between the Resurrection and the Ascension. They recount purely legendary incidents with the aim of exalting Jesus' person. These are often of a fantastic and even rather tasteless nature, such as a series of miracles worked by Jesus as a small boy.

Some of the apocryphal gospels contain sayings of Jesus not found in the Bible, some of which may be authentic. Of particular interest is the "Gospel of Thomas," one of a cache of documents discovered in 1945 at Nag Hammadi in Egypt. Probably dating to ca. 140 CE, it is a collection of one hundred fourteen sayings of Jesus, supposedly delivered to the apostle Thomas after the Resurrection. While the sayings may contain some genuine reminiscences of Jesus' teaching, the great majority represent the doctrines of a religious movement known as Gnosticism that was widespread in the Mediterranean world. Like a number of other non-canonical gospels, therefore, the Gospel of Thomas was the product of a Christian circle that held views different from those of the mainstream Church.

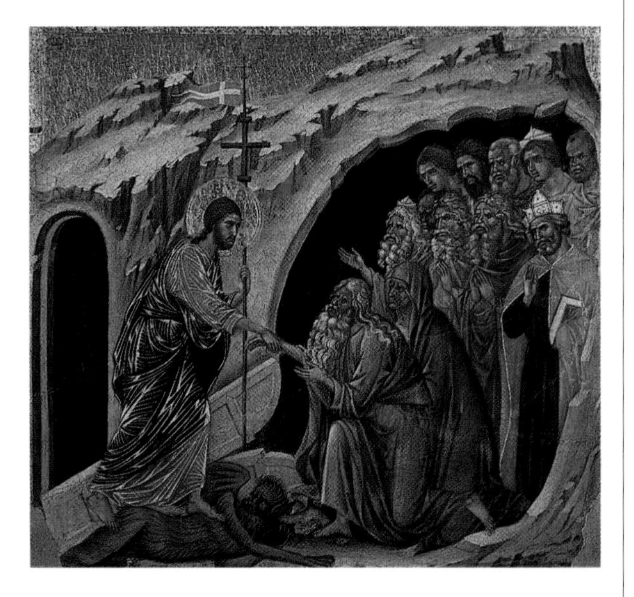

Christ's Descent into Hell *(ca. 1310), by Duccio di Buoninsegna (ca. 1278–1318). This episode does not appear in the canonical gospels, although it was suggested in Matthew 27.52, 1 Peter 3.19, and Psalm 24.7. It is found in the 4th-century apocryphal* Gospel of Nicodemus *or* Acts of Pilate, *itself partly derived from an earlier Latin apocryphal text. According to the story, after his death, Jesus descended into the region of the dead (Hell, or Limbo) and liberated the righteous from sin and death. As here, artists often depicted Adam and Eve as the first to be freed.*

apparently common form of the first three gospels. Today, the most widely accepted theory (with numerous variations) is that Mark was the earliest gospel. According to this idea, Matthew and Luke drew upon both Mark and a second, lost, document which scholars call "Q" (from the German *Quelle*, "source"). "Q" was not a narrative, but consisted almost entirely of sayings of Jesus.

However, this view has been widely challenged. There is no hard evidence to support one chronological ordering of the gospels against any other, and the idea that because Mark's gospel is relatively unsophisticated and untheological, it must therefore be the earliest, cannot be sustained. The existence of "Q" is only hypothetical, and other explanations for the resemblances between the sayings of Jesus in Matthew and Luke are equally possible. A renewed emphasis on the individuality of the gospels, and on the need to study each in its own right, may lead scholars to see that each author used a much wider variety of documents or oral traditions. This in turn may lead to a greater understanding of the particular motives that lay behind each account. Until then, the "Synoptic Problem" is likely to remain unsolved.

St. Matthew, *from an 8th-century Anglo-Saxon manuscript of the Gospels.*

THE CANONICAL GOSPELS

Matthew's gospel, the first book of the New Testament, enjoyed particular prominence among the early Christians, being cited and commented upon more often than its three companions. Its popularity may have been due in part to its attribution to the apostle Matthew, but it is more commonly agreed that the author was an unknown Christian of the last decades of the first century CE. More clearly than any of the other gospels, Matthew sprang from the life and concerns of a particular community, possibly Antioch in Syria. Scholars agree that the gospel was written for a church within which there was apparently considerable tension. Conditions of persecution (Matt. 10.17–18, 23.34) and dissension (Matt. 13.21, 24.10) mandated that the gospel story be told anew. It is also agreed that the church was made up of both gentile and Jewish Christians, which would reflect the population of Antioch.

For Matthew, the history of salvation is divided into two epochs, that of Judaism (the Law) and that of Jesus (the Gospel or Good News). The polemic against the scribes and Pharisees, as in Matthew 23, is designed to stress that Jesus is the Messiah and his followers the true Israel, who can make their own decisions about doctrine and, especially, church order, another of the book's prominent concerns. The many quotations from the Jewish Scriptures highlight the fact that the era of fulfillment has succeeded the time of promise. Christians live by the Law of Moses, but as interpreted by Jesus, who has superseded the great lawgiver. Strikingly, Jesus is identified with the Wisdom of God, which in Jewish circles had by that time come to be equated with the Law (see pp.126–7).

In contrast to Matthew, Mark was, for much of Christian history, the least regarded of the gospels. Toward the end of the nineteenth century, however, critical scholarship came to view it as both the earliest and the most trustworthy record of the life of Jesus, and probably one of the main sources used by Matthew and Luke. Since then, it has increasingly been acknowledged that the author is a theologian rather than a historian, who presents his material in order to convey a particular message.

The opening verses tell the reader who Jesus really is: the Messiah, the one who fulfills the divine promise in the Scriptures, the Son of God who confronts Satan in the wilderness (Mark 1.1–15). But Jesus' contemporaries failed to recognize his true nature: Mark's answer is that this was concealed from them as part of the divine plan and as the result of their own blindness. The teaching, preaching, and miracles of Jesus meet with a response of incomprehension, whether from the people, the religious authorities, or even his own disciples. Only through his suffering, death, and resurrection does his identity become apparent, which accounts for the centrality of the passion narrative in Mark's gospel. The author was probably writing for communities living at the time of the Jewish war (66–73 CE): Mark 13 predicts the fall of Jerusalem in apocalyptic terms. He seeks to prepare them for the even more shattering event of the coming of the Son of Man in judgment. However, as a prelude to this, they must expect to suffer after the pattern of their Master's earthly life.

St. Luke, *from the Lichfield Gospels ("Gospels of St. Chad"), English, ca. 720* CE.

The gospel of Luke is related in content to those of Matthew and Mark, but stands apart from the other two Synoptics in a number of ways. A great deal of material is unique to Luke's work, and he displays considerable literary skill in his use of Greek, suggesting that he was writing mainly for gentile Christians. He has his own distinct historical perspective, because his gospel is the first of two volumes. The second is the Acts of the Apostles, his history of the early Church (see pp.214–15). History, as presented in Luke's two books, is divided into three epochs:

The theology of the Fourth Gospel

John differs from the other gospels in content, in his ordering of events, and, especially, in the long discourses, in which Jesus speaks of his own authority. The author seems to have drawn on a wide range of contemporary thought and speculation. Evidence from Qumran (see p.145) suggests that he may have been influenced by the dualistic belief in the struggle between good and evil, truth and perversity, which would eventually be won by God. Hellenistic Judaism (see p.132) and wider religious currents of the Greco-Roman world, such as Gnosticism, may also have influenced the evangelist. The sharp polemic faced by Jesus suggests that John's audience faced a crisis of faith because of attacks by the synagogues, which threatened to expel anyone who believed that Jesus was divine (John 9.22, 12.42). The author responds by emphasizing that faith in Jesus is the way to salvation and eternal life, and that Jesus is still accessible to those who seek him. A central theme of Jesus' great farewell address to the disciples (John 14–16) is to reassure Christians that without his departure from the earth, the revelation of God would be unfinished: the Spirit, which Jesus bestows as he leaves, continues that revelation and mediates the reality of Jesus to the Church in all ages.

St. John the Evangelist Writing The Gospel, *from the Lambeth Apocalypse (English, ca. 1260).*

the period of Israel, which ends with John the Baptist baptizing the crowds and the designation of Jesus as the Son of God (Luke 1–3); the period of Jesus (Luke 4–21); and the period of the Church, which begins with the descent of the Spirit in Acts 2 (see pp.216–17). The epoch of Jesus is thus the "middle of time," a time of peace when the devil has departed and the true nature of the kingdom of God is manifest in the goodness of Jesus. Luke is concerned to correct the belief of many Christians that the second coming of Jesus and the end of the world are imminent. Rather, he focuses on the nature of the kingdom and its reality in the life of the Church. The end of the world is indeed to be expected, but the time of its arrival cannot be known. Hence Jesus, in Luke, stresses the virtue of patience to his followers.

John's gospel stands apart from the three "Synoptics." It belongs to the "Johannine" tradition (from the Greek *Iohannes*, "John") also represented by the letters of John (see pp.252–3) and, perhaps, Revelation (see pp.254–6). The date of the Fourth Gospel is uncertain: it has often been considered a relatively late work, but scholars now tend to ascribe more authentic historical value to its portrayal of Jesus than they did in the past. Whether or not the author of the gospel of John – commonly referred to simply as the Fourth Gospel – was acquainted with the three Synoptics, his work relies on different sources and traditions, and displays a distinct theological perspective (see box, above).

THE GREAT PRECURSOR

The Visitation, *from the* Heures du Maréchal de Boucicaut *(northern French or Flemish, ca. 1410). On the occasion of her visit to Elizabeth (the mother of John the Baptist), Luke relates that Mary uttered a great prayer (Luke 1.46–55). Known as the* Magnificat *in the West from its first word in Latin, it became part of the liturgy of the Church.*

At the beginning of each gospel story, the figure of John the Baptist appears as the precursor of Jesus Christ, in fulfillment of the prophecy of Isaiah: "A voice cries out: 'In the wilderness prepare the way of the Lord'" (Isa. 40.3; each of the gospels quotes it slightly differently). John introduces a new rite, baptism, as a way to wash away one's sin and prepare for God's imminent judgment (see pp.160–61). According to the Bible, people flocked to him to be baptized in the River Jordan (Matt. 3.5–6; Mark 1.5). Among them was Jesus who, on being baptized, was declared Son of God by the Spirit that descended upon him from heaven "like a dove" (Matt. 3.16; Mark 1.10; Luke 3.22; John 1.32).

Luke's gospel makes a distinct contribution to the understanding of the figure of John the Baptist with its substantial narrative of his conception and birth. Luke opens with an introduction of a common Greek type and in good classical idiom, addressed to a certain Theophilus. This is followed by a group of stories in a very different style peculiar to Luke, like that of the Septuagint, the Greek translation of the Hebrew Bible (see p.132), with its characteristic Hebraic forms of speech. Two similar accounts of the births of John the Baptist and of Jesus are linked by the episode of the visit of Jesus' mother, Mary, to John's mother, Elizabeth (Luke 1.39–56). The Baptist is said to be of priestly parentage: his father, Zechariah, is of "the priestly order of Abijah," and his mother, Elizabeth, is "a descendant of Aaron," Israel's first priest (Luke 1.5). The narrative accurately reflects the arrangements and ritual practice of the Temple of Jerusalem in the first century CE: the angel Gabriel announces the birth of a son, John, to Zechariah on the occasion of one of the twice-daily offerings of incense in the Temple sanctuary (Luke 1.11–17). Zechariah questions the news brought by Gabriel and is rendered mute by the angel as a consequence (Luke 1.18–20). This is a typical outcome of an encounter between a human and a celestial messenger: Ezekiel and Daniel are similarly dumbstruck by their visions (Ezek. 3.15; Dan. 10.15).

The account also follows the familiar pattern of stories in the Jewish

John and Elijah

When Gabriel announces the forthcoming birth of John the Baptist (Luke 1.13–17), he says that John will be possessed by "the spirit and power of Elijah" and "turn the hearts of parents to their children" (Luke 1.17). He is quoting the last words of the canonical prophets, where Malachi foretells the coming of Elijah before the great day of Yahweh (Mal. 4.5–6). There was a belief among the Jews that the final age of history would be ushered in by the appearance of the figure of Elijah from heaven, where he had ascended (see p.98). This was apparently how the Baptist was perceived by his followers and also how he is represented in the Synoptic Gospels. Significantly, Mark's description of John's clothing in the desert – "camel's hair, with a leather belt around his waist" (Mark 1.6) – mirrors that of Elijah in 2 Kings 1.8.

All the evangelists bear witness to a close connection between John and Jesus, and see John as the starting point of the gospel, the one through whom Jesus perceived the imminence of the kingdom of God and his own place in its coming. However, in conceiving of John as Elijah, they stress the precursory character of his mission: he is the forerunner and announcer of one who is even greater. Thus John is seen as the voice crying in the wilderness (Isa. 40.3) who prepares the way of the Lord. This is understood most clearly in the Fourth Gospel, in which John says he is not Elijah but the invisible "voice of one crying out in the wilderness" of Isaiah's prophecy, who points everyone in the direction of Jesus Christ (John 1.23) .

Scriptures, in which a son is born to an aged barren woman and her husband, as in the case of Isaac (Gen. 18.9–15, 21.1–7), Samson (Judg. 13), and Samuel (1 Sam. 1). Luke contains many echoes of these stories, for example Zechariah's objection that he and his wife are too old for childbearing (Luke 1.18) parallels a response from Isaac's father Abraham (Gen. 17.17). But perhaps the most interesting parallel is Gabriel's statement that the future child is to abstain from wine and strong drink (Luke 1.15). The same injunction was laid on Samson's mother before Samson's birth (Judg. 13.4); it also applied to priests and to those who, like Samson, were "nazirites," and had taken a special vow to God (Judg. 13.5). Temperance was also a feature of the radical Israelite religious group called the Rechabites, named for Rechab, the father of Jonadab (2 Kings 10.15), their traditional ancestor. Rechabites renounced all trappings of civilization, such as living in houses and practicing agriculture, as related in Jeremiah 35. This recalls the ascetic life led by the Baptist in the desert.

The story of John's birth probably reflects a tradition among John's own disciples of highlighting the Baptist's exceptional character. There is considerable emphasis on his name, which is unusual in Zechariah's family (Luke 1.61). John (*Yohanan* or *Yehohanan* in Hebrew) means "Yahweh has been gracious," so John is understood to represent a special act of God's grace to his people. For Luke, however, the birth of John is, above all, the prelude to the birth of Jesus. This is the purpose of the visit of Jesus' mother Mary, to her pregnant relative, Elizabeth (Luke 1.39–45). Both Elizabeth and the infant in her womb acknowledge the greater authority of Mary's unborn son. This is the only meeting of Jesus and John in Luke (see Luke 2.18.22, in which John is arrested *before* Jesus' baptism).

The Baptist and Qumran

Since the discovery of the Dead Sea Scrolls at Qumran in the late 1940's and early 1950's (see p.145), it has often been suggested that John the Baptist may have been a member of the Qumran community, or at least of the Essene sect of which it formed a part. Qumran was governed by a priestly hierarchy and this may explain the tradition of John's priestly lineage. Both John and Qumran had a strongly eschatological outlook, that is, they anticipated the end of the world and final judgment, for which they prepared themselves in the wilderness. Ritual washings were prominent in the life of the sect, a feature that may be related to the importance of baptism in John's ministry.

The wilderness of Judea, near the Dead Sea.

However, these and other resemblances can be pressed too far. There were many apocalyptic movements within first-century CE Judaism, which may well have had much in common and influenced one another. John's style of open, public, preaching to all who would listen, contrasts with the closed and secretive character of Qumran, and there is no evidence that he shared the community's distinctive beliefs, such as its particular messianic doctrines. The Qumran washings were repeated every day in order to preserve levitical purity: baptism as practiced by John, at least according to the gospels, occurred only once and was aimed at securing the forgiveness of sins in the face of the imminent doomsday. However, the first-century Jewish historian Josephus states that baptism, like the Qumran washings, was "not to beg off from sins committed but for the purification of the body." While there was not necessarily any direct association between Qumran and the Baptist, both exemplified a widespread climate of apocalyptic expectation.

The Annunciation, *an anonymous Greek icon of the 17th century.*

THE ANNUNCIATION

In the New Testament there are two accounts of the Annunciation, the announcement by an angel of the imminent birth of Jesus. The motive behind these narratives is to make it clear that the birth was a miraculous event, in which God was directly involved. Essentially, they follow the same form as several episodes in the Bible, which recount a promise from God, given through an angelic messenger or a holy man, of a child to a woman who is apparently incapable of childbearing. Examples are the "annunciations" regarding Sarah (Gen. 18.9–15; see p.42); the mother of Samson (Judg. 13.2–5; see p.75); and Hannah (1 Sam. 1.4–20; see p.76). In each case, there is no question of a virgin birth, and pregnancy occurs through ordinary human sexual intercourse. The miracle lies in God's alteration of the normal course of nature, because these women are either infertile or past the usual age of childbearing. Their conceptions demonstrate that nothing is "too wonderful for the Lord" (Gen. 18.14). This pattern is closely reproduced in the gospel story of the birth of John the Baptist to Mary's aged relation, Elizabeth. It is announced by the archangel Gabriel in Luke 1.7–20, and later recounted by him to Mary to give her reassurance in Luke 1.36–37.

Both annunciations of the birth of Jesus are strongly colored by themes from the Hebrew Scriptures. In Matthew's gospel, an angel of the Lord appears in a dream to Joseph and tells him that his fiancée, Mary, is pregnant through the Holy Spirit (Matt. 1.18–20). Dreams and the appearances of angels are customary biblical devices for reporting revelations, as, for example, in Genesis 19 and 37. For Matthew, the chief significance of the account lies in the symbolic names of Mary's child: he is to be called Jesus, the Greek form of the Hebrew Joshua, which means "Yahweh saves." Matthew then quotes Isaiah 7.14: " 'Look, the virgin shall conceive and bear a son, and they shall call him Emmanuel,' " which means 'God is with us' " (Matt. 1.23). Neither the Hebrew word rendered in Greek as

THE ARCHANGEL GABRIEL
The angels, God's heavenly messengers, are headed by superior beings known as "archangels" (see p.119). The number of archangels varies, but Gabriel (see picture on opposite page) is always included among them. His name means "Strength of God," or "Mighty man of God," and he first appears in the book of Daniel, where he is both a messenger from God and an interpreter of Daniel's apocalyptic vision (Dan. 8.16; see p.136). Gabriel is also the instrument of the final destruction of the wicked, and he is said to be the angel whose trumpet blast will herald the end of the world (Rev. 11.15). He is traditionally identified with the archangel who is to announce the return of Christ in 1 Thessalonians 4.16.

The winged deity Ahura Mazda is depicted in this relief from the ancient Persian royal palace of Persepolis. His divine manifestations, the Amesha Spentas, may have served as models for the archangels in the Bible (see above and also p.119).

The Annunciation *(ca. 1445), one of five painted panels from the Magnoli (or St. Lucy), altarpiece, by Domenico Veneziano (ca. 1400–61).*

"virgin" nor its Greek equivalent necessarily imply someone who has never had intercourse, but simply a girl who has reached sexual maturity. However, the evangelist does appear to intend that Mary has never had sexual intercourse: her conception does not occur through human agency, while Joseph, with his strict Jewish background, would have expected his fiancée to be a virgin. He could have broken off the engagement if she were not, as indeed he planned to do before the angel appeared (Matt. 1.19).

The Annunciation narrative in Luke (Luke 1.26–38) foretells Mary's conception as well as the birth of her son, and again the influence of the Bible is prominent. The archangel Gabriel appears in human form, and announces to Mary that Jesus is to fulfill the Messianic promises of the royal line of David (Luke 1.32). But Joseph, not Mary, is descended from David (Luke 1.27), and Mary's statement "I am a virgin" (Luke 1.34) makes it clear that Joseph was not Jesus' father. To account for this genealogical discrepancy, later apocryphal gospels (and the early Church) claimed that David was the ancestor of Joseph *and* Mary.

In Luke, Jesus is explicitly the "Son of God." He is conceived, as in Matthew 1.20, by the Holy Spirit. Luke's narrative leaves no doubt as to Mary's virginity and the virgin birth of Jesus. Scholars have pointed out that the accounts in Matthew and Luke are comparatively late, and that earlier New Testament authors do not appear to know of the virgin birth. For this reason, it has been argued that the early Church created the story from prophecies in the Hebrew Scriptures (such as Isaiah 7.14), in order to magnify the person of Jesus, and to explain the idea of God made flesh.

Any final judgment on the authenticity of the virgin birth must depend on whether it is possible to accept belief in the miraculous. The notion of a deity impregnating a woman was widespread in ancient times: it is common in Greco-Roman mythology, where the offspring of such unions were often mythical or legendary figures, and it is found in Genesis 6.1–4 (see p.32). As Christianity moved out into the wider Mediterranean world, the concept may well have influenced the presentation and understanding of Christ's birth. However, the idea of conception without a human father would have been rejected by most Jews, and the New Testament is careful to avoid presenting the divine fatherhood in the frankly human terms that characterize myths about the lustful couplings of, say, the god Zeus. Mary conceives by the Holy Spirit – a far more refined theological concept.

THE LILY AND THE DOVE

There is a long tradition of representing the Annunciation in art, marked by the use of regularly recurring symbols. One of the most common of these symbols is the lily. A universal emblem of purity, the flower is often depicted flourishing in a pot, usually placed between the figures of Mary and the angel Gabriel. The particular association of the flower with Mary is derived from Song of Solomon 2.1–2: "I am a rose of Sharon, a lily of the valleys. As a lily among brambles, so is my love among maidens." One traditional Christian interpretation of the Song of Solomon (see p.131) has been to understand it as an allegory of the love of God for Mary, his bride, or the love of Christ for the Church. In the painting by the Italian artist Domenico Veneziano (above), Gabriel holds a Madonna lily (*Lilium candidum*). He is also sometimes shown holding a scepter surmounted by a trefoil lily (fleur-de-lys).

The dove, the most frequently mentioned bird in the Bible, also serves a powerfully symbolic function. In Christian illustrations of the Bible, it is used to depict the Holy Spirit, and in representations of the Annunciation, it is commonly portrayed hovering over Mary's head or with its beak inclined to her ear – a sign of her miraculous conception of Jesus (see picture on opposite page).

THE NATIVITY

Only one of the evangelists, Luke, relates the birth of Jesus in detail, but his beautiful and moving narrative poses some problems. He dates the Nativity to the time of a census of the entire Roman empire ordered by the emperor Augustus (27 BCE–14 CE), which is said to have taken place while Quirinius was governor of the Roman province of Syria (Luke 2.1–7). Nothing is known of a universal census under Augustus, and such a monumental undertaking would surely have been noted in other sources. However, the first-century CE Jewish writer Josephus records that Quirinius conducted a census of Syria and Judea, and he agrees with Luke that it was an innovation. Even if this was the census meant by Luke, there are other difficulties: Quirinius' term of office began only in 6 CE, during the reign of Augustus, but ten years after the death of Herod the Great (4 BCE; see p.143), in whose reign Luke places the events leading up to Jesus' birth (Luke 1.5). Luke is probably seeking to lend his story greater authenticity by linking it with historical events in the wider world. In both his gospel and Acts (see pp. 214–15), Luke's aim seems to be to give the Christian story a wider significance and a more universal relevance as part of general human history. It is worth noting that the genealogy of Jesus in Luke 3 goes beyond Matthew's genealogy (Matt. 1), to link Jesus with Adam, the ancestor of all humanity.

In Luke, the census requires all people to register in their ancestral towns,

The Nativity *(ca. 1430), by an anonymous German artist. The birth of Christ was first represented in the early 4th century* CE*, and since then it has inspired more images than perhaps any other New Testament subject with the exception of the Crucifixion. The ox and the ass are traditional details, but they are not mentioned in Luke. They owe their presence ultimately to a passage of Isaiah: "The ox knows its owner, and the donkey its master's crib" (Isa. 1.3). In this painting, unusually, Joseph is absent.*

so Joseph and Mary must travel south from their hometown of Nazareth in Galilee to Bethlehem in Judea, the birthplace of Joseph's supposed ancestor, King David (Luke 2.3–4). However, a Roman census took account only of residence and not ancestral home, so Luke's motive is probably theological. The early Church attached considerable importance to Jesus' descent from David, but, more significantly, Bethlehem was the expected birthplace of the Messiah, according to the prophet Micah (Micah 5.2, quoted in Matthew 2.6 and referred to in John 7.42). The gospels frequently say that Jesus came from Nazareth (Matt. 26.69–71; Mark 1.9; John 7.41 and so on). Luke's story may well be an attempt to reconcile Jesus' Galilean origins with the tradition that the Messiah would be born in Bethlehem.

Luke places the birth of Jesus in a humble setting: there was "no place for them in the inn," so Mary laid Jesus in a manger (Luke 2.7), suggesting that she gave birth in a stable. Since the mid second century, Christian tradition has held that Jesus was born in a cave, and, since at least the early third century, a cave at Bethlehem has been venerated as his birthplace; the present Church of the Nativity was built over it in the reign of the emperor Justinian (527–565 CE). Caves were often used as animal shelters: one cave stable, with its manger, has been dated to ca. 1200 BCE. However, legends of the birth of a divine being in a cave may have influenced the Christian tradition.

The idea that divine revelation is made to unsophisticated people, such as the shepherds in the gospel story, is in keeping with the general outlook of Luke's gospel. But there are also many pagan stories of gods visiting countryfolk, and shepherds are associated with the birth and infancy of several heroes, such as Cyrus and Romulus and Remus. The shepherds are the first to witness Jesus' nativity, having been sent to Bethlehem by an angel bringing "good news of great joy" concerning a "savior" sent by God to bring perpetual peace to his people (Luke 2.8–20). Luke's language here, in particular the expressions "bring good news" and "savior," echoes inscriptions from Asia Minor concerning the birthday of the emperor Augustus.

The Presentation

The story of the Presentation of Jesus to God in the Temple concludes Luke's birth narrative (Luke 2.22–39). It forms a parallel to the story of the shepherds: the very center of Jewish religious life now bears witness to the nature and mission of Jesus. Mary and Joseph present Jesus for the traditional rite of purification. Luke sees the rite as a dedication and a spiritual fulfillment of the legal rule that the firstborn male child belongs to God (Luke 2.22–24). It is probably a deliberate echo of the dedication of Samuel in the temple at Shiloh (1 Sam. 1.24–28). Simeon and Anna, who witness Jesus in the Temple, are both prophetic figures. The Holy Spirit has promised Simeon that he will not die before seeing the Messiah, and guides him to the Temple (Luke 2.26–27). Simeon's song of praise (Luke 2.29–32), known as the *Nunc Dimittis* from its first words in Latin, is composed of reminiscences of Second Isaiah (see p.112), and strikes a note that runs through all of Luke's gospel when it tells of Jesus bringing salvation to the gentiles as well as to Israel. Simeon also foretells the rejection and passion of Jesus, and Mary's heartbreak (Luke 2.34–35). Simeon's moving words to Mary illustrate Luke's interest in women. This is particularly marked in the gospel's early chapters, with the prominence there not only of Mary, but also of the Baptist's mother, Elizabeth, and the female prophet Anna (Luke 2.36–39).

The Presentation in the Temple, *a mosaic (1291) by Pietro Cavallini (ca. 1250–ca. 1350) in the church of Santa Maria in Trastevere, Rome.*

THE THREE WISE MEN

Matthew's account of the "wise men from the East" (Matt. 2) serves the same purpose as the stories in Luke of the shepherds and of Simeon and Anna (see box on p.157). As foreigners who pay homage to Jesus, they foreshadow the future spread of the Christian faith to the gentile world. It is possible that the story of the star that led the wise men to Jesus is based on some unusual natural phenomenon of the time (for example, it has been pointed out that Halley's Comet was visible ca. 10 BCE), but that is hardly the evangelist's prime concern.

Like other peoples, the Jews believed that the stars and the planets could serve as signs and portents revealing divine will. Matthew's account probably has its background in the Bible: in Numbers, the seer Balaam prophesies that "a star shall come out of Jacob" (Num. 24.17). This sign came to be interpreted in Judaism as a reference to the Messiah: a Jewish messianic movement against Rome (132–135 CE) named its leader Simon Bar Kochba, "Simon son of the star." Matthew states that the star, the rising of which the kings observed, was that of the king of the Jews, and various psalms understood by this time in a messianic sense speak of kings bringing gifts to an Israelite monarch. Particularly significant is a passage in Isaiah that came to be interpreted as an address to the Messiah: "Nations shall come to your light, and kings to the brightness of your dawn." It goes on to speak of camels bringing gold and frankincense with which to honor God (Isa. 60.6). Such passages were the origin of the belief that the visitors were kings and that there were three of them.

In the original Greek, the visitors are termed *magi,* a word describing practitioners of Eastern magical arts. The Magi were originally a clan of the Medes (see p.108) who eventually formed the priestly caste in Persia. By the time of Jesus, *magi* had come to indicate professional practitioners of various kinds of magical knowledge, such as dream interpretation or astrology. Such practitioners were widespread in the Mediterranean area, but an awareness of the origin of the *magi* in the remote East always remained, as Matthew's gospel shows. Elsewhere in the New Testament, the *magi* are not portrayed in such a positive light. In the Acts of the Apostles, the magical arts of the magician Simon (Acts 8.9–24) and the

THE MASSACRE OF THE INNOCENTS AND THE FLIGHT TO EGYPT

According to Matthew, King Herod ordered the slaughter of all infants under two years old (the "Massacre of the Innocents") in Bethlehem in an attempt to kill the Messiah whose birth he has heard of (Matt. 2.16–18). Jesus' family had meanwhile fled to Egypt, where they remained until Herod's death (Matt. 2.13–15).

Matthew cites a prophecy of Jeremiah in relation to the Massacre of the Innocents (Matt. 2.18), but the episode is more directly derived from Exodus 1.15–22, where another king threatens male infants with death (see pp.56–7). The tale is part of Matthew's attempt to present Jesus as a new Moses (see also Matt. 2.20 and Exod. 4.19). The massacre cannot be regarded as historical, but Herod the Great (see p.142) was certainly a ruthless monarch, and Matthew's narrative would have fitted with what was remembered of his rule.

The flight of Joseph and his family into Egypt is another story recorded only by Matthew (Matt. 2.13–15). The tale is based on Hosea 11.1, and shows Jesus going into Egypt as the representative of Israel. Again, the story of people being forced to flee royal oppression would have been compatible with memories of Herodian rule.

The record of the family's settlement in Nazareth on their return (Matt. 2.23), appears to be an attempt to reconcile Jesus' Galilean background with the tradition of his birth at Bethlehem. Matthew quotes an obscure and untraced prophecy (Matt. 2.23), apparently to show that the term "Nazorean" or "Nazarene," by which Christians were later sometimes known, indicated that Jesus came from Nazareth.

Matthew relates (Matt. 2.22) that Joseph was afraid to live in Judea under Herod's son, Archelaus, and took his family to Galilee, which was ruled by another son, Herod Antipas (the Herod of the John the Baptist story). Archelaus was a notorious tyrant: appointed ruler of Judea by the Romans on his father's death in 4 BCE, his rule proved so harsh that he was deposed in favor of direct Roman rule in 6 CE (see map on p.143).

The present-day hilltop town of Nazareth.

Jewish magician and false prophet Bar-Jesus, or Elymas (Acts 13.6–12), are defeated by Christ's apostles. The contemporary Jewish philosopher Philo expresses a high regard for the genuine *magi*, but attacks charlatans who pervert their practices. Matthew reflects Philo's positive view of the *magi*: their wisdom enables them to recognize the significance of the star and the royal status of Jesus (Matt. 2.2); and they have the power to receive a message from God in a dream (Matt. 2.12).

Unlike Luke, Matthew sees Bethlehem as the native town of Joseph and Mary: he traces Jesus' descent through King David's royal progeny (Matt. 1) and indicates that his parents have a house there (Matt. 2.11). The astrologers ask in Jerusalem where the king of the Jews has been born (Matt. 2.2), and the chief priests and the scribes tell them that prophecy has designated Bethlehem as the place from which Israel's true ruler, the "shepherd," will come (Matt. 2.4–6). Herod thus knows the child's town but not his actual family: only the Eastern visitors are guided by the star to the house where Jesus lies.

A coin issued by the Jewish revolutionary leader Simon Bar Kochba during the revolt against Rome in 132–135 CE. Bar Kochba was hailed by his followers as the long-awaited Messiah. The coin depicts the façade of the Temple of Jerusalem, destroyed in 70 CE.

The Adoration of the Kings *from one of the most magnificent of all medieval European manuscripts, the* Très Riches Heures du Duc de Berry *(1413–1416), illuminated by the Limbourg brothers (active ca. 1400–1416).*

II THE MINISTRY OF JESUS

According to the gospels, Jesus' adult career began with his baptism by John, who proclaimed the arrival of the Messiah and God's coming judgment. After forty days and nights in the wilderness, Jesus returned to Galilee, where he began his ministry. He traveled in Galilee and to Jerusalem, teaching in synagogues and in the countryside. In preaching to his ever-growing audience, he relied largely on parables, at least according to the Synoptic Gospels. Miracles, which involved healing the sick, exorcising demons, and overriding the power of nature, also formed an important part of his ministry.

THE BAPTIZER

The gospel writers shaped and added to the traditions about John the Baptist (see pp.152–3) for their own literary and theological purposes. However, the three Synoptic Gospels all agree that the central element in his work was his advocacy of baptism for the remission of sins: Mark gives him the explicit title "the baptizer" (Mark 1.4). John's practice of baptism has been linked with the Jewish baptism of proselytes as a rite marking initiation into a new community, although it is doubtful whether this occurred as early as the period of John's activity. Scholars have also speculated on the relationship between John and the Qumran community (see p.153), which also used washings as a means of attaining ritual purity. Other baptist movements seem to have flourished in the Jordan valley in the late centuries BCE and early centuries CE, in opposition to the official Jerusalem priesthood. One descendant of these may be the Mandaeans, a Gnostic sect which claims the Baptist as a member and, to this day, baptize in streams which they call "Jordan."

A distinctive feature of John's baptism as recorded in the gospels was its eschatological tone, that is, it anticipated the end of the world. He proclaimed an imminent universal judgment, from which people could escape only by repentance and cleansing from sin. According to Matthew, the Jews, and especially their religious leaders, the Pharisees and

The Baptism of Christ, *from the workshop of Bicci di Lorenzo (1375–1452). John baptizes Jesus as the dove descends from Heaven (compare Luke 3.22).*

Sadducees (see pp.184–5), could no longer rely on their traditional claim of descent from Abraham to justify their status as the chosen people.

In Luke, the crowds, tax collectors, and soldiers ask John how they should avoid God's judgment. He tells them to lead a life founded on justice and social responsibility, in accordance with their professions (Luke 3.10–14). Luke's gospel shows a special interest in these particular social groups, who also feature prominently in his account of Jesus' ministry.

According to all the gospels, John taught that he was not the Messiah, as people thought he might be (see, for example, Luke 3.15). The gospels all quote Isaiah 40, which speaks of the one who will "prepare the way of the Lord" (Isa. 40.3; compare Matt. 3.3; Mark 1.3; Luke 3.4, John 1.23). This precursor will be followed by a greater figure, who will bring judgment on the wicked, but will deliver the righteous, through baptism in the Holy Spirit. For the Fourth Gospel, the representation of the Baptist as the great forerunner of Jesus eclipses everything else about him: his baptizing is mentioned only incidentally (John 1; see sidebar, right).

John's baptism of Jesus might suggest that Jesus was originally a disciple of John, and the gospels seem concerned to deny this. In Matthew, John recognizes Jesus' superiority and baptizes him only when he insists (Matt. 3.13–15). Luke places John's imprisonment (Luke 3.20) before Jesus' baptism (3.21), and does not state explicitly that John baptized him. The Fourth Gospel has no direct account of a water baptism of Jesus at all.

The baptism acknowledges Jesus as God's son and as the Messiah. In Mark, this seems to be a personal experience in which Jesus recognizes his true nature: Jesus alone sees "the heavens torn apart" and hears the divine voice proclaiming "You are my Son, the Beloved" (Mark 1.10–11). In Luke, the proclamation is described as a physical phenomenon: "the Holy Spirit descended upon him in bodily form like a dove" (Luke 3.22). In Matthew, the divine words are spoken in the third person ("This is my son"), thus addressing the public who stand as witness to Jesus' divinity (Matt. 3.16–17). God's utterance is drawn from Psalm 2.7 and other messianically interpreted passages from the Hebrew Scriptures. It designates Jesus as the Messiah, God's son, and "Beloved," a word which in Greek can equally mean "Unique One."

Jesus with the Temple teachers

Before his baptism, there is only one story about the youth of Jesus in the New Testament. According to Luke (the one gospel to record the event, at the age of twelve), Jesus stayed behind in Jerusalem after a visit there with his parents. They later found him in the Temple, "sitting among the teachers, listening to them and asking questions" (Luke 2.41–50). As with accounts of his birth and early infancy, it reflects the Christian tendency to see Jesus' messianic status as something that was realized not only at his baptism, but from the very beginning of his life. The central theme of the narrative appears when Jesus tells his parents that "I must be in my Father's house" (Luke 2.49): he is conscious that his sonship to God may override family ties. However, Luke is also careful to record Jesus' submission to parental authority (Luke 2.51).

The story was probably inspired by the desire to know more of Jesus' early life. Tales about the prodigious childhood of great figures occur elsewhere in ancient literature: for example, the first-century CE Jewish historian Josephus relates that, when he was only fourteen, he was praised for his learning in an assembly of the chief priests and rulers of his city.

The traditional site of Jesus' baptism by John in the Jordan River.

THE PROLOGUE TO THE FOURTH GOSPEL

The Fourth Gospel has no account of Jesus' birth, baptism, or temptation (see p.162). It begins with a picture of the pre-existence of the incarnate Jesus Christ as the Word of God: "In the beginning was the Word, and the Word was with God, and the Word was God. He was in the beginning with God" (John 1.1–2). The description points (John 1.3) to the creative divine word seen in the first chapter of Genesis (see pp.26–7), and is also strongly influenced by the concept of divine Wisdom (see pp.126–7) that developed in Judaism. In the gospel, this Wisdom is uniquely present in the person of Jesus Christ.

The Greek term for "word" is *logos*, and the fourth evangelist's teaching resembles that of the first-century CE Jewish author Philo of Alexandria, who saw the *logos* as a divine entity alongside God himself. It can also be compared with the view of the *logos* expressed by the Stoic philosophers, who saw it as the ordering principle of the cosmos. (However, this is unlikely to be a direct source for the Fourth Gospel.)

This gospel, like the others, is greatly concerned with John the Baptist, but here the sole purpose of the Baptist's message is to witness to the true nature of Jesus. In his dialogues with the Jewish religious authorities, John asserts that he is only a voice announcing the savior to come (John 1.19–23), and the descent of the Spirit on Jesus is the sign to John that this is indeed the one he has foretold (John 1.32–34). He calls Jesus "the Lamb of God"(John 1.29, 1.36), no doubt a reference to the sacrificial lamb (Exod. 29.38–42) and to Jesus as the bringer of redemption by his death. But the lamb may also have overtones of the language of apocalyptic (see pp.136–7), where the animal is a conquering figure who overcomes evil beasts, the symbols of sin, as in Revelation 5.6–12.

THE TEMPTATION

The Synoptic Gospels all record Jesus' temptation by the devil in the desert as occurring between his baptism and the beginning of his active ministry. Matthew and Luke describe three temptation scenes which have strong similarities: the devil tempts Jesus to create bread from stone, to fly from "the pinnacle of the Temple," and to worship him in return for riches (Matt. 4.1–11; Luke 4.1–13). Matthew and Luke differ in the order of the second and third temptations, and it is generally accepted that Matthew preserves the more original sequence, with the temptation of "all the kingdoms of the world and their splendor" (Matt. 4.8) coming last. In both gospels, the whole episode has features of a form of Jewish biblical interpretation known as *midrash*, where a Scriptural passage is elucidated by comparison with verses from elsewhere in the Bible (for example, Matt. 4.4; 4.7; and 4.10).

Jesus' fast of forty days and nights (Matt. 4.2; Luke 4.2) recalls the sojourn of Moses on Mount Sinai when he received the tablets of the Law (Exod. 24.18–31.18), and it is likely that the number forty also alludes to Israel's forty years in the wilderness, which Deuteronomy calls a time of testing (Deut. 8.2). The Scriptural verses with which Jesus counters the devil's temptations are all spoken by Moses in Deuteronomy, and each temptation scene is related to an episode in Israel's wilderness wanderings. The statement "one does not live by bread alone" (Matt.

The Temptation in Mark

Mark's version of the temptation is brief, recording that Jesus was "in the wilderness forty days, tempted by Satan; and he was with the wild beasts; and the angels waited on him" (Mark 1.13). In this cryptic verse may lie an interpretation of the temptation which sees Jesus as the new Adam in a restored paradise. In Eden (see pp.28–9), the first man lived at peace with the animals, and according to Jewish tradition angels served Adam and Eve in Eden. (Angels are also mentioned in the context of the temptation story in Matthew 4.11.) The prophets frequently claim that the new messianic age will be marked by the restoration of the paradisial harmony between humans and wild animals. For example, the Syriac *Apocalypse of Baruch*, which dates in its original form from ca. 100 BCE, states that when the new age arrives "wild beasts will come from the forest and minister to people." A fourth-century Latin prologue to Mark's gospel makes explicit the friendly character of the wild beasts referred to in Mark 1.12.

The three temptations that are put before Jesus by the devil, as described by Matthew and Luke, do not occur in Mark, who says simply that Jesus was "tempted by Satan." This also points to Adam and Eve: in the garden of Eden they were tempted by the serpent, which by Jesus' time had come to be identified with the devil. Unlike Adam, Jesus overcomes Satan and inaugurates the new age, when wild animals are no longer a menace and the angels serve humanity. Sometimes the Bible views Israel's years in the wilderness as a time when it enjoyed a perfect relationship with God. When the new age dawned, this relationship would be renewed.

4.4; Luke 4.4) refers, in Deuteronomy, to the divine gift of manna, nourishment supplied to the Israelites on their journey through the wilderness (Deut. 8.3). Putting God to the test (Matt. 4.7; Luke 4.12) was the sin committed by the Israelites at Massah (Deut. 6.16). In showing Jesus all the kingdoms of the world from a mountain (Matt. 4.8–10; Luke 4.5–8), the devil may be parodying God showing Moses the whole Promised Land from Mount Nebo (Deut. 32.49; 34.1). In a way that is typical of apocalyptic thought (see pp.136–7), this part of the Temptation episode presents the modern age as being entirely under the power of evil.

At his baptism, Jesus was designated Son of God (see p.161); the narratives of the Temptation show the character of this sonship, and also reject popular messianic conceptions. Jesus is implicitly contrasted with the numerous wonder-workers in the Palestine of his day and later, such as Simon Magus (see pp.224–5), who is credited with the ability to turn stones into bread and to fly through the air, just as Jesus was tempted to do. But the true Messiah will not use his powers for self-preservation or world dominion: his work can only be accomplished through his Passion, death, and Resurrection. Luke's final words that the devil departed "until an opportune time" (Luke 4.13) point to a future renewal of the conflict.

Left: The Temptation of Christ in the Wilderness, *by Juan de Flandres (active ca. 1496–1519). In the foreground, the devil in the guise of a monk tempts Jesus to make bread from stones (Matt. 4.3; Luke 4.3). In the background (left) are the temptations on the high mountain (Matt. 4.8; Luke 4.5) and on the "pinnacle of the Temple" (Matt. 4.5; Luke 4.9).*

Below: *The southeast corner of the Temple Mount in Jerusalem. The large blocks (bottom and right) are from the original south wall of Herod's Temple (see picture on p.91). Because this corner was the highest part of the Temple remains, in Byzantine times it was identified as the "pinnacle" of the Temptation story.*

THE GALILEAN MINISTRY

A view of the landscape west of Tiberias, with the Sea of Galilee in the distance.

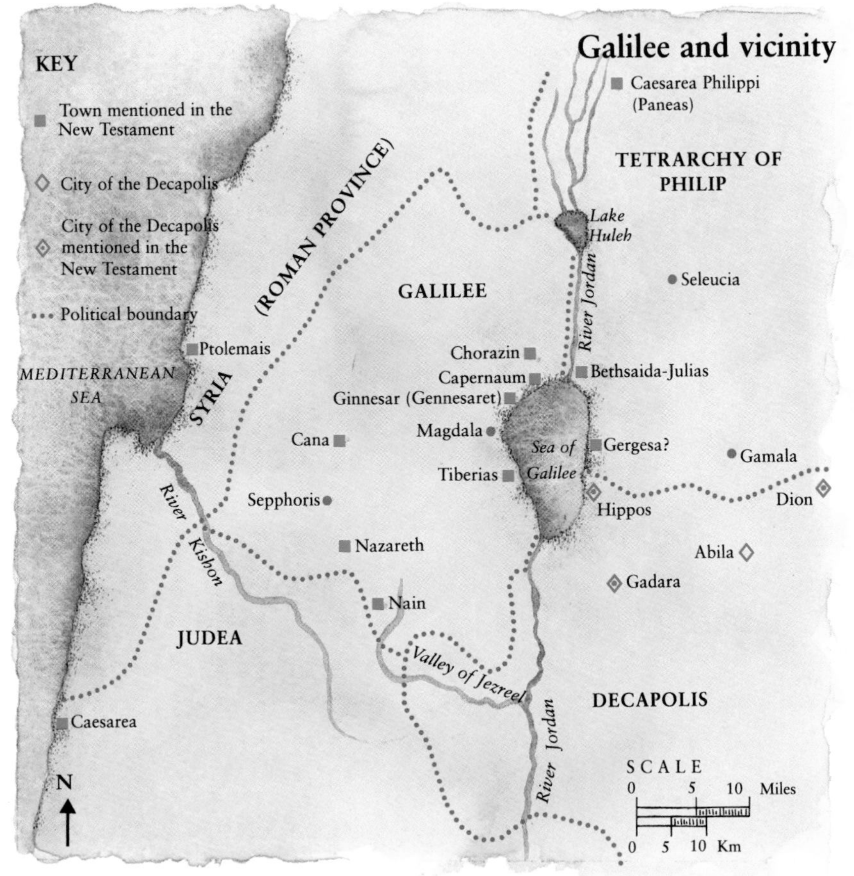

After his contact with John the Baptist (see pp.160–61), Jesus returned to Galilee and began to preach. The Synoptic Gospels relate that until the final journey to Jerusalem, his ministry was virtually confined to this region, and in particular to a restricted area along the western shore of the Sea of Galilee (also called the Sea of Tiberias). In Mark, there sometimes appears to be a degree of tension between Galilee and the religious authorities in Jerusalem (Mark 3.22 refers to critical "scribes who came down from Jerusalem"; compare Mark 7.1–2). However, this is primarily a literary device, followed by Matthew (compare Matt. 15.1) but absent in Luke. The supposed opposition between Jerusalem and Galilee that has sometimes been inferred from it has little historical basis.

The region has a particular significance in the gospels of Mark and Matthew. For Matthew, it is the place of revelation where the coming of Jesus brings light to the people "who sat in darkness" (Matt. 4.15–16). This foreshadows the Christian mission to the gentiles, a theme that is taken up again when, after the Resurrection, Jesus appears on a mountain in Galilee and commissions his followers to "make disciples of all nations" (Matt. 28.16–20). The Fourth Gospel (John 21.1) records an appearance of Jesus to the disciples at the Sea of Galilee after the Resurrection (see p.211). In Mark, Jesus tells his disciples that, after the Resurrection, he will go ahead of them to Galilee (Mark 14.28): the women at the tomb receive the same message (Mark 16.7).

It is impossible to construct a coherent account of Jesus' Galilean

Nazareth and Capernaum

Two towns, Nazareth and Capernaum, are given particular significance in the account of Jesus' time in Galilee. Nazareth was then little more than a village. It is a symbol of the Jewish rejection of Jesus, all the more poignant because it was his home. All the Synoptic Gospels record Jesus' teaching in the synagogue there, the unbelief that greeted him, and his response that "a prophet is not without honor except in his own country" (Matt. 13.54–57; Mark 6.1–6; Luke 4.16–24), but the fullest and most vivid account of the episode is to be found in Luke.

Luke places Jesus' visit to the Nazareth synagogue at the very beginning of the ministry, and presents it as the basic revelation of Jesus' person and mission. Jesus reads out a passage from Isaiah announcing the advent of "the year of the Lord's favor" and then says that, even as he speaks, the text has come true (Luke 4.16–21). But he continues that the miracles of Elijah and Elisha were performed not on Israelites but on foreigners (Luke 4.25–27). Here Luke is foreshadowing the rejection of the gospel by Judaism and its proclamation instead to gentiles, a major theme of his second volume, Acts (see pp.214–15). As a result of these words, the congregation attempts to lynch Jesus by throwing him from the steep hill on the edge of the town, but the mysterious authority of his personality means that he is able to walk through the crowd and escape (Luke 4.28–30).

Part of the excavated remains of Capernaum; in the background is the synagogue of ca. 350 CE that replaced the building that stood in Jesus' time.

The town of Capernaum is generally presented as a counterweight to Nazareth, the scene of success, although it is finally cursed for its failure to accept Jesus. As the obduracy of the people of Nazareth prevented Jesus from performing many miracles there, so Capernaum is the setting for several spectacular healing acts. Jesus settled at Capernaum and it was here that his ministry had its beginning and center. In the gospels, the town is the setting of many incidents, such as the curing of a centurion's servant (Matt. 8.5–13; Luke 7.2–10), and is implied in others. Particularly important was Jesus' visit to the synagogue there (Mark 1.21–28; Luke 4.33–37), paralleling his visit to the synagogue at Nazareth. On this occasion, his teaching is recognized as of unique authority, as shown by his exorcising of a man possessed by a demon. Through this man, the whole company of demons speaks as one, acknowledging Jesus as "the Holy One of God" (Mark 1.24; Luke 4.34). The healing symbolizes that the kingdom of God, which Jesus proclaims, vanquishes the kingdom of evil. Jesus goes to the home of his disciples Simon Peter and Andrew, and cures Simon Peter's sick mother-in-law (Matt. 8.14–15; Mark 1.29–31; Luke 4.38–39). The episode ends with a triumphant scene in which Jesus heals crowds of sick people and drives out numerous demons (Matt. 8.16; Mark 1.32–34; Luke 4.40–41).

ministry from the accounts in the gospels. They do not provide a systematic chronology, and each records events in a different order and often with no indication of their geographical setting. They present a picture of an itinerant preacher, constantly moving about between a few towns and the countryside. The gospels speak of Jesus enjoying great success and attracting large crowds (for example, Matt. 4.24–25; Mark 3.7–8), but they also sound a note of failure, especially when Jesus denounces several towns for not accepting his message (as in Mark 11.21). Hence he seems increasingly to have preached his mission to those who followed him from the towns into the open country.

THE CALL OF THE DISCIPLES

The Sea of Galilee, sometimes called Lake Tiberias and Lake Gennesaret, figures prominently in Jesus' ministry. He crosses it regularly by boat (Mark 4.35–41, 5.21; Luke 8.22–25) and teaches from a ship there (Mark 4); several important episodes also take place on its shores.

In the Synoptic Gospels, the lake is the scene of the call of the first disciples. Walking by the lake, Jesus approaches two pairs of fishermen (Simon Peter and his brother Andrew, and James and his brother John) who abandon their occupations to follow him (Matt. 4.18–22; Mark 1.16–20). In Luke, the event occurs farther into Jesus' ministry, after his preaching at the synagogues of Nazareth and Capernaum (see p.165). Jesus enters Simon Peter's boat and instructs the fisherman to put down his nets, whereby he grants a miraculous catch of fish: "so many ... that their nets were beginning to break." On witnessing Jesus' power, Simon and his partners, James and John, leave everything to follow him (Luke 5.1–11). In all three Synoptics, Jesus promises the fishermen that, instead of catching fish, they will now catch people. The first four disciples inaugurate the Christian community and symbolize its evangelizing mission, the success of which is foreshadowed in Luke by the miraculous catch of fish. Peter, James, and John come to form an inner core among Jesus' disciples, accompanying him on occasions such as the healing of Jairus's daughter (Mark 5.22–48; Luke 8.41–55) and the ascent of the mountain of Transfiguration (Matt. 17.1–8; Mark 9.2–8; Luke 9.28–36; see box on p.189). Andrew plays little part in the Synoptic narrative, but he is more prominent in the Fourth Gospel where he is represented as originally a follower of John the Baptist.

There are accounts of the choice of other disciples. The fourth evangelist relates the call of Philip, who plays a considerable role in his gospel, and Nathanael (John 1.43–51), and the Synoptics relate the call of the tax collector Matthew, also known as Levi (Matt. 9.9; Mark 2.14; Luke 5.27). Later, Jesus nominates a closed band, the "twelve," whom he empowers to heal, to cast out demons, and to carry his message to a wider audience (Matt. 10; Mark 3.14–19; Luke 9.1–6). There were twelve Israelite tribes (see map on p.69), and the choice of twelve disciples indicates that the Christian community is to be the New Israel. Matthew explicitly states that the mission of the twelve is to Israel (Matt. 10.6), and Jesus promises that, in the coming new age, the disciples will sit on twelve thrones to judge the twelve tribes of Israel (Matt. 19.28). The twelve are also called apostles, from Greek *apostolos*, "one who is sent out," a term corresponding to the Hebrew term *shaliach*, meaning one

Cana of Galilee. The churches commemorate the wedding at Cana, where Jesus turned the water into wine, according to the story in John's gospel.

out," a term corresponding to the Hebrew term *shaliach*, meaning one with full powers to represent the person who appoints them. When Jesus sends out the twelve, he says: "Whoever welcomes you welcomes me, and whoever welcomes me welcomes the one who sent me" (Matt. 10.40). Luke records the commissioning of seventy disciples (Luke 10.1–16) in addition to the twelve, while other ancient authorities mention seventy-two. This echoes the number of the nations of the world recorded in Genesis (seventy in the Hebrew Bible, seventy-two in the Greek translation), so the additional disciples may symbolize the Church's mission to all nations.

For the Fourth Gospel, the key place in Galilee for the early career of Jesus after the call of the disciples is Cana, the home of one of the first disciples, Nathanael. The evangelist sets there the first two miracles performed by Jesus. The first is the turning of water into wine (see sidebar, right), the second the healing of the royal official's son (John 4.46–53), which parallels the healing of the centurion's servant in Matthew 8.5–13 and Luke 7.2–10. Matthew and Luke set the episode in Capernaum, but in the Fourth Gospel, Jesus is in Cana while the sick child is in Capernaum. Jesus' ability to effect such a miracle at a distance stresses the greatness of his power, and causes those involved to believe in him.

THE WEDDING AT CANA

In the Fourth Gospel, Jesus' first miracle is to turn water into wine at a wedding feast at Cana (John 2.1–11). The miracle reveals his "glory" – his divine nature – to the disciples, as it had previously been revealed to the Baptist. As so often in this gospel, the episode is highly symbolic. Many expected a great banquet to accompany the coming of the Messiah (Rev. 19.9 refers to a "marriage supper"). At the feast, Jesus is the bringer of new life (wine symbolizes vitality) and the huge quantity of wine he provides, over one hundred and twenty gallons (John 2.6), shows the fullness received by all who believe in him. The transformation of the water, which was for Jewish purification rites, indicates how Jesus' coming supersedes the ancient rituals of Judaism (see also p.177).

Above: *Large terracotta jars for storing water or wine, from Jerusalem; Roman period, ca. 100* CE.

The Marriage Feast at Cana, *by Juan de Flandres (active ca. 1496–1519). Jesus (left) raises his hand in a gesture of blessing to perform the miracle. Next to him is his mother, Mary.*

THE SERMON ON THE MOUNT

The "Seat of Moses," where rabbis sat to teach, in the 4th-century CE *synagogue at Chorazin in Galilee. Chorazin is mentioned in Matthew 11.21.*

JESUS AND MOSES

It has often been claimed that the Sermon on the Mount represents Jesus as the new Moses, teaching a new Law to the people. There are parallels between the five discourses of Jesus in Matthew's gospel (Matt. 5–7; 13; 18; 23–25) and the five Mosaic books of the Law. The mountain setting and the fact that Jesus spoke while seated, the regular Jewish posture for teaching, could also suggest Moses: the bench in the synagogue from which the Law was expounded was known as "The Seat of Moses" (see picture, above). Even if all this is admitted, it must be recognized that, in the Christian view, Jesus is a much greater figure than Moses, and that his Sermon is not just promulgating a new Law. The Mosaic Law remains in force, although this is only for the present age. The Sermon is the Messiah's proclamation of the coming of the kingdom, which makes demands on the disciples of Jesus that go beyond the requirements of the Law.

According to Matthew, after the call of Simon Peter and Andrew, large crowds came to Jesus from all over the region to hear "the good news of the kingdom" (Matt. 4.23) and to be healed. The high point of this period was a great discourse delivered by Jesus on a mountainside and therefore known as the Sermon on the Mount (Matt. 5–7). The sermon in Matthew is very similar to a discourse in Luke, delivered "on a level place" (Luke 6.17) and therefore commonly called the Sermon on the Plain (Luke 6.20–49). Both sermons begin with a list of beatitudes (Latin *beatitudo*, from *beatus*, "blessed"), promises of blessing to those who belong to the kingdom (Matt. 5.3–12; Luke 6.20–23). Next comes a list of rules for human conduct (Matt. 5.17–7.12; Luke 6.27–48), and each discourse ends with parables (Matt. 7.13–27; Luke 6.39–49). Matthew's version is significantly longer than Luke's, but much of his Sermon appears at other points in Luke (for example, Matt. 7.7–11 parallels Luke 11.9–13). This suggests that Luke preserves the basic outline of a sermon, which Matthew expanded by incorporating other teachings of Jesus.

In the Hebrew Scriptures, a reward for right conduct is promised in this present world; in the gospels, rewards are related to the kingdom of salvation, which may be either present or future (see p.171). The beatitudes are the terms of admission to the kingdom for those to whom it represents the only hope. There are differences between the beatitudes that open the Sermon on the Mount and the Sermon on the Plain. Matthew has nine, Luke only four; most of Matthew's are in the third person (such as: "Blessed are the poor in spirit"), while Luke's are in the second person ("Blessed are you who are poor"). It is often claimed that while

Judaism

There are significant links between the Sermon on the Mount and certain aspects of contemporary Judaism. The saying, "Do to others as you would have them do to you" (Matt. 7.12; also Luke 6.31), known as the "Golden Rule," was stated in very similar terms by Hillel, a rabbi of the first century CE: "What is hateful to you, do not do to your fellow creature." Hillel expresses the rule in a negative form, but the positive form, as in the gospels, also occurs in Jewish literature of the period.

There are similarities between the Sermon and some of the Dead Sea Scrolls (see p.145). In one of them, the phrase "poor in spirit" occurs, as in Matthew 5.3. This suggests that both the people listening to the Sermon and the Qumran group thought of themselves as "the poor," each following its own authoritative interpreter of the Law, either Jesus or, at Qumran, the "Teacher of Righteousness." Matthew 5.43 refers to a command to "hate your enemy." Such a command is not found in the Hebrew Scriptures or rabbinical literature, but it occurs more than once in Qumran writings, which may have provided Matthew's source.

Another suggestion is that the Sermon was constructed as a response to the codification of the oral Law (see p.118) that was undertaken by the rabbis after the destruction of Jerusalem in 70 CE (see p.145). As such, the Sermon sought to highlight the ethical and moral aspects of Jesus' teaching as a guide for living, such as the rabbis aimed to provide.

The Sermon on the Mount, *by Joos de Momper the Younger (1564–1635).*

Matthew speaks of the "poor in spirit" and "those who hunger and thirst for righteousness," Luke is referring literally to disadvantaged groups when he addresses the "poor" and "hungry." However, in postexilic Judaism, the "poor" and the "righteous" can be synonymous, denoting faithful members of the community who seek to trust completely in God, rather than in political schemes, high position, or wealth. Hence, Luke may well understand "poor" and "hungry" in the same sense as Matthew. This is further implied by the fact that both Luke and Matthew were conscious of Isaiah 61, a postexilic composition (see p.117), as a source of Jesus' inspiration. Matthew's first four beatitudes echo Isaiah (compare, for example, Isa. 61.1–2 and Matt. 5.4), and Luke records that Jesus read out Isaiah 61 in the synagogue at Nazareth (Luke 4.18–19).

Following the beatitudes in the Sermon on the Mount are two sets of demands for right conduct. The first (Matt. 5.21–48) consists of a set of antitheses, contrasting what is commanded in the Law with the more searching conduct required of Jesus' followers. For example: "You have heard that it was said, 'An eye for an eye and a tooth for a tooth.' But I say to you, Do not resist an evildoer. But if anyone strikes you on the right cheek, turn the other also" (Matt. 5.38–39). The second set of demands (Matt. 6.1–7.11) deals with the worshiper's proper relationship to God: "When you give alms, do not let your left hand know what your right hand is doing, so that your alms may be done in secret; and your Father who sees in secret will reward you" (Matt. 6.3–4). This section, which includes the Lord's Prayer (see pp.170–71), ends with an assurance that prayer will be answered (Matt. 7.7–11).

The final part of the Sermon presents the theme of the coming judgment by means of three parables, each offering a set of alternatives. The first presents the choice between entering "through the narrow gate," Jesus' way, where "the road is hard that leads to life," or through the wide gate, where "the road is easy that leads to destruction" (Matt. 7.13–14). The second is directed against false prophets, who are like trees bearing bad fruit, unlike Jesus, the tree that bears good fruit (Matt. 7.15–20). The third parable, about two builders, offers the choice between building on rock and weathering the storm of judgment, or building on sand and being swept away (Matt. 7.24–27).

THE BEATITUDES IN MATTHEW

"Blessed are the poor in spirit, for theirs is the kingdom of heaven.
"Blessed are those who mourn, for they will be comforted.
"Blessed are the meek, for they will inherit the earth.
"Blessed are those who hunger and thirst for righteousness, for they will be filled.
"Blessed are the merciful, for they will receive mercy.
"Blessed are the pure in heart, for they will see God.
"Blessed are the peacemakers, for they will be called children of God.
"Blessed are those who are persecuted for righteousness' sake, for theirs is the kingdom of heaven.
"Blessed are you when people revile you and persecute you and utter all kinds of evil against you falsely on my account. Rejoice and be glad, for your reward is great in heaven, for in the same way they persecuted the prophets who were before you."

Matthew 5.3–12

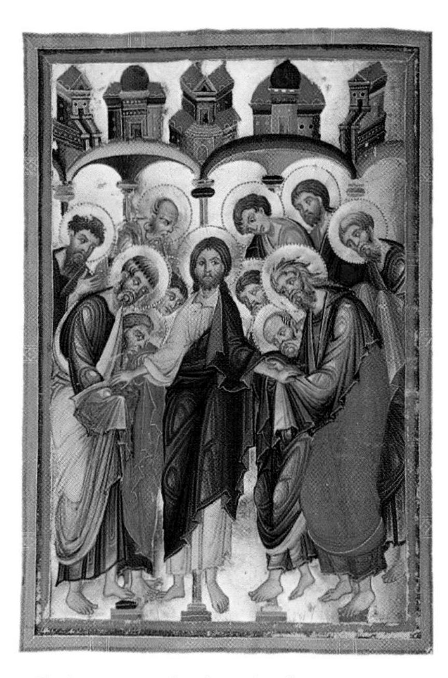

Christ among the Apostles *from a Byzantine Psalter of ca. 1250.*

THE LORD'S PRAYER

Matthew and Luke record that Jesus gave his disciples their own prayer as a model of how to pray (see box on opposite page). It is known as "the Lord's Prayer" and is deliberately brief, in contrast to the "many words" (Matt. 6.7) that characterized the prayers of gentiles. The most familiar version of the Prayer is contained in Matthew 6.9–13 in the Sermon on the Mount (see p.168). Luke's version is even shorter and occurs at a later stage of Jesus' ministry (Luke 11. 2–4), when Jesus' disciples ask him to teach them to pray, as John the Baptist taught his followers.

Luke's text differs from Matthew in both wording and length, and is generally believed to be closer to the original words of Jesus. The Lord's Prayer in Matthew may be an elaboration of the original, in accordance with the liturgy of the church that the evangelist was addressing (see p.150). The Prayer was assimilated into synagogue practice, acquiring more rhythm, a more repetitive pattern, and, in some sources, the concluding words of praise ("For the kingdom and the power and glory are yours forever. Amen."). Matthew's version came to be the one generally used by Christians, and manuscripts of Luke's gospel were increasingly brought into conformity with it.

There is evidence that the Prayer was originally in Aramaic, which Jesus would have spoken. The Prayer's opening address, "Father" (Luke 11.2) or "Our Father" (Matt. 6.9), translates the Aramaic *Abba*, a word that seems to have characterized Jesus' own prayers: it occurs, alongside its Greek equivalent, in Mark 14.36 and in two of Paul's letters (Rom. 8.15; Gal. 4.6). Religious groups of Jesus' time, such as the Pharisees and the Qumran community, had their own special forms of prayer, and Jewish prayers of the period also sometimes address God as "Father." Matthew's addition of the words "in heaven" reflects Jewish piety, with its stress on the transcendence of God. There is overlap in both form and content between the Lord's Prayer and Jewish synagogue prayers, especially the Kaddish, which calls for the sanctification of God's name and the establishment of his kingdom, and the Eighteen Benedictions, which the Lord's Prayer summarizes in many respects. In their present shape, these Jewish prayers date from after the time of Jesus, but there is evidence that they derive from much older originals that may well have been used in the synagogues of Jesus' day.

The Prayer has the same basic three-part structure in both Matthew and Luke, again suggesting that both versions are based on a single source, presumably ultimately from Jesus himself. It begins with an address to God (Matt. 6.9; Luke 11.2), followed by "you" petitions (Matt. 6.9–10; Luke 11.2), and then "we" petitions (Matt. 6.11–13; Luke 11.3–4). The petition "hallowed be your name" (Matt. 6.9; Luke 11.2) is again characteristic of Jewish worship: the Talmud says that "a benediction which contains no mention of the Divine Name is no benediction." The petition for the coming of God's kingdom (Matt. 6.10; Luke 11.2) represents the central element in Jesus' teaching: the Prayer looks forward to the end of the kingdoms of this world and the establishment of God's royal rule over all the nations (see sidebar on opposite page).

By contrast, the petition that occurs only in Matthew, "Your will be done, on earth as it is in heaven" (Matt. 6.10), expresses the demands made by Jesus on his followers in the present as they wait for the impending kingdom. This liturgical addition reflects the common Jewish concept of God's reign as an eternal and existing reality. His will is unchallenged in the heavenly sphere, and God is called upon to exercise it with equal freedom from opposition on earth.

In the second half of the Lord's Prayer, the "we" petitions move the focus from God to the personal needs of the disciples. These clauses may appear straightforward, but in fact their interpretation presents a number of problems, not least because of the differences in wording between Matthew and Luke. In the first of the "we" petitions, a request for "daily bread," the Greek word that is usually translated "daily" has not been found anywhere else in Greek literature before the New Testament. It is the only adjective in the entire Prayer, and its meaning is uncertain. The presumed Aramaic original could simply mean our bread "for the day," as is suggested in Luke, where the petition asks for the gift to be given "each day" (Luke 11.3). This would make the petition a prayer for each day's sufficiency. However, the petition as it occurs in Matthew 6.11 is for the bread to be given "this day." Here, what is normally rendered "daily bread" could in fact mean "bread for the next day." If this is the correct interpretation, the petition in Matthew is therefore a request for reassurance that our existence will continue.

In the next petition, Matthew asks for forgiveness of "debts" (Matt. 6.12) while Luke asks for forgiveness of "sins" (Luke 11.4). Here, the difference may turn simply on a variant understanding of an Aramaic word meaning either "debt" or, in a specifically religious context, "sin" or "guilt." Paul described the pious Jew as "a debtor to the whole Law," and some scholars have suggested that Matthew has the Jewish Law in mind when God is asked to forgive, that is, cancel, our debts to him. However, earlier in Matthew's account of the Sermon on the Mount, Jesus emphatically denies that he wants to abolish the Law (Matt. 5.17–18). It is therefore unlikely that Matthew's Jewish Christian church would have prayed to be delivered from the obligations of Judaism.

The final petition, "do not bring us to the time of trial" (Matt. 6.13; Luke 11.4), given in some sources as "do not bring us into temptation," may appear difficult to interpret. It has often been suggested that "trial" refers to the woes which were expected as an immediate precedent to the messianic age. However, it may simply refer to temptation, which is always a test or trial for the righteous, as Jesus' encounter with the devil in the wilderness shows (see p.162). A number of Jewish texts ask God not to lead the petitioner into temptation in words very similar to those in the Lord's Prayer.

The two versions of the Lord's Prayer

Our Father in heaven,
hallowed be your name.
Your kingdom come.
Your will be done,
on earth as it is in heaven.
Give us this day our daily bread.
And forgive us our debts,
as we have forgiven our debtors.
And do not bring us to the time of trial,
but rescue us from the evil one.

Matthew 6.9–13

Father, hallowed be your name.
Your kingdom come.
Give us each day our daily bread.
And forgive us our sins,
for we ourselves forgive
everyone indebted to us.
And do not bring us to the time of trial.

Luke 11.2–4

GOD'S KINGDOM

For the Synoptic Gospels, the central theme of the preaching of Jesus is the kingdom of God (Matt. 4.17, 23; Mark 1.15; Luke 4.43). Some fifty of Jesus' sayings and parables deal with the kingdom. Matthew uses the expression "kingdom of heaven" but this is only a reverent, characteristically Jewish, paraphrase, and there is no difference in meaning.

The teaching of Jesus reveals two main elements in its presentation of the concept of the kingdom. It is eschatological in tone, that is, it refers to the end of time, when God will decisively establish his royal rule over the universe and vindicate the righteous. This is the urgent note struck by Jesus when he appears proclaiming "the kingdom of God has come near" (Matt. 4.17; Mark 1.15).

However, in rabbinic thought, the kingdom is a present, moral, reality: a person can be said to assume the "yoke of the kingdom" by reciting Israel's confession of faith in the one God. This is the second element present in Jesus' teaching about the kingdom. He was a teacher of ethics, and as such emphasized the necessity for his followers to prepare themselves here and now for the time when the kingdom would be fully realized (Luke 10.25–28). This demands repentance (Matt. 4.17; Mark 1.15), a wholesale change of heart, through which people will accept the arrival of the kingdom as good news, or gospel.

THE PARABLES

The term "parable" (Greek *parabole*) is used to describe some thirty stories attributed to Jesus in the gospels, and there are about another fifty tales in which the word is not used, but which are clearly the same type of material: one of these is the tale of the Good Samaritan (Luke 10.29–35). Despite the fact that Jesus regularly made use of parables in his teaching, it is not easy to give a precise definition of the term. In classical Greek, the word primarily means a comparison, and the Septuagint, the Greek translation of the Jewish Bible, employs the word to translate the Hebrew *mashal*, which has the same basic sense. A parable could therefore be broadly defined as a metaphor, in which case it could also refer to a wide range of literary forms that occur in the Bible, such as proverbs, riddles, taunts, aphorisms, and allegories. These and similar linguistic devices characterize the teaching of Jesus to a very great degree, and so it is understandable when Mark's gospel says that Jesus did not speak to the people "except in parables" (Mark 4.34).

However, the term parable is most commonly applied to a realistic but metaphorical story, which has a deeper meaning than that which appears on the surface. The parable has two basic forms. The first is the brief

Above: *Present-day fishing boats on the Sea of Galilee.*

St. Peter and St. Andrew Fishing on the Sea of Galilee, *a 12th-century Spanish wall-painting.*

The social background

The parables in the Bible almost certainly represent an authentic element of the preaching of Jesus. They are full of scenes drawn from ordinary life which reveal a distinctively Palestinian background, above all in Galilee of the first century CE. The parable of the great net of fishes, where the good are separated from the bad (Matt. 13.47–50), reflects what would have been the practice of Galilean fishermen. Many features of the parables reveal their Palestinian background: for example, the parable of the sower (see p.174) reflects Palestinian farming practice, in which sowing took place *before* plowing. The parable of the laborers in the vineyard, who are paid at the end of the day in accordance with Jewish Law (Matt. 20.1–16), shows how paid workers, rather than slaves, were the norm in Palestine.

King David and Nathan, *from the* Hours of Nicolas Le Camus; *French, 15th century.*

Recent scholarship has done much to explain the character and background of the New Testament parables. In composing his parables, Jesus appears to have developed and extended the biblical tradition of telling stories that bring home to the hearer the realities of his own situation. A good example is the parable of the rich man and the poor man, through which the prophet Nathan brings King David to condemn himself (2 Sam. 12.1–15).

However, the closest parallels to the gospel parables are probably the parables of the rabbis. Jesus himself was often addressed as "Rabbi," which means "Teacher" (for example Matt. 26.25; Mark 9.5), and was clearly regarded as one, even if his message sometimes differed from that of other Jewish teachers of the time. The rabbis regularly used arresting stories to clarify some aspect of the Law, and drew on the social and economic conditions of ordinary life, involving characters such as landlords, tenants, and laborers. Many of their typical settings – the field, the vineyard, the banquet, the king's presence – occur in the gospel parables.

Not surprisingly, it is in Matthew's gospel, with its Jewish links, that the closest resemblances to the rabbinic style of teaching are found. More than the other evangelists, he presents the parables as teaching aids and as a means to inculcate moral duties, much as the rabbis did.

straightforward narrative, such as that of the pauper Lazarus and the rich man (Luke 16.19–31). Without stating whether either man was sinful or virtuous, Jesus says that the rich man goes to Hades and Lazarus to heaven "to be with Abraham." Abraham ignores the rich man's appeal that his living brothers should be warned of what awaits them. The patriarch tells him: "They have Moses and the prophets; they should listen to them." The parable makes a point about obedience to the Law that the hearers, or readers, are implicitly challenged to grasp and apply to themselves. There are also parables which call explicitly for the comparison by using an introductory formula such as "it is like," or "it is as if," for example Mark 4.26–32. These two main types of parable highlight the chief aspects of Jesus' teaching. Many of them serve to explain the kingdom of God (see p.171), while others are concerned with moral duties and obligations toward fellow human beings, which are required as preparation in the present for those seeking to enter the kingdom.

Jesus' parables are confined to the Synoptic Gospels; the Fourth Gospel presents Jesus' teaching in a very different way (see p.176), and the term usually translated as "parable" in John 10.6 represents a different Greek word from that used in the Synoptics. The first three gospels often group parables with a similar form and content, such as the "seed" parables in Mark (see p.174). Sometimes the evangelists differ in the form and setting that they give to the same parable, in order to convey their own particular understanding of how the story illustrates the teaching of Jesus.

THE SOWER

The main purpose of Jesus' parables was to illustrate the principal themes of his preaching to a popular audience. However, the meaning of the parables is not always obvious, and Mark says that they were in fact intended to conceal from the people at large the real secret of his message, which was to be made known only to the innermost circle of disciples (Mark 4.10–12). The early Church clearly felt that the parables needed to be explained, and the gospels themselves show Jesus explaining some parables. The most notable instance of this is the parable of the sower, which occurs in each of the Synoptic Gospels, together with an explanation given by Jesus to the disciples in private (Matt. 13.3–23; Mark 4.3–30; Luke 8.5–15). Jesus' interpretation was probably a later addition, stemming from the teaching activity of the early Christians. The apocryphal Gospel of Thomas (see p.149), which is believed to contain authentic teachings of Jesus, has the parable but not the explanation.

In Mark, the parable of the sower is followed by the parables of the

The parables in Luke

Some of the best known of Jesus' parables are found in Luke. They are often very realistic and human, and strike a note of joy and optimism, in contrast to the doom-laden, apocalyptic tone of many of the parables that Luke shares with Matthew and Mark, such as the parable of the sower (see main text, above).

In the famous story of the Good Samaritan (Luke 10.30–35), Jesus tells how a priest and a Levite ignore the distress of an injured man who has been attacked and robbed on the road to Jericho. The man is eventually saved by a Samaritan, who tends his wounds and carries him on his animal to an inn. In Luke, Jesus takes a special interest in the Samaritans (see p.125) that reflects his concern for outcast groups in general. The lawyer to whom Jesus relates the story understands its truth and is told to behave in the same way. The parable recalls the account in 2 Chronicles of how some Samaritans tended the wounded of Judah and carried them on donkeys to Jericho (2 Chron. 28.15).

The Prodigal Son, *by the Master of the Female Half Lengths (Low German, ca. 1550). The return and repentance of the prodigal son was celebrated with a feast.*

Three parables in Luke 15 clearly represent the evangelist's concern for sinners and how they can be restored to God through repentance. These are the parables of the lost sheep (Luke 15.3–7), the lost coin (Luke 15.8–10), and the Prodigal Son (Luke 15.11–32), which echoes the story of Joseph (see pp. 48–51). They are probably critical of the scribes and Pharisees (see pp.184–5), whose attitude may be reflected by the prodigal's brother. However, the three stories mainly serve to stress the love of God, who always welcomes back those who repent and return to him. When the son, the coin, and the sheep return, there is great rejoicing, even in heaven.

The vineyard and the wedding banquet

Two parables that appear in more than one gospel have as their theme the rejection of the old Israel. The first is the parable of the vineyard and its evil tenants (Matt. 21.33–43; Mark 12.1–11; Luke 20.9–17). It is based on the opening verses of Isaiah 5, which Mark and Matthew virtually quote. The story is strongly allegorical: the vineyard is Israel; the tenants are the Jewish nation; the owner is God; the servants that he sends are the succession of prophets (whom Matthew and Luke describe elsewhere as all having been killed by the Israelites); and the owner's son is Jesus. The outcome is the rejection of the Jews, and the transfer to the righteous gentiles of the kingdom promised to them.

The watchtower of a vineyard in Judea. A watchtower is mentioned in Isaiah 5.2.

Matthew appends the parable of the great wedding banquet to the parable of the vineyard (Matt. 22.1–10). Luke also includes the wedding parable, but puts it in a different setting (Luke 14.15–24), and gives it the straightforward message that nothing should stand in the way of a person's response to God's gracious summons. In Matthew, however, the parable becomes a sophisticated allegory. The host is a king – God himself – and the banquet is the messianic wedding-feast for his son. The servants sent to summon the guests are killed by those whom they invite. In response the king sends his troops to destroy those who have so brutally rejected his invitation and burn their city. Here, Matthew is probably referring to the destruction of Jerusalem by the Romans in 70 CE. The guest without a wedding garment appears only in Matthew (22.11–14). It does not fit well with the preceding story and is probably a moralistic addition, rebuking unworthy members of the community for whom the evangelist was writing.

seed growing secretly (Mark 4.26–29), and of the mustard seed (Mark 4.30–32). The three parables are linked by the word "seed," and the last two, at least, refer to the coming of the kingdom of God (see p.171). Mark sets the parable of the sower against a background of repeated rejections of Jesus, perhaps as a warning that Jesus' message, "the word of the kingdom," might not be universally accepted. In the interpretation of the parable, Jesus gives each detail of the story its own individual significance: the birds are Satan; the heat of the sun is persecution; and the thorns are worldly cares and pleasures. The parable becomes more of a moralistic instruction for the daily life of all Christians than a call for Jesus' followers to prepare for the arrival of God's kingdom, which is the essence of Matthew's understanding of the parable.

Matthew follows his account of the sower with a different group of parables from that in Mark. He leaves out the secretly growing seed, but adds to the parable of the mustard seed (Matt. 13.31–32) the parables of the wheat and weeds (Matt. 13.24–30), the leaven (Matt. 13.33), the hidden treasure (Matt. 13.44), the pearl (Matt. 13.45–46), and the net of fishes (Matt. 13.47–50). In this string of parables Matthew appears to stress the fact that the kingdom is hidden at present, but will finally triumph, at which time sinners will be separated from the righteous. This interpretation is supported by the later parable of the sheep and goats, which occurs only in Matthew (Matt. 25.31–46): when the kingdom comes, the "Son of Man" will set the righteous on his right hand and the sinners on his left, "as a shepherd separates the sheep and the goats."

JESUS' TEACHING IN JOHN

A bronze commercial weighing scale from Roman Judea. The traders in the Temple may well have used a similar utensil.

It has often been thought that the Fourth Gospel represents a later interpretation of the three Synoptic Gospels (see p.151). Nowadays, the Synoptics are still believed to preserve the most authentic record of Jesus' speech, but it is questioned whether the author of the Fourth Gospel drew on them at all. It seems more likely that his work is based on independent, and possibly very early, traditions of Jesus' life and teaching. This would account for the marked difference between the Synoptics and the Fourth Gospel in the presentation of Jesus' message. In the Synoptics, Jesus' teaching consists mainly of short sayings and parables (see pp.172–5), but in the Fourth Gospel he delivers lengthy discourses (see box on opposite page) and has extended debates with his opponents. The first half of the gospel (John 2–12) is structured around six discourses which explain the significance of six miracles. This leads up to the point where Jesus withdraws from the world to be with his disciples. The second half of the gospel (John 13–20) includes a further series of speeches, especially Jesus' farewell address to the disciples (John 14–17).

The content of the discourses is very different from that of the sayings of Jesus in the Synoptic Gospels, and they serve to further the stated aim of the Fourth Gospel, that its readers "may come to believe that Jesus is the Messiah, the Son of God, and that through believing you may have life in his name" (John 20.31). In the Synoptics, Jesus mainly proclaims the advent of the kingdom of God (see p.171) and gives moral and religious instruction about the demands entailed by acceptance of the kingdom. The Jesus of the Fourth Gospel speaks more of his own role and person, his claim to be the Messiah and the divine Son. His discourses develop the description of him as "the Word" in John 1, where the principal themes of the whole gospel are introduced (see also p.161).

The Fourth Gospel is notable for six sayings beginning "I am", which serve as a starting point for each discourse. In them, Jesus depicts himself

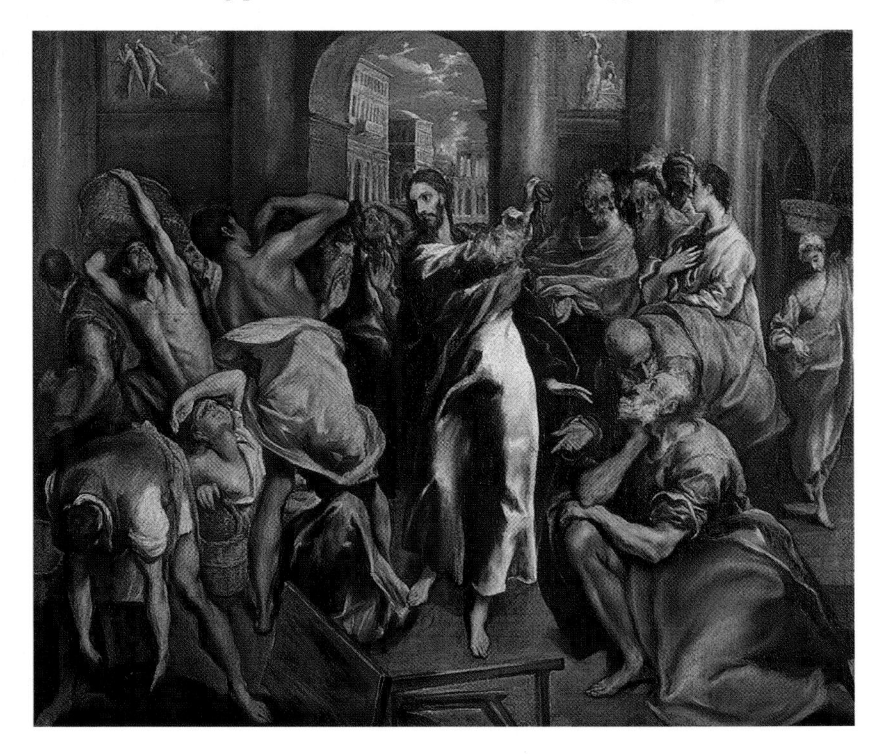

The Cleansing of the Temple, *ca. 1600, by Domenikos Theotokopoulos, alias El Greco (1541–1614). This is El Greco's third painting of the scene.*

as, for example, the bread of life (John 6.35–51), the light of the world (John 8.12–14), and the true vine (John 15.1–8). These "I am" sayings have a special authority, because they echo the revelation to Moses of the name of God in Exodus 3 (see p.56).

John recounts a number of events that are not in the Synoptics, and when he relates incidents that are in the other gospels, he often gives them a different setting and significance. These features are clearly illustrated by the episodes that he places at the very beginning of Jesus' career, in chapters 2 to 4, which form a distinct section. They consist of two narratives and two discourses, all on the same basic theme: the new order of life that Jesus brings. The first narrative, about the wedding feast in Cana (John 2.1–11), shows Jesus replacing the water of Judaism with the new wine of his own teaching (see also p.167). This story is immediately followed by the narrative of the cleansing of the Temple in Jerusalem (John 2.13–22), which the other gospels place at the end of his ministry. (In John, Jesus pays regular visits to Jerusalem on the occasion of the major Jewish festivals; in the Synoptics he makes just one visit, which represents the climax of his ministry.) The story is placed at the beginning of the gospel because it introduces two of John's principal themes: the faith implied by discipleship (John 2.22) and Jesus' controversy with "the Jews." In the Fourth Gospel, Jesus is not so much a figure within Judaism, as he clearly is in the Synoptic Gospels, but a unique being sharply distinguished from "the Jews." Typically for John, the term "Jews" has above all a symbolic meaning as those who reject Jesus as Messiah.

John gives his interpretation of the cleansing episode in a debate between Jesus and "the Jews" (John 2.18–21): the Temple is not just to be purified but, as Jesus predicts elsewhere in the gospels, destroyed. However, John adds the promise of a new Temple in which God will dwell: the risen body of Jesus (John 2.19–21).

Nicodemus and the Samaritan woman

The first two discourses in John clearly present one of the Fourth Gospel's central ideas, that of a new beginning through Jesus. The first (John 3) is addressed to the Pharisee Nicodemus, a Jewish leader and skilled theologian who comes to Jesus by night. He receives illumination from "the light" that has "come into the world" (John 3.19). Jesus teaches that entering the kingdom involves a complete new birth, symbolized by baptism (John 3.5). Baptism conveys the gift of "the Spirit," a new realm of existence that contrasts with "the flesh," the life of this world (John 3.6). Only Jesus can know heavenly truth, having been "born from above" (John 3.3). His saving power, to those who accept him, will be revealed when he is lifted up, "as Moses lifted up the serpent in the wilderness" (John 3.14): this statement foreshadows the crucifixion.

The second discourse (John 4.7–26) is set in the context of a dialogue with a Samaritan woman at Jacob's well. Jesus tells the woman that the water of Jacob's well will be superseded by the living water of eternal life which he brings. The Samaritans (see p.125) must accept the Messiah who fulfills God's promises to the Jews, but this does not entail acceptance of the exclusive claims of the Jerusalem Temple. Jesus inaugurates a new era, in which worship is not bound to particular sites, such as the Temple of Jerusalem or the Samaritan cult center on Mount Gerizim.

A simple village well, typical of those found in Judea.

The healing of the paralyzed man (Matt. 9.2–7; Luke 5.18–25; compare Mark 2.3–12; John 5.5–9). A wall painting from the 3rd-century church excavated at Dura Europos in Syria.

THE MIRACLES

The New Testament relates numerous occasions on which Jesus, and later the apostles, perform miracles. A miracle is often described as an event that suspends the laws of nature, but this definition is inappropriate when applied to the cosmological outlook of the Bible, which does not view the universe as a closed system operating by observed cause and effect. A miracle in the biblical sense is an occasion in which God acts in a remarkable or exceptional way to produce a particular end. In the New Testament, there are three Greek words for miracle: "wonder" (meaning something out of the ordinary); "sign" (evidence that God is at work); and, most commonly, "power" (a demonstration of divine might and control).

Miracles were taken for granted in the ancient world, not least in Judaism. The Bible contains many records of God's miraculous interventions on behalf of his people, above all in the deliverance from Egypt, with the plagues of Egypt, Moses' parting of the Red Sea (Exod. 14), and the provision of nourishment in the desert (Exod. 16) serving as three examples (see pp.52–67). There are also numerous tales of miracles performed by holy men, such as the prophets Elijah and Elisha (see pp.96–9), who possess the power to intercede on behalf of God. These accounts have influenced many of the miracle stories in the New Testament. The New Testament miracles offer reassurance that, just as he was active at the beginning of the nation's history, God will intervene again to deliver his people from the grip of evil and usher in the new age of salvation.

Jesus' miracles demonstrate to unbelievers that he is the expected Messiah and that the promised new age of God's kingdom (see p.171) is about to dawn. When he is asked by the disciples of John the Baptist whether or not he is the Coming One, Jesus points to the visible evidence

Miracle workers of Jesus' time

Both Jesus and his disciples are portrayed as itinerant preachers who often performed miracles in the course of their missions. As such, they would not have seemed unfamiliar or exceptional in the Roman empire: a close analogy is to be seen in the Cynic-Stoic teachers who wandered throughout the Middle East during the first two centuries CE.

There were many Jewish wonder-workers of a similar kind, who healed by using magical names, as Jesus' disciples healed in his name (as at Acts 3.6). Miracles are also attributed to a number of rabbis. Legends of their power to produce rain occur in the Talmud, particularly about Honi ha-Meaggel, a rabbi of the first century BCE. But the most famous rabbinic miracle-worker was probably Hanina ben Dosa, of the late first century CE. Among the stories told of him is that of his miraculous healing of the son of another rabbi, Gamaliel II. Gamaliel's son was sick, so the rabbi sent two of his disciples to Hanina, who declared that the fever had left the boy. The disciples noted the hour of Hanina's announcement and returned to Gamaliel, who confirmed that the fever had left his child at that precise time. There is a striking resemblance between this story and the gospel story of Jesus' healing of the son of a royal official at Capernaum (John 4.46–54; see p.167).

The pool of Siloam, where Jesus cured the blind man (John 9.7). It is at the southern end of King Hezekiah's tunnel between the Gihon spring and the City of David (see map on p.190).

of his miracles as proof of his messianic role (Luke 7.18–23). The list of healings in these verses recalls several passages in Isaiah, where such miracles also appear as portents of the establishment of God's rule (Isa. 29.18–19; 35.5–6; 61.1). The extraordinary phenomena which accompany Jesus' birth and death would have been taken for granted in Jesus' day, when people expected the birth and career of a great person to be marked by unusual signs and portents.

Jesus endowed his disciples with the authority to perform miracles as he himself did, and Acts records that "many signs and wonders were being done by the apostles" in the first days of the Church (Acts 2.43). It is through Jesus' name and power that the apostles are able to heal, and some of the miracle stories in Acts closely parallel those of Jesus in the gospels, such as Peter's healing of the crippled beggar at the Beautiful Gate of the Temple of Jerusalem. Peter says to the beggar: "In the name of Jesus Christ of Nazareth, stand up and walk" (Acts 3.6), which recalls the words with which Jesus healed the cripple (Matt. 9.6; Mark 2.11; Luke 5.24; John 5.8) and raised Jairus's daughter from the dead (Mark 5.41; Luke 8.54). Just as Jesus' own miracles mark significant moments in his life, the miracles carried out on his authority mark significant developments in the growth of the Church. Peter's miracle causes a considerable stir and his subsequent preaching results in his imprisonment; but it also wins five thousand new converts to Jesus (Acts 4.3–4).

Miracles in the Fourth Gospel

The fourth evangelist goes further than the Synoptic Gospels in his presentation of miracles as records of Jesus' remarkable powers and evidence of his character, mission, and unique relationship with God. John describes the miracles as "signs" and the six signs he records correspond to, and are sometimes directly linked with, the six discourses and six "I am" sayings in the first part of the gospel (see p.176). This half of John has often been called the "Book of Signs."

In most cases, the stories in the Fourth Gospel have parallels in the Synoptic Gospels, but John lends them his own particular significance. The healing of a sick man on the sabbath (John 5.2–16) is a violation of Jewish law, as it is in the Synoptics, but in the following discourse, Jesus claims that the miracle confirms him as God's son and equal (John 5.17–47). The account of a blind man receiving his sight (John 9) also has parallels in the Synoptics, but in the Fourth Gospel its purpose is to reveal Jesus as the light of the world: true blindness belongs to those who cannot see this truth.

The miracle stories that are found only in John, such as the episode at Cana (see p. 167), are of the same nature. Thus the raising of Lazarus from the dead (John 1; see p.181) serves to awaken faith and is a sign of the eternal life which is assured through faith in Jesus.

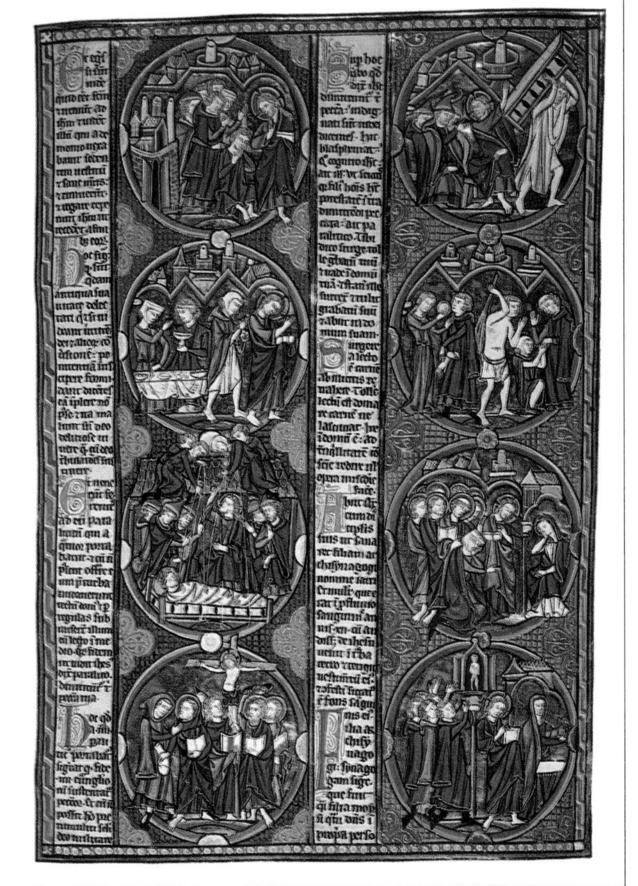

Christ's Miracles of Healing, *from the* Bible Moralisée *produced ca. 1240 for King Louis IX of France (St. Louis).*

THE EXORCISM OF DEMONS

A herd of swine, from a relief in the Assyrian royal palace at Nineveh, ca. 690 BCE.

OTHER HEALING MIRACLES

The accounts of illnesses cured by Jesus are not all directly related to demonic possession. The narrative of the paralyzed man (Matt. 9.2–8; Mark 2.3–12; Luke 5.18–26; John 5.5–9) takes the common view of the time that illness was the result of sin, and shows that Jesus has divine power to forgive sin and cancel its effects. The stories of the man with a withered hand (Matt. 12.9–13; Mark 3.1–5; Luke 6.6–10), and of the blind man (John 9) serve, like other episodes in the gospels, to prove that Jesus has the messianic authority to override the strict legal regulations for the sabbath.

Special importance is attached to the account of the cure of a leper (Matt. 8.1–4; Mark 1.40–41; Luke 5.12–15). It should be noted that in the Bible, the word "leprosy" does not necessarily refer to Hansen's Disease but may refer to one of a variety of skin conditions. In Mark, when Jesus is confronted with the leper, he reacts with strong and indignant emotion at his supremely pitiful condition. The Jewish Law provides no means for curing leprosy: if the disease disappeared of its own accord, the leper had to be inspected by the priest and then undergo a purification ritual. But Jesus supersedes the old Law and can cleanse the leper of his disease. Luke records the cure of ten lepers when Jesus was in Galilee and Samaria (Luke 7.11–19). The lepers are thought to symbolize the ten Israelite tribes who inhabited that region. The only leper who acknowledges Jesus is a Samaritan, a non-Jew, and the episode probably symbolizes the carrying of the gospel to the gentiles. The message is clear: Jesus is the healer who lavishes his benevolence on all in need, whether they are Jewish or not. All can be saved through faith in his power.

The exorcism of evil spirits played an important role in the ministry of Jesus, as is recorded in the Synoptic Gospels. In the ancient world, it was commonly accepted that evil spirits could cause both disease and natural disasters, such as storms and floods, and that their influence had to be repelled through exorcism and other practices designed to ward off dark forces. The Fourth Gospel has no record of such activity, and outside the Synoptics, exorcism is mentioned only in Acts (Acts 5.16, 8.7, 16.16–18, 19.11–19). When Jesus and his disciples are recorded as having cast out demons from individuals, it is impossible to know exactly what medical conditions are being cured, because such information was not available to the narrators. The vivid picture of the boy who was convulsed by a demon, so that he rolled on the ground foaming at the mouth (Mark 9.14–27; Luke 9.37–43), suggests that he was suffering from an epileptic fit, and the various stories of Jesus restoring speech to the dumb or sight to the blind may indicate the removal of some sort of psychological or psychosomatic condition.

The New Testament accounts of the curing of demonic possession generally follow a fixed pattern: the intractable and long-lasting nature of the illness; the method of the cure; and the observable consequences. This pattern is also found in Greek and Jewish popular tales of the first century CE, and Greek magical texts dealing with exorcism have a number of features in common with those attributed to Jesus. For example, the demon often recognizes the healer's supernatural character: Mark records that whenever the "unclean spirits" saw Jesus, "they fell down before him and shouted, 'You are the son of God!' " (Mark 3.11; compare also Mark 8.29). The Greek texts, like the Bible, also describe the casting out of the demon by the power of the divine name, and remark on the importance of forcing the spirit to disclose its own identity, as in the version of the story of the Gerasene (or Gadarene) demoniac recounted by Mark and Luke. The demons declare that their name is "Legion," because there are so many of them (Mark 5.9; Luke 8.30). The name of Jesus soon came to be employed to cure demonic possession, not only by his own followers but also by other exorcists.

For the gospel writers, Jesus is much more than just one among other wonder-workers. The evangelists reflect the common world view of their day, according to which the demons inhabited the sphere between heaven and earth, constituting a realm of evil under a leader, Satan or Beelzebul. (Compare Matt. 12.24–27, where Jesus denies that he acts on the authority of Beelzebul, a figure derived from the Canaanite title Baal-zebul, "Lord Baal." The Hebrew form, Baalzebub, means "Lord of the Flies" and is probably a deliberate corruption.) The arrival of God's kingdom in the person of Jesus brings about the overthrow of the kingdom of evil, and this is anticipated by his ability to master the evil spirits. In the narrative of the Gerasene demoniac, Jesus expels an entire company of such spirits from a possessed man (two men in Matthew). The demons beg to be cast out and Jesus gives them permission to enter a herd of swine, unclean animals and therefore a suitable environment. As soon as they have taken possession of the swine, the animals rush down a steep bank to their deaths (Matt. 8.28–33; Mark 5.2–13; Luke 8.27–33).

The gospels' view of Jesus' exorcisms is best summed up in his own saying, as recorded in Luke: "if it is by the finger of God that I cast out the demons, then the kingdom of God has come to you" (Luke 11.20). The image of the finger of God is also used by the Egyptian magicians describing God's actions to liberate Israel from Egypt (Exod. 8.19).

The raising of the dead

Above: *The traditional "Tomb of Lazarus" at Bethany.*

Left: The Raising of Lazarus, *ca. 1309, a fresco by Giotto (ca. 1266–1337) in the church of St. Francis at Assisi, Italy.*

The gospels record three accounts of Jesus bringing a dead person back to life. Women feature prominently in each of these stories, which also illustrate the defeat of the forces of evil, since, as the Letter to the Hebrews states, it is the devil who has the power of death (Heb. 2.14).

In Mark's account of the healing of Jairus's daughter (Mark 5.22–24, 35–43), Jesus' words to the girl (*Talitha cum*, "Little girl, get up") are preserved in the original Aramaic (Mark 5.41), probably because the phrase had come to be used by Christian exorcists. (Between the two parts of this episode is the story of the woman cured merely by touching Jesus' clothes (Mark 5.25–34). She is healed by faith, but the story preserves the idea of a mysterious power within the healer.) At Nain in Galilee, Jesus restores a widow's son to life. Luke's account (Luke 7.11–17) recalls similar miracles by the prophets Elijah (1 Kings 18) and Elisha (2 Kings 4). When the dead man is revived, Jesus "gave him to his mother," a quotation from the Elijah story (1 Kings 18.23). Jesus, like Elijah, is hailed as a great prophet (Luke 7.15–16). The story of Lazarus (John 11) is the final, climactic, "sign" in the Fourth Gospel (see p.179). The evangelist emphasizes Jesus' emotional response to Lazarus's death (John 11.35, 38), and, as elsewhere, heightens the miraculous nature of the event. Lazarus is said to have been dead four days when Jesus arrives. In popular Jewish belief, the soul was said to leave a corpse on the fourth day after death, when decomposition set in. The raising of Lazarus prefigures the promise of Jesus' own death and resurrection, as Jesus tells Lazarus's sister, Martha: "I am the resurrection and the life. Those who believe in me, even though they die, will live" (John 11.24–25).

THE MIRACULOUS FEEDINGS

THE FEEDING OF THE FIVE THOUSAND IN JOHN

In the Fourth Gospel, the symbolic meaning of the feeding of the five thousand (John 6.1–14) is greatly developed by the accompanying discourse (John 6.25–59), which includes one of Jesus' key "I am" sayings (see pp.176–7): "I am the bread of life" (John 6.35). John interprets the event in the light of Moses and the manna in the wilderness. The feeding takes place on a mountain, hinting at an association with Moses. The true "grain of heaven," the expression in Psalms 78 for the manna, was not provided by Moses, but is given by God in the person of Jesus (John 6.49–51).

In the Synoptics, when Jesus takes the loaves and the fish, he blesses, breaks, and distributes them, foreshadowing his actions at the Last Supper (Matt. 26.26–29; Mark 14.22–25; Luke 22.15–20). In John, as in the feeding of the four thousand (see main text, right), instead of "blessing" the food, Jesus "gives thanks" (Matt. 15.36; Mark 8.6; John 6.11), a phrase recalling the Eucharist (Greek *eucharistia*, "thanksgiving"). If the word "fish" already has its later Christian significance as a symbol for Jesus, the miracle may also be a reference to the body and blood of Christ in the eucharist. The Fourth Gospel has no account of the Last Supper, and its narrative of the feeding appears to be its equivalent. The discourse concludes with Jesus saying that only those who eat the flesh and drink the blood of the Son of Man will receive eternal life and the assurance of resurrection on the last day (John 6.54). (See also pp.192–5.)

The majority of Jesus' miracles are cures of individual illnesses, and it is not difficult to accept the reality of these, whatever their religious significance; similar healings are attested in all centuries down to the present. Nature miracles, involving natural objects or phenomena, may seem a different matter, but the New Testament writers were living in a world where it was taken for granted that deities could, and did, intervene directly to affect the course of nature. They would have found plenty of scriptural evidence for God's control of the weather, the quality of harvests, and so on.

In the gospels, Jesus' feeding of the multitude from a tiny amount of food is treated as a nature miracle of special importance. All four evangelists record the feeding of the five thousand (Matt. 14.13–21; Mark 6.30–44; Luke 9.12–17; John 6.1–14), and Matthew and Mark have a second and very similar story of the feeding of the four thousand (Matt. 15.32–38; Mark 8.1–10). The latter are generally thought to be variants of the same episode, but it is not easy to see the relationship between them. There are some grounds for thinking that the story of the four thousand may be the original. The inner meaning of the narratives is not always clear, beyond the obvious fact that they demonstrate Jesus' divine power. Matthew and Mark record a discussion between Jesus and the disciples about bread, when the disciples discover that they have only one loaf between them (Matt. 16.5–12; Mark 8.14–21). Jesus refers to both feedings and complains of the disciples' failure to understand the significance of what had happened. Mark gives no indication of what this significance might be, but Matthew, in a very obscure way, introduces a warning against "the yeast of the Pharisees and Sadducees" presumably the teaching of these two groups (see pp.184–5).

Jesus Blessing the Loaves and Fishes, *a mosaic in the early 5th-century church of St. Apollinare Nuovo in Ravenna, Italy.*

Various nature miracles

Other nature miracles ascribed to Jesus have a distinctively Galilean setting. The gospels record regular journeys of Jesus and his disciples by boat across the Sea of Galilee, and two of these are the occasion for a demonstration of Jesus' powers. Once, a sudden storm threatens to sink the vessel and the disciples appeal to Jesus, who is sleeping in the stern, to save them: he rebukes the wind and the waves and restores calm (Matt. 8.23–27; Mark 4.35–41; Luke 8.22–25). Sea and tempest, as often in the Bible, are hostile forces which the word of Jesus defeats. His stilling of the "raging waves" shows his divine power, as when, in Psalm 89, God is said to "rule the raging of the sea" and, "when its waves rise ... still them." On a similar occasion, Jesus reveals his supernatural character – his disciples think he is a ghost – by walking on the water (Matt. 14.22–33; Mark 6.47–51; John 6.16–21). In Matthew, he is contrasted with Peter who tries, and fails, to do the same. In rescuing the disciple, Jesus is to be viewed as the savior of all in distress and worshiped as the Son of God (Matt. 14.28–33). Another story set in Galilee is that of the miraculous catch of fish. In Luke, this accompanies the call of Peter and foreshadows the success of his mission to catch people (Luke 5.4–10; see also pp.166–7). In John, the episode is placed after the Resurrection (John 21.4–14) and, in the evangelist's usual manner, its symbolism is heightened. The net contains one hundred fifty-three fish (John 21.11): ancient Greek zoologists believed that there were this number of species of fish, so the catch depicts the universal nature of the Church, embracing all humanity.

The Miraculous Draught of Fishes, *by Duccio di Buoninsegna (ca. 1278–1318), which illustrates John's account of the story (John 21.7–8).*

The feeding stories are full of allusions and symbolism. As God had provided food for Israel in the wilderness (Exod. 16), so Jesus feeds the multitude in a remote or desert setting. The messianic age would be a time of abundance and Jesus may be seen, as at Cana, as the giver of life and fertility at the messianic banquet (see p.167). However, the closest parallel to the feedings in the Bible is the story of Elisha feeding a hundred men from twenty loaves and some ears of grain (2 Kings 4.42–44). Like Jesus, Elisha commands that the people be fed, and also like Jesus, he is met with incredulity. Elisha's story ends with the statement that some food was left over, as in the gospel accounts: twelve baskets full of broken pieces remained after the feeding of the five thousand, seven baskets after the feedings of the four thousand in Matthew and Mark.

JESUS AND THE PHARISEES

The opposition of various Jewish parties to Jesus' words and actions plays an important role in the gospels' definition of his person and message. In places, Jesus is made to denounce each opponent in extremely harsh terms (for example, Matt. 23; Luke 11.39–52). This may represent a projection onto Jesus' lifetime of a situation that existed only later, when early Christians were denounced as heretics and expelled from synagogues. In the Fourth Gospel, the blanket term "the Jews" serves to denote the whole opposition to Jesus (see p.177).

All the same, there is no reason to doubt that confrontations with other religious parties within Judaism were an authentic element in Jesus' career. His main opponents are depicted as the Pharisees and the "scribes," the latter, especially in Luke, are sometimes called "lawyers." Whatever their origins, in the New Testament, scribes and Pharisees are virtually indistinguishable and their identity is confirmed from Jewish sources. They are the authoritative exponents of the Jewish Law, in its twofold aspect: the written Law of Moses, as found in the Pentateuch, and the unwritten, oral traditions handed down from former generations (see pp.118–19). The gospels acknowledge the Pharisees as both those who "sit on Moses' seat," the place in the synagogue from where the Law was taught (Matt. 23.2; see p.168), and those who uphold the "tradition of the elders" (Matt. 15.2; Mark 7.3, 7.5).

Many of Jesus' disputes with the scribes and Pharisees turn on points of Jewish Law and religion, and frequently they take the form of a rabbinical argument. Mark records a series of conflicts at the beginning of Jesus' ministry, in which Jesus' personal authority is contrasted with that of the recognized religious teachers (Mark 2.5–12; 15–17; 23–28; 3.22–30). He is able to reject their rules for the strict observance of the sabbath (Mark 2.23–28), though in one story he appeals for the Pharisees' permission to carry out healing on the sabbath, when a life is in danger. Elsewhere, the question of divorce revolves around the issue of

Jesus in the House of Simon the Pharisee, *by a follower of Dierick Bouts (ca.1420–1475). This episode is also discussed on pp.192–3.*

Jewish sects in the time of Jesus

A number of groups existed within Judaism of the first century CE, each vying to uphold its particular understanding of the Law, and each, according to the gospels, coming into contact with Jesus. The Pharisees were the heirs of the Hasideans, who emerged as defenders of the Law during the Maccabean revolt of the mid-second century BCE (see pp.140–41). Originally, they were very active in the affairs of the Jewish state which sprang from the revolt, and gave their support to the ruling Hasmonean dynasty. Increasingly, however, they viewed these rulers as departing from the strict observance of the Law, and they were persecuted under John Hyrcanus (134–104 BCE) and Alexander Jannaeus (103–76 BCE). Under Salome Alexandra (76–67 BCE) a compromise was reached which distinguished between state and religious authority. From this time, the Pharisees ceased to play an active part in public affairs, devoting themselves instead to strengthening their popular support through local courts and synagogues, with much success. The Pharisees were also particularly active in the Sanhedrin, the senate or council in Jerusalem, which served in part to uphold and disseminate their understanding of the Law. But their prime achievement, through their control of the synagogues and their concentration on the rules of purity for the individual, was to transfer the task of preserving the nation's holiness from the Temple of Jerusalem to the family and the local community. Pharisaic Judaism was able to survive the destruction of the Temple in 70 CE and the disappearance of any hope of the restoration of a Jewish state.

The Pharisees Disputing with Jesus in the Cornfield, *by Abel Grimmer (1570–ca. 1619). The episode is related in Matthew 12 and Mark 2–3.*

Another party, the Sadducees, has often been viewed as a priestly group. It has been suggested that their name derives from Zadok, one of David's two priests, but their name in Hebrew simply means "righteous ones." They should probably be seen as similar to the Pharisees, a scholarly group with their own particular interpretation of the Law. They rejected the oral Law and regarded written Law as the sole authority. They did not believe in the resurrection of the dead (Matt. 22.23; Mark 12.18; Luke 20.27), since it is not attested in the Pentateuch, and clashed on this issue with both the Pharisees and Jesus. They also rejected Jewish angelology and demonology, which the Pharisees embraced (Acts 23.8). The Sadducees were basically traditionalists and, with the fall of the Temple and the ascendency of Pharisaic Judaism, they rapidly disappeared.

Virtually nothing is known about the Herodians, mentioned only in the Bible and then only three times (Matt. 22.16; Mark 3.6, 12.13). The name suggests that they supported the Herodian dynasty, in contrast to the Sadducees who, according to the historian Josephus, were pro-Hasmoneans. The Bible links them with the Pharisees in plotting Jesus' death and trying to trap him over the question of paying tax to Caesar. The Herodian dynasty (the rulers of Galilee in Jesus' day; see p.144) depended on Roman sufferance and the Pharisees opposed resistance to Rome, so both groups may well have seen Jesus as a dangerous agitator.

how Scripture is to be interpreted (Mark 10.2–9), and the legality of paying taxes to Rome is settled by an authoritative decision of Jesus as one who teaches "the way of God in accordance with truth" (Matt. 22.16–22; Mark 12.14–17).

Much of Jesus' teaching is, in fact, in line with that of the Pharisees, but his followers were predominantly ordinary Jews, who found it difficult to observe the elaborate Pharisaic regulations governing daily life, especially those concerning ritual purity. The Pharisees are condemned for imposing "heavy burdens" on people (Matt. 23.4), in contrast to Jesus' lighter load, and for being concerned with petty legal details at the expense of the real purpose of the Law, the establishment of justice, mercy, and faith (Matt. 23.23). Matthew's favorite term for the Pharisees is hypocrites (Matt. 23). He sees them as people who do not live up to their own pretensions and seek only to enjoy the prestige, flattery, and acclaim which they receive as teachers and scholars.

III

CRUCIFIXION & RESURRECTION

•

The climax of Jesus' ministry was reached when he traveled from Galilee to Jerusalem shortly before Passover to face the death which, the gospels record, he had already predicted. After a final meal with the disciples Jesus was arrested, tried, and crucified. According to the gospels and Acts, this was not the end of the story: Jesus rose from the dead and, after issuing further guidance and reassurance to his followers, ascended to heaven.

•

THE DEATH OF THE BAPTIST

Herod Antipas, whom the Romans had appointed tetrarch (ruler) of Galilee and Perea after the death of his father Herod the Great in 4 BCE (see table on p.144), feared that John the Baptist's great influence on the masses could provoke a revolt. According to the Jewish historian Josephus, Herod Antipas (commonly referred to as Antipas, but called "Herod the tetrarch" in the gospels) jailed and executed John at Machaerus, a palace-fortress east of the Dead Sea. When the tetrarch was subsequently defeated in battle, some saw this as divine retribution for John's execution.

The circumstances of John's death are described somewhat differently in the gospels of Mark and (in briefer form) Matthew. The court and the leading men of Galilee are attending a banquet in honor of Antipas's

The Death of John the Baptist, *a 19th-century Russian icon.*

birthday (Mark 6.21). John has been imprisoned for offending Herodias, the wife of Antipas, for whom Antipas had set aside his previous wife, the daughter of the Nabatean king Aretas IV (9 BCE–40 CE). Herodias herself had previously been married to Antipas's half-brother, called Philip in the Bible (Mark 6.17; Matt. 14.3), but more usually known as Herod or Herod Philip, and not to be confused with another brother, Philip the tetrarch (see p.144). Under Jewish law the marriage of Antipas and Herodias was prohibited, and John denounced it, incurring Herodias's wrath. In contrast to Herodias, Antipas is represented by Mark as being impressed by John and his teaching, and unwilling to go beyond imprisoning him (Mark 6.20).

Herodias's daughter Salome, who asked for the head of John the Baptist in the gospel account. A granddaughter of Herod the Great, Salome's name is not given in the Bible, but is known from the Jewish historian Josephus. She first married an uncle, Philip the tetrarch, then another relation, Aristobulus of Calchis in Syria (ruled 70–92 CE). This portrait is from one of her second husband's coins.

The climax of the account of John's demise is the story of the dance performed before Antipas by Salome, Herodias's daughter by her first marriage (Mark 6.22–28; Matt. 14.6–11). The tale has a certain folktale quality: it tells of a king beguiled by a beautiful young woman, to whom he rashly vows to give anything she desires (Mark 6.23). The king is horrified when, prompted by her mother, Salome asks for the head of John on a plate. The story was probably preserved among the Baptist's disciples, who are recorded as giving his corpse a decent burial (Mark 6.29; Matt. 14.12). In Mark, the account serves to link John's fate with that of Jesus: as the Baptist had to suffer, so must Jesus, the true Messiah of whom John is the precursor.

The Baptist's movement did not end with his death. In the Bible, Christian missionaries twice encounter followers of John (Acts 18.24–19.7). These meetings took place at Ephesus in Asia Minor on one of the apostle Paul's evangelizing missions (see pp.232–3), so the Baptist's followers, like the early Christians, were presumably active missionaries. The followers of John and Jesus differed, among other things, over the nature of baptism (Acts 18.25–26; 19.2–3). In Acts, twelve disciples of John receive Christian baptism (Acts 19.5–6), and it seems that most of John's followers were soon absorbed into the Church. However, assertions that John was the Messiah continued into the early third century CE. The *Clementine Recognitions* of that date tells of one of the Baptist's disciples who "declared John to be the Messiah and not Jesus." Christians took care to refute such claims (for example, John 1.20).

The remains of the Herodian palace of Machaerus (modern Mukawir, Jordan).

Jesus and the Zealots

Like John the Baptist, Jesus may have been regarded by the authorities as a dangerous agitator: Luke records an episode where friendly Pharisees warn him to go into hiding because Herod Antipas wanted to kill him, as he had killed John. It has sometimes been suggested that Jesus was a political revolutionary, who advocated armed resistance to Rome, and even that he was associated with the "Zealots," a term that covers a variety of disparate rebel movements in Judea in the first century CE. Some have supported this view by drawing attention to the fact that one of Jesus' disciples was "Simon the Zealot." Other supposed evidence has included Jesus' violent cleansing of the Temple (see p.191); sayings such as "Do you think that I have come to bring peace to the earth? No, I tell you, but rather division" (Luke 12.51); the popular desire to make Jesus king; and the fact that, if we may trust the inscription on the cross, Pilate may have regarded him as one seeking to seize political power.

However, it seems highly unlikely that Jesus was associated with violent revolutionaries. His statement that violent men take the kingdom by force seems to be aimed at them, and his own description of the coming and spread of the kingdom is very different. His comments on paying tax to the Romans (Luke 20.25) would not have won the approval of those committed to the overthrow of Rome. Certainly, the mission of Jesus must be seen against a background of social unrest. But the general tenor of his teaching, not least the Sermon on the Mount (see pp.168–9), shows advocacy neither of militant resistance nor of armed revolution.

PORTENTS OF THE PASSION

The traditional site of the Transfiguration of Jesus is Mount Tabor, in the south of Galilee. However, Mount Hermon (right) in the far north of Israel, is a more probable location; it is very close to the site of Caesarea Philippi, the scene of Peter's confession, which precedes the Transfiguration.

In the Synoptic Gospels, a momentous change occurs when Jesus' ministry in Galilee comes to an end and he sets out on his journey to Jerusalem. There are now explicit predictions of the Passion, the suffering (Latin *passio*) that Jesus will undergo during the final days of his life on earth. Jesus explains to his disciples that the journey to Jerusalem will end with his death and resurrection (for example, Matt. 16.21; Mark 8.27; Luke 9.22). The climax of this period of Jesus' career is the extraordinary episode of the Transfiguration (see box on opposite page).

At Caesarea Philippi, north of Galilee, near one of the sources of the Jordan (Matt. 16.13–21; Mark 8.27; Luke 9.18–20), Jesus asks his disciples what people think of him. They reply that he is believed to be one of the great figures who will precede the Messiah. Then Jesus asks the disciples how they themselves view him and Simon Peter, in his typical role as spokesman, replies that Jesus is the Messiah himself. According to Matthew, Simon Peter's confession (used here in the sense of "statement of belief") earns him a great reward, based on the name Peter, in Greek *Petros*, from *petra*, "rock." Jesus promises Peter that he will be the rock on which the Church will be built (Matt. 16.17), and in effect confers on him the power of a Jewish doctor of the Law to decide what is forbidden and what is allowed: "whatever you bind on earth will be bound in heaven, and whatever you loose on earth will be loosed in heaven" (Matt. 16.19). This points to Peter's central role in the early Church (see pp.218– 19).

Immediately after Peter's confession, Jesus utters the first prediction of the Passion. He makes it clear that the Messiah will not be the triumphant figure of Jewish nationalist expectation, but one who must suffer and be killed, yet who will rise from the dead. Peter objects vehemently to this, and is rebuked by Jesus with the words "Get behind me, Satan!" (Matt. 16.23; Mark 8.33). Here, Peter appears in a way that is typical of him in the gospels: an impetuous speaker who embodies the failure of mortals, even disciples, to understand the true nature of Jesus

The Transfiguration

The mysterious episode of Jesus' Transfiguration (Matt. 17.1–8; Mark 9.2–8; Luke 9.28–36) is closely connected with the preceding confession of messiahship (see main text), and contains several themes.

The episode reveals the divine nature of Jesus and his intricate relationship with his heavenly father: the climax of the narrative is when, as at the Baptism, God proclaims Jesus as his unique son. In the Bible, mountains are often the scene of revelation and it is perhaps significant that both Moses and Elijah, who appear in the story, experienced a divine revelation on a holy mountain. The cloud is a regular sign of the presence of God, and Matthew makes this explicit when he describes the cloud as "bright."

During the Transfiguration, Jesus' whole physical being is transformed. His clothes become glistening white, like the dress of supernatural beings, such as angels. The statement in Matthew that Jesus' face shone, recalls the description of Moses after his meeting with God (Exod. 34.29–30). The presence of Moses and Elijah during the Transfiguration is of great significance. They are the representatives of the Law and prophecy, the cornerstones of the old Israel, whose destiny Jesus now fulfills. Luke gives the substance of their conversation with Jesus. It concerns his "departure" (the Greek word here is *exodos*) and certainly refers to his forth-coming death. However, by hinting at the Exodus of the Israelites from slavery in Egypt, it may be saying that this death also means redemption.

The event takes place for the benefit of three disciples. Peter wishes to secure the permanent presence of the three celestial beings by making a tent for each of them: "dwelling in a tent" was a Jewish expression for the sojourning of a divine being among humans. But Moses and Elijah vanish, leaving Jesus alone. Only he can now claim the allegiance of the three, and the whole story affirms the earlier teaching about the way of the cross, and the divinity of Jesus.

The Transfiguration *(top), from the Floreffe Bible (Flemish, ca. 1156). Jesus is flanked by Moses and Elijah. The apostles John, James, and Peter are astounded; only Peter (typically depicted with gray hair and beard) turns to look directly at Jesus.*

and his mission on earth. Not to accept the necessity of the Messiah's Passion is to side with the world and with Satan. Similarly, when Jesus shortly afterwards again foretells his death, the disciples can only react to his words with incomprehension, fear, and distress (Matt. 22.23; Mark 9.31–32; Luke 9.44–45).

Nonetheless, only the disciples learn of Jesus' messiahship and its consequences for him, and they are warned not to make it generally known until its truth is manifested by the Resurrection. At Caesarea Philippi, Jesus instructs them, as he continued to do on other occasions, on the conditions of their discipleship: to follow Jesus means taking up the cross and being prepared to meet death on his behalf. But that death will be the gateway leading to eternal life and salvation (Matt. 16.24–28; Mark 8.34–9.1; Luke 9.23–27).

JESUS IN JERUSALEM

All the Synoptics record a considerable number of events in the course of Jesus' journey to Jerusalem, but with some notable differences. Matthew and Mark speak of his traveling through the border region between Judea and Perea, eventually reaching Jericho (Matt. 19–20; Mark 10). John also briefly mentions a ministry east of the Jordan. By contrast, Luke appears to suggest that Jesus went through Samaria (Luke 17.11), but he also has him visiting Jericho. Luke may have been influenced by the fact that the regular route for Galilean pilgrims to Jerusalem passed through Samaritan territory, but he had a special interest in Samaria, and perhaps wished to include a mission there to foreshadow the future expansion of the Church (see Acts 1.8). In any case, Luke's account of the Jerusalem journey forms a distinct section of his gospel, and contains most of Luke's well-known parables.

A note of increased urgency and gravity marks the episodes and sayings common to more than one of the three gospels at this point. There is a third, more detailed, prediction of the Passion (see pp.188–9), with the explicit statement that Jesus will be handed over to the gentiles (Matt. 20.17–19; Mark 10.32–34; Luke 18.31–33). There are a number of episodes concerned with the demands of discipleship and the conditions for entry into the kingdom (see p.171). Only those who have the unquestioning trust that characterizes a little child can belong to God's kingdom. A rich young man, who is unable to accept that he must renounce everything to follow Jesus, is warned that eternal life is only for those prepared to abandon all earthly ties and concerns (Matt. 19.25; Mark 10.24). The

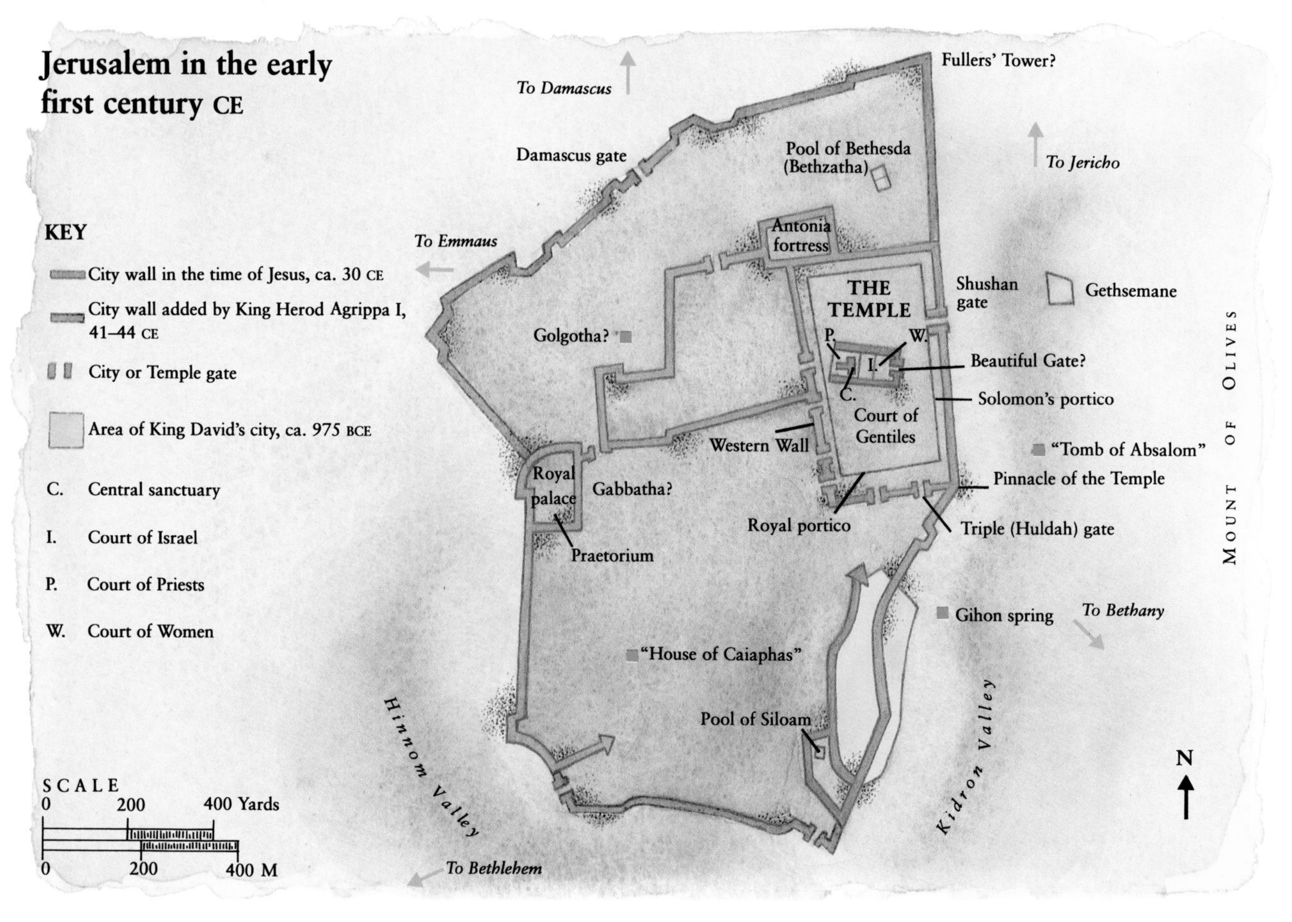

sons of Zebedee ask for a position of honor in the coming kingdom, but Jesus says that this is not in his gift: all he offers is the cup of suffering and the baptism of death. He stresses that discipleship is a calling to serve others, after the pattern of the Son of Man, who gives up his life to deliver many. The restoring of sight to one or two blind men at Jericho is no doubt intended as a typical healing miracle, although it emphasizes the fact that Jesus is hailed as "Son of David" (Matt. 20.29–35; Mark 10.46–52; Luke 18.35–43).

Jesus' arrival in Jerusalem is marked by three dramatic episodes. Arriving at the Mount of Olives, Jesus himself arranges for his triumphant entry into the city (Matt. 21.1–11; Mark 11.1–11; Luke 19.29–40). Certainly his reception is shown as an outburst of popular nationalistic fervor, which naturally arouses the alarm of the authorities; their plot to destroy him hints that his triumph will be short lived. But Matthew's quotation of the old prophecy (Matt. 21.5) gives the real significance of the event: Jesus is the expected king of Zion. The stress on the humility of the ruler points to the true meaning of his kingship.

Again, prophecy is the clue to understanding the episode of the cleansing of the Temple, when Jesus drives money changers and traders from the Temple precincts (see also p.176–7). According to Isaiah, the Temple in the messianic age will become "a house of prayer for all nations." Jesus realizes this, and combines the Isaiah verse with one from Jeremiah, in which the prophet denounces the Temple as a bandits' cave (Matt. 21.13; Mark 11.17). The Temple has failed to achieve its `divine purpose. There follows in the gospels a string of events and teachings that, among other things, demonstrate Jesus' authority over the official religion, and predict the destruction of the existing Temple.

The strange episode of Jesus' cursing of the fig tree (Matt. 21.18–19; Mark 11.12–14) is to be understood in the same way. The tree is a symbol of Israel, which has not produced the fruit God desired and so must be destroyed. Mark's addition "it was not the season for figs" has often been seen as showing the unreasonableness of Jesus' action. However, Mark may mean that it is now too late for Israel to fulfill its destiny.

Christ's Entry into Jerusalem, *painted ca. 1310 for Siena cathedral, Italy, by Duccio di Buoninsegna (ca. 1260–1318).*

The Synoptic apocalypse

Hearing Jesus predict the destruction of the Temple (Matt. 24.1–2; Mark 13.1–2; Luke 21.5– 6), his disciples ask how they are to know when this will happen. As Matthew says, it will mark the second coming of Jesus and the end of the present world.

Jesus' reply is a solemn sermon on the end of the world, the "Synoptic apocalypse," given on the Mount of Olives (Matt. 24.4–25.46; Mark 13.5–37; Luke 21.7–36). A passage couched in the typical language and imagery of Jewish apocalyptic writings (see pp.136–7), it falls into three main sections: the woes that precede the time of the end; the final consummation; and the need to be vigilant and observe the signs of the times. The evangelist's discourse was probably colored by the actual events of the fall of Jerusalem in 70 CE, but it is possible that Jesus himself foretold an imminent catastrophe for the nation: in his day there was already a simmering climate of revolt in Judea.

Before the end, Jesus says, there will be widespread warfare and natural disasters, together with severe persecution of his disciples. But the key sign that believers must be awake for is the profaning of the Temple by the "desolating sacrilege," a reference to the desecration of the Temple by the Seleucid king Antiochus Epiphanes (see pp.138–9). When this occurs, the faithful must flee Judea – Jewish Christians do in fact seem to have been dispersed after Jerusalem's fall. There will be many false hopes, many false Christs and prophets to lead the Church astray, but finally the Son of Man will return in glory to gather his flock from every corner of the world. The apocalypse is imminent – "this generation will not pass away till all these things have taken place" – but no one can know the precise time of its happening. The discourse ends with two parables urging constant watchfulness for the approach of the last day.

THE LAST SUPPER

THE ANOINTING AT BETHANY

All the gospels know of a tradition that Jesus was anointed by a woman (Matt. 26.6–13; Mark 14.3–9; Luke 7.37–50; John 12.1–8). Every account, apart from Luke's, sets the episode at Bethany, a village outside Jerusalem, in Jesus' last days. The woman's act foreshadows the regular Jewish anointing of the dead, for, as Jesus says, "you will not always have me" (Mark 14.7). In Matthew and Mark, she pours the richly perfumed oil on Jesus' head: this is probably intended to recall the anointing of kings and high priests with precious oil at their consecration (for example, Exod. 40.13–15; 1 Sam. 10.1, 16.13; 1 Kings 1.39). As well as anticipating the burial-anointing, her act is a recognition of Jesus as the Messiah or Christ, God's "anointed one."

In John, the woman (here identified with Mary, the sister of Lazarus) anoints Jesus' feet rather than his head, because John probably intends to focus less on the royal overtones of the anointment than on its funeral aspect. Jesus says that the unction is "for the day of my burial" (John 12.7), and, indeed, it is only in the Fourth Gospel that Jesus' corpse is actually anointed (John 19.40).

Luke places the anointing episode in the context of the Galilean ministry and gives it a quite different interpretation. The woman is a sinner, a characteristic Lukan motif: she has traditionally been identified with Mary Magdalene, but the narrative gives no justification for this. She pours oil on Jesus' head, and humbly washes, kisses, and anoints his feet. Jesus contrasts her actions with those of his host, the Pharisee Simon, who has not welcomed him with the usual courtesies of kissing the cheek and washing the feet. The woman's action is a sign of love and trust, for which she receives forgiveness (Luke 7.47). The story illustrates two great themes in the Gospels: Jesus' authority to forgive sins, and his concern for outcasts and sinners.

The gospels record that Jesus and his disciples ate one final meal together on the evening of his arrest, around the time of the feast of Passover. The narratives of this Last Supper are immediately preceded by the accounts of the plans of the religious authorities to arrest and kill Jesus, and of his betrayal by Judas. Jesus is presented as one who knows of, and accepts, his coming suffering and death, "for the Son of Man is going as it has been determined" (Luke 22.22). In each gospel, he says that someone present will betray him (Matt. 26.21; Mark 14.18; Luke 22.21; John 13.11).

There has been much discussion about whether or not the Last Supper was a Passover meal, eaten on the eve of the Passover sabbath. According to the Fourth Gospel it was not, because the crucifixion took place on the day before Passover. But John does say, in typically symbolic manner, that the crucifixion happened at the time of Preparation, when the paschal (Passover) lambs were sacrificed (John 19.14), thus in a way making Jesus himself into the paschal lamb (John 19.36, quoting Exod. 12.46). The Synoptic Gospels imply that the supper included the paschal lamb, the principal constituent of the Passover meal (Matt. 26.17; Mark 14.12; Luke 22.7–8 and 22.15), but do not mention it in connection with Jesus' impending sacrifice. However, the Synoptics did not intend to describe in detail what was eaten at the meal, but only Jesus' distinctive words and actions on that occasion, especially his eucharistic utterances (see box on opposite page). It may be that the Synoptics, too, saw Jesus taking the place of the lamb, in giving himself as food to the disciples. In John (see pp.194–5), the consumption of Jesus' body and blood are associated with the narrative of the loaves and fishes (John 6, especially verse 45–58).

Jesus and his disciples may have constituted a "fellowship group," who regularly met for formal religious meals. At these meals, which were an important part of the life of, for example, the Qumran community (see p.153), blessings were pronounced over bread and wine, in accordance with Jewish custom. The Last Supper may have been the last time Jesus and his disciples shared such a fellowship meal. If this theory is true, the supper may well have been a paschal meal, because Jewish texts testify to a close connection between fellowship groups and the Passover.

At their final meal together, Jesus and his disciples may well have used utensils very similar to this domestic crockery from Roman Judea, from a monastery at Martirius, east of Jerusalem (ca. 350 CE).

An early Byzantine fresco of the Last Supper from Göreme in central Turkey. The fish, a symbol of Jesus Christ, here represents the Christian Eucharist; the beardless figure of Jesus blesses the chalice containing the Eucharistic wine.

The eucharistic words of Jesus

Blessing the bread, a Christian painting of the late 3rd century CE *in the cemetery of SS. Paul and Marcellius, Rome.*

Jesus' words at his final meal with the disciples, as recorded in the Synoptic Gospels, may be said to transform the Last Supper into the Lord's Supper, the Holy Communion or Eucharist (from the Greek *eucharistia*, "thanksgiving"). When Jesus says that he will not again eat the Passover or drink the fruit of the vine until the coming of the kingdom of God (Luke 22.15–18), the meal becomes a foretaste of the great messianic feast at the end of time. The Qumran community (see p.153) apparently understood its own religious meals in the same way: one of the Dead Dea Scrolls speaks of the presence of the Messiah at the communal table.

In all three gospels, Jesus takes bread and wine, saying respectively: "this is my body" and "this is my blood." These words have provoked a great debate over the centuries about the nature of Christ's presence in the Eucharist. But the evangelists see Jesus as one of the old Israelite prophets, performing symbolic acts that point to his coming death and its significance. The "blood of the Covenant" (Matt. 26.28; Mark 14.24) refers to Sinai, where Moses sealed the Covenant with sacrificial blood (Exod. 24.8). The first Covenant centered on the book of the Law. The new Covenant, spoken of, according to Christian understanding, in Jeremiah 31.31–34, is made through the redemptive death of Jesus. His blood is "poured out for many" (Mark 14.24), a reference to the servant who "bore the sin of many" in Isaiah 53.12. Paul explains the breaking of the bread: it represents "my body that is broken for you" (1 Cor. 17.24). Luke's gospel probably preserves the original text: "This is my body, which is given for you"– that is, given up to death on the cross (Luke 22.19).

THE LAST SUPPER IN JOHN

The Fourth Gospel places the Last Supper before the Passover and records only one event during the meal: Jesus' washing of his disciples' feet. This follows the regular pattern in John of a "sign" followed by an explanatory discourse. Washing of the feet was a lowly duty and a mark of exceptional deference, and Jesus' action therefore shows what his incarnation means: the Word descends to the humblest position for the salvation of humanity. Above all, the narrative teaches the nature of true discipleship and how the Church will function on the basis of mutual service, following the model of its Lord and Teacher.

Following on from this episode, the Synoptics' motif of the traitor at the table is taken up by the fourth evangelist, but with his own interpretation. Judas is not just someone who betrays his master for money, but the agent of the devil and the representative of the powers of evil (John 13.27). The little sentence "And it was night" (John 13.30), which accompanies Judas's departure to carry out his grim task of betrayal, is highly symbolic. Judas is one who, in Jesus' earlier words, loves darkness rather than light because his deeds are evil.

The most distinctive elements in John's account of the Last Supper (John 13.1–17.26) are the three lengthy discourses in which Jesus takes leave of his disciples. This follows John's usual pattern, by which an action of Jesus is accompanied by teaching to show its significance, but on this occasion the teaching precedes rather than follows the event. The farewell discourses expound to the puzzled disciples the meaning of their master's forthcoming crucifixion and resurrection. Jesus' first speech proceeds in three stages, in response to three questions. Thomas asks about the road that Jesus is to take and is told that Jesus alone is the way of life that will lead his followers to the heavenly father (John 14.5–6). Then Philip asks to see the father and receives the reply that whoever has seen Jesus has seen the father, because the father is at work in him in perfect unity. The world cannot know this ultimate secret, which will be disclosed only to those who love and follow Jesus. Another disciple inquires why only the disciples and not the world in general have this revelation. He receives no direct answer, because his question shows that the disci-

Jerusalem in the Fourth Gospel

In contrast to the Synoptic Gospels, John centers the ministry of Jesus on Jerusalem. He makes five visits in all, while the Synoptics record only one, at the end of his ministry (apart from the Presentation; see p.157). In John, Jesus goes to the city for the great Jewish festivals, and the author may wish to show him as both the true heir of Judaism and its transformer.

In John 2, Jesus goes up to Jerusalem at Passover, when he cleanses the Temple. Later, he visits for "one of the Jewish festivals," possibly Pentecost (John 5.1), which by this time may have become a commemoration of the giving of the Law at Sinai (see pp.60–61). When Jesus heals a cripple he is accused of breaking the Law by working on the sabbath, and his comment at John 5.17 heralds the advent of the new Law through Jesus. His following visit is for the feast of Booths or Tabernacles, which was marked by a daily ritual drawing of water and a brilliant illumination of one of the Temple courts: Jesus teaches openly in the Temple, proclaiming himself the water of life (John 7.37–39) and the light of the world (8.12). Jesus is next at Jerusalem for the festival of Hanukkah or Dedication (John 10.22; see also p.140). This is his last visit before the Passion, which is foreshadowed when "the Jews" accuse him of blasphemy and attempt to kill him (John 10.31–39).

ples still do not understand what the promised revelation implies or that its condition is Jesus' death. Instead, Jesus states clearly that he must leave them; their earthly companionship with him has finally ended, but he will send them the Advocate, or Helper, who will be with them for ever (John 14.15–17, 14.25–26). The Advocate (often called the Paraclete) is the Holy Spirit, the spirit of truth through whom the disciples will learn the full and glorious significance of Jesus' life and mission which, at the present moment of crisis and tragedy, they are unable to comprehend.

The second discourse extends several of the themes of the first but deals more directly with the disciples and, through them, the future Church. In the last of John's "I am" sayings (John 15.1–11; see also p.176–7), Jesus describes himself as the vine, a symbol for Israel: he and the disciples, the branches, form a new community, and, following the teaching of some Synoptic parables, the disciples are warned of the fate that awaits them if they do not produce fruit. They must expect, like their Lord, persecution and suffering from the people of this world, but when Jesus has conquered the world their final reward will be peace and joy.

Jesus' final discourse is a moving prayer (John 17). He prays first to be allowed to complete the mission given to him by God, then for the disciples who are to represent him in the world, and last for all those who, in the future, will come to faith through the disciples' teaching.

Above: *1st-century* CE *sandals from Masada (see p.145). The open footwear of Jesus' day meant that people washed their feet frequently, and it was a courtesy to enable guests to do so (cf. Luke 7.44).*

Below: Jesus Washing the Disciples' Feet, *School of Arezzo, Italy, ca. 1390. Here, Jesus is shown washing the feet of Peter.*

GETHSEMANE

Above: *Ancient olive trees in the traditional Gethsemane on the Mount of Olives (see map on p.190). The exact location of Jesus' agony and arrest cannot be known for certain, but this olive grove is widely accepted as the most likely place.*

Right: The Arrest of Christ, *painted ca. 1310 for Siena cathedral, Italy, by Duccio di Buoninsegna (ca.1260–1318).*

According to the gospels, after the Passover meal (see pp.192–5), Jesus and his disciples crossed the Kidron valley to the Mount of Olives and a place called Gethsemane, a name meaning "oil vat" or "oil press," in other words it was an olive grove. Before reaching it, Jesus warned his disciples that they would abandon him. Typically, Peter protested his loyalty, but Jesus predicted that even he would disown him in the hours to come (Matt. 26.33–34; Mark 14.29–30; Luke 22.33–34; see p.199). Luke adds a passage, the meaning of which is not clear, in which Jesus apparently warns his disciples to prepare for persecution, even to the extent of arming themselves (Luke 22.36). The disciples produce two swords, and Jesus says, perhaps with heavy irony, "It is enough."

The scene in Gethsemane (Matt. 26.36–56; Mark 14.32–42; Luke 22.39–46) begins with a moving picture of Jesus' agony in the face of his approaching Passion – particularly vivid in Luke – in which Jesus subjects himself to God's will. John omits the episode of the agony because he has already provided its equivalent in Jesus' great prayer at the Last Supper (see p.195). The disciples fall asleep, emphasizing Jesus' isolation. John says that Judas knew where to find Jesus because the band of disciples often met at Gethsemane, which he does not name but calls "a garden" (John 18.1). The force sent to arrest Jesus consists mainly of Jewish Temple police, but John also mentions "soldiers" (that is, Roman troops) and their officer. This has led some scholars to suggest that the Romans took the lead in arresting Jesus, but it is more likely that it only represents the particular outlook of the Fourth Gospel, which appears to stress the role of the occupying Roman power: for example, John gives no details of a trial before Caiaphas.

The motif of the sword recurs when one of the disciples (named in John 17.10 as Peter) cuts off the ear of the high priest's servant. Jesus calls for no further resistance: nothing must stand in the way of Jesus' prescribed fate. Mark's gospel rounds off the account with an odd scene describing a young follower of Jesus dressed only in a linen cloth, which he leaves behind when the soldiers try to seize him (Mark 14.51–52). Whether or not the young man was Mark himself, as is often conjectured, the story may well be a nugget of authentic information.

Annas and Caiaphas

After his arrest, Jesus is taken to the palace of the high priest (Matt. 26.57; Mark 14.53; Luke 22.54; John 18.13). The New Testament is not clear about the high priest's identity: Mark and Luke do not name him, and for Matthew he is Caiaphas, whom John also describes as high priest (John 18.24). But in John, Jesus is taken first to Annas, the father-in-law of Caiaphas, and in the following passage the term "high priest" appears to refer to Annas. It is possible that John 18.24 has been dislocated and belongs before John 18.15, although Acts, speaking of a time shortly after the Ascension, also calls Annas the high priest (Acts 4.6).

By the early first century the high priest was appointed from among a restricted group of families by the Roman authorities. According to the Jewish historian Josephus, Annas or Ananus was appointed to the post in 6 CE by Quirinius, the Roman legate of Syria. Annas was removed in 15 CE by the procurator of Judea, Valerius Gratus (15 CE–26 CE), the predecessor of Pontius Pilate (26 CE–36 CE). Valerius appointed Josephus Caiaphas to the post in 18 CE, and Caiaphas was deposed by Vitellius, the legate of Syria, in 36 CE or 37 CE. Annas had therefore ceased to be high priest well before any likely date for the crucifixion (ca. 30 CE), it seems certain that Caiaphas was the one before whom Jesus was tried. But Annas was an influential figure – no less than five of his sons, as well as his son-in-law, attained the high-priesthood. Popular memory may thus have credited him with a longer tenure than he actually enjoyed. Also, before and after Caiaphas the high priests changed so frequently – there were eighteen in the sixty years after 6 CE – that the exact succession is unlikely to have been remembered by the population at large.

Judas Iscariot

Judas Iscariot is a mysterious figure. The derivation of his name is uncertain; it may be a patronymic, indicating "man from Kerioth," perhaps a town in southern Judea, or it could have an Aramaic root meaning "false one." Judas means "Judean" or "Jew," and he has sometimes been viewed as an archetype, representing all the Jews, who rejected Jesus. But it was a common name – there was another apostle called Judas – so this seems unlikely. There is no indication of Judas's motives for betraying Jesus. It has been suggested that he had either lost faith in his master, or was seeking to force him to assert the power that was popularly expected of the Messiah.

Silver shekels of the mid-1st century CE.

What the New Testament says of Judas is full of biblical allusions. "Thirty pieces of silver," his payment for the betrayal, was a fairly small sum, but it is derived from an obscure passage of Zechariah (Zech. 11.12–13), also the source of the scene where the repentant Judas throws down the money in the Temple. The two versions of Judas's death – in Matthew he hangs himself (Matt. 27.5), but in Acts he falls, spilling his entrails (Acts 1.18–19) – derive from Bible stories of the deaths of evil men, such as Ahithophel (2 Sam. 15.23) and Antiochus Epiphanes (2 Macc. 9.7–9). In Acts, Judas's death explains the name of a place in Jerusalem, *Hakeldama*, the "Field of Blood." In Matthew, it is so called because Judas's "blood money" paid for it (Matt. 27.6–7). The purchase of the field is said to fulfill a prophecy of Jeremiah, although it actually conflates verses from Jeremiah and Zechariah.

JESUS ON TRIAL

The traditional "House of Caiaphas" in Jerusalem, where Jesus is said to have been taken after his arrest (Matt. 26.57).

All the gospels record two trials, or at least interrogations, of Jesus, one by Jewish authorities, the other by Pontius Pilate, the Roman governor (see pp.200–201). The Jewish part of the proceedings has often been regarded as unhistorical, an attempt by the evangelists to transfer the real blame for Jesus' death from the Roman governor to the Jews. This certainly accords with the generally favorable attitude found elsewhere in the New Testament toward the Roman power, especially in Acts (see pp.234–7). There are discrepancies between the gospel accounts of the trial before the high priest and what is known of Jewish legal procedure. But the evangelists were not court reporters giving a verbatim account of events: their concerns are primarily theological.

The Synoptics speak of Jesus being brought before the "council" (Matt. 26.59; Mark 14.55; Luke 22.66), a reference to the Sanhedrin, the body of Jewish elders in Jerusalem which the Romans allowed to regulate, within limits, the internal affairs of the Jews. The history of this body in the first century CE and earlier is very obscure, and it probably underwent considerable development. An important question turns on how far its powers extended, in particular whether, as John's gospel states, it was unable to carry out the death penalty (John 18.31), thus requiring Jesus to be taken to Pilate for execution. It is now increasingly accepted by scholars that Roman provincial government reserved the death penalty to itself, and that there is therefore nothing improbable in Jesus facing two trials, one before Jewish authorities and a second before the Romans.

It has also been claimed that the trial before the Sanhedrin would have infringed rabbinic laws, such as the prohibition of trials at night, on the eve of a sabbath or festival, or on the sabbath or festival itself. Trials were also banned anywhere other than in a formal court building. However, such legal requirements are laid down in sources much later than Jesus' time, and their provisions may not have been in force then. They represent the rulings of the Pharisees (see p.185), who played no direct role in the condemnation of Jesus, and whose regulations might not have been followed by the priestly authorities of Jesus' day. Besides, although Jesus was taken at night from Gethsemane to the high priest's house (see pp.196–7), no tradition states that his trial actually took place there, and a careful reading of the gospels does not prove that it took place during the night rather than in the early morning.

Jesus was condemned by Jewish authorities as a blasphemer for claiming to be the son of God (Matt. 26.65; Mark 14.62; Luke 23.2b; John 19.7), but rabbinic legal sources state that a person could not be found guilty of blasphemy unless he spoke the divine Name (Yahweh) itself. Nothing in the gospels indicates that Jesus did this, although in Luke he says "the son of man will be seated at the right hand of the power of God" (see also Matt. 26.64, Mark 14.62). The phrase "power of God" could be a euphemism, and Jesus may actually have used the Name. However, the law against uttering the Name may also be of later origin, and blasphemy may have been interpreted more widely in Jesus' own day: for example, the gospels record elsewhere that he outraged his opponents by claiming to be able to forgive sins.

There are indications in Matthew and Mark that the crucial charge against Jesus, supported by the two witnesses that Jewish law required, was that he had threatened to destroy the Temple (Matt. 26.60–61; Mark 14.57–58). If this were true, Jesus was liable to condemnation by both the authorities he faced: by the Sanhedrin as a flouter of religion, and by the Romans as a dangerous agitator.

Peter's denial

As recounted on p.197, Jesus was first taken to the house of the high priest, which is very likely to have been situated in the wealthy quarter on Jerusalem's western hill. Recent excavations there have revealed large and handsome houses from the time of Jesus, built around an inner courtyard and approached by a pillared gateway. Such was the setting of Peter's denial of Jesus (Matt. 26.69–75; Mark 14.66–73; Luke 22.54–62; John 18.25–27), which fulfilled Jesus' prophecy in Gethsemane (see p.196).

The episode shows Peter in a poor light, but it is significant as the final stage in the progressive isolation of Jesus. On his arrest, all his disciples had abandoned him and fled, apart from Peter, who was determined to carry out his promise "to see how this would end" (Matt. 26.58). According to John, Peter was accompanied by "another disciple," who was acquainted with the high priest and able to gain him admittance.

The story is carefully crafted, with Peter's three denials each increasing in intensity. The accounts in the Synoptics end with Peter breaking down after the third denial and weeping bitterly. John seems to soften the depiction of Peter, omitting both his recollection of Jesus' prophecy and his bitter repentance. However, this also weakens the pathos of the story. In contrast, Luke heightens the emotional intensity, with his usual narrative skill, by recording the meaningful look that Jesus casts on Peter after his third and final denial (Luke 22.61).

Peter's Sorrow *(Mark 14.72; Matt. 26.75; Luke 22.61); Ethiopian, 17th century.*

THE TRIAL BEFORE PILATE

Jesus' trial before the procurator of Judea, Pontius Pilate (Matt. 27.11–14; Mark 15.1–5; Luke 23.1–16; John 18.28–38), essentially conforms with Roman provincial practice. It took place at the praetorium, the governor's residence in Jerusalem, which may have been the old Herodian royal palace or the Antonia fortress (see map on p.190). The hearing was an official one, held at the tribunal (public speaking platform), on the paved court before the praetorium. It was here that Pilate's judgment seat was placed.

In their judicial function, Roman provincial governors were mainly concerned with public order which they had substantial power to enforce. A trial began with the submissions of the accusers – in the case of Jesus, the chief priests, elders, and scribes. Their charge is not about the breaking of a specific law but about certain disruptive actions: that Jesus has been subverting the nation, opposing the payment of tax to the emperor, and claiming to be the Messiah, a king – implying that Jesus is a leader of resistance to Rome. Initially Pilate tries to put off making a decision. In John, he attempts to pass responsibility to the Jewish courts – "judge him according to your own law" (John 18.31) – which it was in his power to do. Governors sometimes avoided tiresome cases by sending the accused to be tried in their place of origin. In Luke, when Pilate discovers that Jesus is from Galilee, he sends the case to the client ruler of that region, Herod Antipas (see p.186), who is in Jerusalem for Passover (Luke 23.7).

Pilate treats Jesus in a way not untypical of Roman legal custom. Those who, like Jesus, did not offer a defense, were given three opportunities to change their minds, as is known from a famous letter by the Roman author Pliny about the Christians. When Jesus made no defense, Pilate was forced to convict him. Jesus' adversaries put further pressure on Pilate by accusing him of treachery toward Caesar (John 19.12), a charge that could not be taken lightly. It could well have been used against Pilate

Pontius Pilate

Pontius Pilatus (Pilate) was governor of Judea from 26 to 36 CE. His name, with his correct title "prefect," rather than the more popular "procurator," was found in 1961 on an inscription at Caesarea (see picture). Pilate would have gone from Caesarea to Jerusalem to keep an eye on the city during Passover, a time when insurrection was often in the air. (see pp.144–5). Jewish sources generally view Pilate unfavorably. Two first-century writers, the historian Josephus and the scriptural commentator Philo, relate two occasions when he offended Jewish sensibilities by bringing into Jerusalem shields and standards bearing idolatrous Roman images. He suppressed a riot sparked by his use of funds from the Temple treasury to build an aqueduct. Following a disturbance in Samaria he executed some leading Samaritans, for which he was dismissed and sent back to Rome.

The inscription bearing the name ...tius Pilatus.

The fact that Pilate held office for ten years suggests that he was not entirely unsuccessful as an administrator. The gospels show a tendency to exonerate him and to blame the Jews for Jesus' death, a development seen most clearly in John. Mark portrays him as a reasonably impartial judge, and this perhaps comes nearest to reality. In later Christian writings Pilate became something of a hero: the Coptic Church honors him as a saint and a martyr.

by his enemies: the unscrupulous manipulation of the treason law for political ends was a common feature of Roman public life.

Matthew, Mark, and John state that it was the custom at Passover for the governor to release a prisoner of the people's choice (Matt. 27.15; Mark 15.6; John 18.39). No such custom is known, although Barabbas, who is freed instead of Jesus, may have been a well-known figure. Mark and Luke say that he was in prison for insurrection (Mark 15.7; Luke 23.19), which fits well with the disturbed conditions in Palestine at the time. His name means "Son of the Father." Some versions of Matthew give his name as Jesus Barabbas, so the episode may have been included in the gospel story to illustrate the contrast between the fate of Jesus the false Messiah, who is spared suffering, and Jesus the true "Son of the Father," who must suffer at the hands of his enemies.

Matthew heightens his narrative of Jesus' last hours with various signs and portents, one of which is the dream of Pilate's wife (Matt. 27.19), who acknowledges that Jesus is an "innocent man." The purpose of this scene is to show that Jesus was recognized by an important figure among the gentiles. It corresponds to the scene in Luke where, at Jesus' death, a Roman centurion says: "Certainly this man was innocent" (Luke 23.47).

Above: *Part of the Roman pavement discovered in the early 20th century on the site of the Antonia fortress in Jerusalem. The pavement of the fortress is the traditional site of Jesus' trial before Pilate.*

Left: Ecce Homo (Behold The Man), *by Quentin Massys or Metsys (1466–1530). Pilate (left) hands over Jesus to be crucified as a mocking soldier places a scarlet cloak over the shoulders of Jesus, who already wears a crown of thorns. The Latin title of this scene and its traditional English translation echo Pilate's words in John's gospel (John 19.5; translated in the NRSV as "Here is the man").*

THE WAY OF THE CROSS

THE WOMEN OF JERUSALEM
According to Luke, among the crowd that followed Jesus on his way to execution were many women who, in accordance with Jewish mourning rites, "were beating their breasts and wailing for him" (Luke 23.27). Jesus addresses them as "daughters of Jerusalem" and tells them to weep not for him but for themselves and their children. In the Hebrew Bible, the phrases "daughter of Zion" and "daughter of Jerusalem" mean the city itself, since town names are feminine in Hebrew. But Luke uses the phrase literally of real women who will mourn the disaster that is to befall their city, which Jesus foretold in Luke 21. The blessing of the barren (Luke 23.29) recalls the woe pronounced on pregnant women on the day of disaster (Luke 21.23), and the quotation from Hosea (Luke 23.30) perhaps echoes his warning (Luke 21.21) to people to flee to the mountains on that day. The statement about the green and dry wood (Luke 23.31) may be a popular proverb. Its significance is not obvious, but it could mean that if the innocent Jesus suffers such a fate, that of guilty Jerusalem will be so much greater.

It has often been noted how Jesus shows a special concern for women, both in his teaching and in the stories about him, and this aspect is particularly emphasized in Luke. All the Synoptic Gospels mention the presence at the Crucifixion of women who had looked after Jesus in Galilee. Luke alone records them in their actual Galilean context and depicts them as a group of well-to-do women who were able to support Jesus financially. There is now considerable evidence that in some Jewish communities at least, women had a degree of financial independence and in some cases were even leaders of synagogues.

The route taken by Jesus to the place of crucifixion depends largely on the location of the praetorium, Pilate's Jerusalem residence, where it began (see p.200). If it was the Antonia fortress, the journey could have been along the traditional Via Dolorosa, the "Way of Sorrow," that pilgrims follow to this day. However, the praetorium was identified with the Antonia no earlier than the fourth century CE, and it is more likely that Jesus was taken from the former royal palace of Herod the Great, in the west of the city. This would have involved a completely different route.

Condemned persons usually carried their own cross, as Jesus does in the Fourth Gospel. But the Synoptists record that a man called Simon was dragooned for this task (Matt. 27.32; Mark 15.21; Luke 23.26). Simon was from Cyrene in North Africa, and was probably in Jerusalem for Passover. He is a model of discipleship: Luke says that he carried the cross behind Jesus, echoing Jesus' sayings about the need to take up the cross and follow him (as in Luke 9.23). Mark says that Simon was "the father of Alexander and Rufus," referring to them without further description in a way that suggests that they were people known to his readers, perhaps members of the church for which he wrote.

The goal of the journey was a site called Golgotha ("The Skull") in Aramaic, translated in Latin as *Calvaria*, Calvary. Why it was so called is unknown, but it seems likely that it was associated with death in some way, either as a place of execution or as a burial ground. The Fourth Gospel and the letter to the Hebrews assume, probably correctly, that Jesus was executed outside the city walls, in accordance with both Roman practice and the provisions of Jewish law. The traditional site of the Crucifixion was first identified in the fourth century CE, and the present Church of the Holy Sepulcher was founded on the spot. In ancient times it was a small quarry, a place suitable for both executions and burials, and in Jesus' time there was a small precipice, on the east side of the quarry. The area has been extensively excavated (see picture, below). Numerous burial caves have been discovered in and around the church, a fact that supports John 19.41 and also indicates that the spot was originally outside the urban area. Modern archaeology may yet confirm the authenticity of the traditional site of the Crucifixion.

This cross-section of the Church of the Holy Sepulcher, Jerusalem, shows how it was built over the low precipice traditionally identified with Golgotha.

The scourging and mocking

The gospels recount that Jesus was scourged (flogged) and mocked before the Crucifixion. Two types of scourging seem to be mentioned in the gospels, and they are in accordance with Roman judicial practice. A judge could administer a light beating before releasing a prisoner, as a warning not to repeat an offense. So when, in Luke, Pilate decides to free Jesus, the phrase translated as "have him flogged" is a fairly mild term that may mean simply "give him a warning" (Luke 23.16; 23.22). In Matthew and Mark, the flogging of Jesus before the crucifixion is expressed by a stronger term: it was a severe beating of the sort that in Roman practice always preceded further punishment (Matt. 27.26; Mark 15.15).

The Roman soldiers' mockery has greater symbolic value. The story parallels the physical abuse of Jesus when he was brought before the Sanhedrin (Matt. 26.67; Mark 15.65; John 18.22). There, he had been accused of claiming to be a king; now he appears as one. In Matthew, Jesus is dressed in a scarlet robe, something that is normally worn only by Roman generals and emperors (Matt. 27.28); in Mark the cloak is imperial purple (Mark 15.17). A crown of thorns is placed on his head, a parody of the imperial crown of laurels. The homage of the soldiers is only a taunt, and a prelude to Jesus' execution as a criminal. But for the evangelists it contains a supreme irony: Jesus, and not the Roman emperor, is indeed the real king of the world.

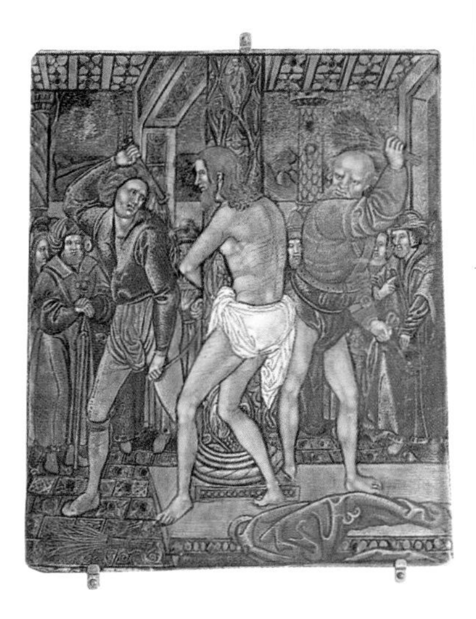

The Flagellation of Christ, *an enameled panel by Jean I Pénicaud (1500–1530).*

The Road to Calvary, *1527, by Albrecht Dürer (1471–1528). Jesus (center) carries his own cross, as in John's account (John 19.17).*

THE CRUCIFIXION

THE SEVEN LAST WORDS
This is the traditional order given to the seven last "words" (utterances) of Jesus on the cross, as reported in the gospels (see box on opposite page).

"Father, forgive them; for they do not know what they are doing."
(Luke 23.34.)

"Truly I tell you, today you will be with me in paradise."
(Luke 23.43.)

"Woman, here is your son" ... "Here is your mother."
(John 19.26–27.)

"My God, My God, why have you forsaken me?"
(Matt. 27.46; Mark 15.34.)

"I am thirsty."
(John 19.28.)

"It is finished."
(John 19.30.)

"Father, into your hands I commend my spirit."
(Luke 23.46.)

The gospels devote a large amount of space to the Passion narrative and this in itself shows the great importance that they attach to it. Another indication of its significance to the evangelists is the way in which, even more than elsewhere in their narratives, they endeavor to show that the events of the Passion were prefigured in the Scriptures. This has led some commentators to conclude that many incidents in the Passion story have been constructed from passages in the Psalms and prophets. But, although the Passion story is clearly colored by the evangelists' own interpretation, it does not follow that it has no historical basis.

Mark relates that before being crucified, Jesus was offered a drink of wine mixed with myrrh, such as would be offered to condemned criminals to deaden the pain they were about to suffer (Mark 15.23). Matthew, on the basis of Psalm 69.21, changes "myrrh" to "gall," thereby making the offering of the drink into a hostile act (Matt. 27.34). The Roman soldiers are then said to have cast lots for Jesus' clothes. John (19.24) sees this as a fulfillment of Psalm 22.18, which he actually quotes (Psalm 22 figures prominently throughout the Passion story). Roman legal sources show that it was the accepted right of an execution squad to divide the minor possessions of the condemned.

Jesus is crucified between two bandits. According to an additional verse in some versions of Mark's gospel (Mark 15.28), this fulfills a verse of Isaiah about how he "was numbered with the transgressors" (Isa. 53.12). The passage of Isaiah may have been in all the evangelists' minds, although the extra verse in Mark was probably originally a marginal note by a copyist, who was influenced by Luke's earlier quotation of the same passage (Luke 22.37). There is nothing improbable in Jesus' being executed alongside "bandits" or "criminals": they too may have been anti-Roman rebels like Barabbas, who is described in John 18.40 as a "bandit." The description of the crowd jeering and shaking their heads at

The death of Yehohanan

Recent archaeological evidence has provided a grim reminder of the horror of death by crucifixion, which was a common penalty for rebels and "bandits" in Jesus' day. The remains of a crucified man have been unearthed from a first-century CE tomb just outside the walls of Jerusalem. The victim may have been crucified for his participation in an anti-Roman revolt that took place in 6 CE. He had been given a proper Jewish burial: he was interred in the family tomb, and subsequently his bones were reburied in an ossuary, together with those of a young child, probably his son. The inscription on the ossuary gives their names: the man is called Yehohanan (John); the child's is difficult to interpret but may read "son of the hanged." Yehohanan's feet remained fixed to a portion of a wooden post by a single nail four and a half inches (11 cm) long through both heels. The nail had bent on entering the wood, and apparently could not be pulled out without further damaging the body. Both legs were smashed, the left one by a single blow: this act, which supports John 19.32, hastened death by causing the victim's full weight to crush the lungs. It was probably carried out so that the body could be buried before sunset, as laid down in Deuteronomy (see p.207). Yehohanan's arms may have been tied to the horizontal beam, or fixed by nails through the wrists or forearms, but not the hands.

The heelbone, with the nail still in place, of the crucifixion victim Yehohanan.

Jesus (as at Matt. 27.39; Mark 15.29) is taken from Psalm 22.7, as are the words of the Jewish authorities in Matthew: "He trusts in God; let God deliver him now, if he wants to" (Matt. 27.43, quoting Ps. 22.8).

The Fourth Gospel relates episodes which it alone records as the fulfillment of Scripture. That Jesus' legs were not broken after his death (John 19.33) is said (19.36) to fulfill a verse from Exodus (Exod. 12.46), which is a regulation about the treatment of the Passover lamb. A Roman soldier pierces Jesus' side, from which flow water and blood, perhaps symbolizing baptism and the eucharist (John 19.34). According to John 19.37, this fulfills a prophecy of Zechariah (Zech. 12.10).

Jesus' words from the cross

The seven "words" (utterances) spoken by Jesus on the cross (see sidebar on opposite page) have long been a traditional element in Christian devotion. They are spread among all four gospels. Matthew and Mark have only one saying, the famous "My God, my God, why have you forsaken me?" (Matt. 27.46; Mark 15.34). These are the opening words of Psalm 22, quoted by Mark in Aramaic, but closer to the biblical Hebrew in Matthew. This has commonly been understood as a cry of abandonment. However, Psalm 22 ends with God's restoration of the sufferer who has appealed to him, and thus the evangelists may be anticipating the triumphant conclusion of their narrative.

Luke has three sayings (Luke 23.34; 23.43; 23.46). The first is Jesus' prayer for the forgiveness of his executioners, very much in keeping with the tone of Luke's gospel in general. The second, a promise to the penitent criminal alongside him, continues the same theme and also showing Luke's typical concern for individual repentance and forgiveness. The third saying is Jesus' final cry, an expression of trust taken from Psalm 31.5.

John also has three sayings (John 19.26–27; 19.28; 19.30), which are all highly symbolic. Jesus' words to his mother and "the disciple whom he loved" may refer to the Church (the mother) and its members (the disciple). The apparently simple "I am thirsty" is said to fulfill the Scripture: several passages in the Psalms speak of a thirst for God (such as Pss. 42.1 and 69.3); John may be spiritualizing the actual thirst of Jesus implied in the Synoptics. The final words of Jesus in John are triumphant: in Jesus' death, the divine purpose is accomplished.

The Crucifixion, *by Lorenzo Monaco (ca. 1370–ca. 1423).* Center left: *Jesus' mother and the two other women (Matt. 27.55; John 19.25; Mary Magdalene traditionally has red hair).* Right: *soldiers divide Jesus' garments (John 19.24).*

DEATH AND BURIAL

According to the Synoptic Gospels, dramatic cosmic portents accompanied Jesus' crucifixion and death. In the Bible such phenomena are often signs of the final apocalypse, as stated in a prophecy quoted by Peter in Acts (Acts 2.17–21, quoting Joel 2.28–32). The Synoptics tell of darkness overwhelming the land (Matt. 27.45; Mark 15.33; Luke 23.44). At the moment of Jesus' death, the curtain of the Temple is said to be torn in two from top to bottom (Matt. 27.51; Mark 15.38; Luke 23.45). The curtain divided the main body of the Temple from the Holy of Holies, where God manifested himself. Only the high priest could pass through the curtain, and then only once a year, with stringent precautions in the face of the divine presence. The tearing of the curtain symbolizes God's accessibility to all people at all times, as implied in the letter to the Hebrews: Christians are free "to enter the sanctuary by the blood of Jesus, by the new and living way that he opened for us through the curtain (that is, through his flesh)" (Heb. 10.19–20). Matthew adds another portent, a great earthquake accompanied by the opening of graves and the resurrection of God's saints, who enter Jerusalem (Matt. 27.51–53). In Jewish apocalyptic thought (see pp.136–7), the resurrection of the dead was expected at the end of the world, as in Daniel 12.2, which Matthew seems to have in mind here.

The Synoptic Gospels record that when Jesus died, he was acknowledged by the Roman centurion on duty at the execution ground (Matt. 27.54; Mark 15.39; Luke 23.47). Matthew makes this event more public by putting the words into the mouths of all the soldiers present, who are filled with awe when they experience the earthquake. In Mark and Matthew the soldiers declare: "Truly this man was God's Son!" The phrase "God's Son" need not necessarily mean "Messiah": the troops were presumably non-Jews, and the idea that an exceptional person was divine was common in the Greco-Roman world. Luke interprets the expression as "Certainly this man was innocent"; one of the meanings of "son of God" in Jewish literature was "righteous person" (see Wisd. of

The Deposition (Descent from the Cross), *ca. 1450, by Petrus Christus (active in Flanders, 1444–1473). Joseph of Arimathea (left) helps lay out the body of Jesus as Mary Magdalene (far left) and the Virgin Mary (center) lament. St. John the Evangelist is traditionally depicted, as here, supporting Jesus' mother.*

Sol. 2.12–20, another text used by the evangelists in their narratives of the death of Jesus.

The presence of a group of women is mentioned at this point, to prepare for their future role as witnesses of the empty tomb (Matt. 27.55–56; Mark 15.40–41; Luke 23.49). Matthew and Mark name the two Marys, who are later the only two to go to Jesus' grave.

It was Roman custom to leave crucifixion victims on their crosses to the mercy of carrion birds, but Jewish law (Deut. 21.22–23) stated that the body of an executed criminal must be buried before sunset. This was particularly important in the case of Jesus and the two men executed alongside him, because sunset on this day marked the start of the sabbath, and moreover the sabbath of Passover (John 19.31). The gospels relate that Jesus' body is handed over for burial to Joseph, a wealthy disciple from the Judean town of Arimathea (Matt. 27.57–58; Mark 15.42–45; Luke 23.50–53; John 19.38; see box, below). If Joseph had not asked for the corpse, it would probably have been thrown into a common grave for criminals. According to Mark, Pilate is surprised that Jesus died so quickly (Mark 15.44): crucifixion was in most cases a slow death, with victims sometimes lingering for more than two days.

In the parallel account in John, before Jesus is buried, Pilate is asked to have the legs of the victims broken (probably to ensure that they were dead; see box on p.204), and the bodies taken down. The soldiers do not break Jesus' legs, because he appears to be dead already; they pierce his side instead (John 19.34). Typically, John loads the incident with heavy symbolism (see p.205). Some groups in the early Church seem to have denied the truth of Jesus' death, which may be why Mark and John seem at pains to verify it. Mark states twice that Jesus "breathed his last" (Mark 15.37 and 15.39). John stresses that his death had been observed by an eyewitness whose testimony was entirely trustworthy (John 19.35).

Joseph of Arimathea

Joseph, the man who gives Jesus' body a proper burial, is variously described in the gospels as a wealthy man, a member of the council (Sanhedrin), and a secret disciple of Jesus. Luke 23.50–51 connects the end of Jesus' life on earth with its new beginning after death, by describing Joseph in much the same manner that he described Simeon, the righteous man who witnessed the Presentation of the infant Jesus in the Temple (Luke 2.25; see p.157).

Jesus' body was placed in a tomb hewn out of rock, according to Matthew the tomb Joseph had planned for himself (Matt. 27.60). He seals the tomb by rolling a large stone in front of the entrance, a common practice at the time. The Galilean women witness the burial, and, in Luke, they make preparations for the anointing (Luke 23.56). In John, the women play no part and Joseph is accompanied by Nicodemus, another secret follower of Jesus. The two men follow Jewish burial customs, embalming the body with spices and wrapping it in linen strips (compare John 11.44). John says that the tomb is close to the place of crucifixion. There are tombs near the traditional Golgotha (see p.202), and it is in "a garden." He also described Gethsemane as a garden (John 18.1), and in both cases he may be recalling the garden of Eden (see pp.28–9): Jesus in the garden can be seen to represent humanity restored, through his death, to the original state of perfection of Adam and Eve in paradise.

A 1st-century Jerusalem tomb, sealed by a stone rolled against the entrance.

THE RESURRECTION

According to Matthew, on the morning after the Crucifixion "the chief priests and the Pharisees" asked Pilate to make Jesus' tomb secure until the third day (Matt. 27.62; the day of the Crucifixion counted as the first day). Jesus, they said, was an imposter – a false Messiah – who claimed that he would rise again "on the third day," and his disciples might steal the body to make it look as if the claim had come true. Pilate's response was "You have a guard of soldiers; go, make it as secure as you can" (Matt. 27.65). This is often taken to mean that Pilate assigned them a Roman guard, but he may be telling the delegation that the matter was out of his hands and that they should deal with the matter themselves by deploying a guard from the Jewish Temple police.

This story is only in Matthew and is not easy to accept as it stands. Jesus predicted his resurrection in private, and the Jewish authorities would probably not have taken his words very seriously even if they had learned of them, say from Judas. The episode looks like a conscious prelude to Matthew's effort, later in the chapter, to dismiss the idea that Jesus' body had been stolen by his followers: he says that the guards at the tomb were bribed by the chief priests to spread this false rumor (Matt. 28.11–15). There may have been disputes about the truth of Jesus' resurrection between Christians and Jews, especially the Pharisees (see pp.184–5), who appear only here in the Passion narrative.

In the Synoptic Gospels, Jesus' women followers come to anoint his

The empty tomb

The author of the Fourth Gospel, John, relates the events at the empty tomb in his own way and gives them his own interpretation. Only one woman, Mary Magdalene, comes to the sepulcher, in the darkness of early morning and for reasons that are not explained. She finds the stone rolled back from the entrance and runs to tell Peter and "the other disciple, the one whom Jesus loved" that the corpse has been taken away (John 20.1–2). As usual, Peter is enthusiastic and impetuous, and he enters the tomb first. But it is the "other disciple," the epitome of truth and faith and always favored above Peter in John's gospel, who first realizes the significance of what he sees (John 20.8).

The Resurrection: The Angels Remove the Stone from the Sepulcher, *ca. 1800, by William Blake (1757–1827)*

The gospel's attention is now concentrated on Mary. Like the Synoptics, John knows the tradition of angels at the tomb (John 20.12). But their words to Mary only lead her to repeat her earlier misunderstanding of what has happened to Jesus (John 20.13). She turns around and comes face to face with Jesus (a scene corresponding to Matt. 28.9–10). She at first takes him to be the gardener – perhaps another hint at the idea of the new Eden (see box on p.207). When she finally recognizes him, Jesus forbids her to cling to him, because he has "not yet ascended to the Father." His followers are no longer to enjoy the fellowship of their earthly Master: from now on their contact with him will be spiritual and not dependent on his physical presence. This is the message that Mary must take to the disciples (John 20.17).

body after the sabbath, and wonder how they will get to it, because of the large stone sealing the tomb (see p.207). To explain why they are not stopped by the guards, Matthew tells of a great earthquake and the appearance of an angel who rolls back the stone, making the soldiers senseless with terror (Matt. 28.2–4). The women find the tomb empty and meet one or more supernatural beings who tell them that Jesus has risen from the dead. The beings are said to be white-robed men (Matt. 28.3; Mark 16.5; Luke 24.4; compare also John 20.12), a typical description of angels in Jewish writings of this era. In Matthew and Mark, the angels command the women to tell Jesus' disciples that he will meet them in Galilee. Luke, however, sets all the appearances of Jesus after the resurrection in or near Jerusalem, and so makes Galilee simply the place where Jesus had foretold his resurrection (Luke 24.6). Matthew was aware of both traditions, and as they leave the tomb, the women meet Jesus and kneel before him in worship (Matt. 28.9), just as the disciples later do in Galilee at the close of his gospel .

In Matthew and Luke, the women are instructed to carry the news of the Resurrection to the disciples (in John, Mary Magdalene has this task). But Mark records that they fled from the tomb in "terror and amazement" and said nothing to anyone "for they were afraid" (Mark 16.8). Very early versions of Mark's gospel end with these words, and most scholars agree that Mark 16.9–20 are an attempt to "complete" the gospel by a different author, probably of the second century CE. Many believe that Mark 16.8 is the intended conclusion of the work, in which case the evangelist may have understood the Resurrection to mean the Messiah's imminent final coming. Jesus had foretold that this would be accompanied by cosmic disasters and portents (as in Mark 13; see also p.191), so it would be no surprise if the women were too afraid to speak of it. Some scholars believe that either the original ending of the gospel is lost, or that the evangelist left the work unfinished, perhaps at his death; a missing ending might have included a post-Resurrection appearance of the risen Jesus in Galilee.

Noli Me Tangere (Touch Me Not), *painted ca. 1310 for Siena cathedral, Italy, by Duccio di Buoninsegna (ca. 1260–1318). This is the traditional title for depictions of the Fourth Gospel's story of Mary Magdalene's encounter with the risen Jesus in the garden; it refers to John 20.17.*

VISIONS OF THE RISEN JESUS

Doubting Thomas, *Hungarian School, 15th century. The story of the apostle who disbelieved that Jesus was really alive is recounted in John 20.24–29.*

The gospels recount a number of appearances of the resurrected Jesus. The nature of these appearances has been much discussed. In his first letter to the Corinthians, the apostle Paul says that Jesus "was raised on the third day in accordance with the scriptures" (1 Cor. 15.4). He goes on to say that he subsequently "appeared to Cephas [Peter], then to the twelve," and then "to more than five hundred" of his followers at once. Last of all, Paul says "he appeared also to me" (1 Cor. 15.8), a reference to his experience on the road to Damascus (see p.226). Paul makes no distinction between this appearance of the risen Jesus and the earlier appearances to the disciples, so these too could have taken the form of visionary experiences, perhaps the result of the highly emotional state into which Jesus' followers had been plunged by their master's death.

This view has found support in the fact that Paul makes no mention of the empty tomb, unlike the gospels (see pp.208–9), which are generally thought to be later than Paul's letters. When they relate the appearances of the risen Jesus, the evangelists often seem at pains to stress that it was his physical person, and not a disembodied spirit, that the disciples saw. John's story of Doubting Thomas is probably the most famous example of this tendency (see box on opposite page). Matthew's account of the two Marys at the tomb tells how the women encountered Jesus and then "took hold of his feet" (Matt. 28.9), which they could hardly have done if their experience had been only visionary. In Luke, the disciples react as if they had seen a ghost when Jesus appears, but he asks them to touch him and see for themselves that he is not a ghost, "for a ghost does not have flesh and bones" (Luke 24.39). To prove that Jesus is not a ghost, Luke has him asking the disciples for something to eat and then eating "in their presence" (Luke 24.41–43): it was commonly believed at the time that ghosts and spirits were unable to eat.

According to the gospels, the disciples did not immediately accept Jesus' appearances: they were unable to understand Jesus' promises that he would rise from the dead (Matt. 16.21–23; Mark 8.31–33), and found it difficult to believe that it had actually happened. The disciples' first reaction to the idea of an individual rising from the dead would probably have been similar to Martha's (John 11.23–24). Matthew says that when the disciples saw Jesus in Galilee some were still doubtful (Matt. 28.17), and, in Mark, Jesus upbraids them for their incredulity (Mark 16.14, although this part of the gospel is a later addition; see p.209). According to Luke, the apostles found the testimony of the women who had encountered Jesus "an idle tale" (Luke 24.11). In John, the disciple Thomas will not believe that Jesus is alive until he receives confirmation of the risen Jesus with his own eyes (John 20.25– 27; see box on opposite page). In conclusion, it can probably be said that the gospel narratives testify to a belief that Jesus was alive. However, the expression of this belief in terms of bodily resurrection, as the authors understand it, may well be a literary device rather than a literal account of what happened.

It has sometimes been suggested that the disciples probably shared the Jewish doctrine of bodily resurrection and that this was how they were able to believe that Jesus had risen from the grave. The belief was common and it was highly topical around this time. It was one of the main points of contention between the Pharisees and the Sadducees (see pp.184–5), and Paul plunged the Sanhedrin into disarray by mentioning the issue (Acts 23.6–10). The doctrine referred to a general resurrection at the end of the world, not the resurrection of a particular person, however remarkable that person may be.

The appearances in Luke and John

Part of the site of biblical Emmaus.

Both Luke and John have stories of the appearances of Jesus after the resurrection that are not in the other gospels. In Luke, they all occur in the vicinity of Jerusalem. Luke's vivid story of how Jesus appears to two disciples walking to Emmaus, a village seven miles (11 km) from the city, shows the evangelist's narrative powers at their height (Luke 24.13–31). At the beginning of the story, the disciples still do not understand the significance of Jesus' death and resurrection: they feel disappointed in their hope that he would be the Messiah, "the one to redeem Israel" (Luke 24.21). Jesus rebukes the disciples (who have not yet recognized him): he *is* the true Messiah, because his suffering and rising to glory have been foretold in the Scriptures, as he explains at length (Luke 24.25–27). The disciples invite Jesus to stay at Emmaus, and at supper he breaks bread, blesses it, and gives it to them. It is only at this point that the disciples know that their guest is Jesus, who at once vanishes (Luke 24.31). The breaking of the bread refers to the Eucharist: from now on, Luke is saying, those who partake of the Communion meal will experience the presence of Jesus. This story is followed almost immediately by Jesus' appearance to all the disciples in Jerusalem. It has several themes in common with the Emmaus story, such as the disciples' initial disbelief (Luke 24.37–41), a meal that was also probably symbolic of the Eucharist, and the proof from Scripture that Jesus is the Messiah (Luke 24.44). The story's main purpose is to show the reality of Jesus' bodily resurrection (see main text on opposite page).

John's stories of the risen Jesus have some themes in common with those of Luke. The theme of the disciples' doubt is focused on Thomas. He is convinced by the physical evidence that the risen Jesus is the man who was crucified, but he goes beyond this to express faith in Jesus' divinity (John 20.28). In his final blessing, Jesus commends those future believers who, unlike Thomas, "have not seen and yet have come to believe" (John 20.29).

Like Matthew, John is aware of a tradition that the risen Jesus made a final appearance in Galilee. John picks up themes which occur in the gospel accounts of the Galilean ministry, such as the miraculous catch of fish (mentioned only here in John's gospel), and a distribution by Jesus of bread and fish (John 21.13; compare John 6.11 and Luke 24.30,42). Another central theme is the restoration of Peter: Jesus grants him supervision of his flock and bids him follow in the way of martyrdom (John 21.15–19; compare John 10.11). But, as always, Peter's insight is limited and he is not to know what is in store for "the disciple whom Jesus loved" (John 21.21–22). The beloved disciple is not, like Peter, to govern the Church, but to be the witness of the truth of the gospel (John 21.24).

Christ among the Apostles, *depicted on a marble Christian sarcophagus of the late 4th century* CE *from Roman Gaul (France).*

THE ASCENSION

The Ascension, *from a manuscript by the Master of Osma, Spain, ca. 1460.*

According to the Acts of the Apostles, the risen Jesus manifested himself many times to his disciples, "speaking about the kingdom of God" (Acts 1.3), before ascending to heaven (Acts 1.9). Matthew's gospel does not record the Ascension. Mark records that Jesus "was taken up into heaven and sat down at the right hand of God" (Mark 16.19). This passage, which is by a later author than Mark (see p.209), is probably derived from the story of the ascension of the prophet Elijah, who "ascended in a whirlwind to heaven" (2 Kings 2.11), and from the opening of Psalm 110: "The Lord says to my lord, 'Sit at my right hand until I make your enemies my footstool.' "

In his preface to Acts, Luke states that his first book included all that Jesus did "from the beginning until the day when he was taken up to heaven" (Acts 1.1–2). The Ascension takes place as a prelude to the descent of the Holy Spirit on the day of Pentecost (see pp.216–17), fifty days after Passover. The risen Jesus is said to have made appearances to the disciples for forty days, a period found frequently in the Bible, for example in the stories of the Flood (Gen. 7.12) and the Exodus (Exod. 24.18). It is also found in a Jewish writing about the ascension of Abraham (see box on opposite page). Typical biblical imagery is prominent in the Acts narrative. The cloud into which Jesus enters out of the apostles' sight (Acts 1.9) is a regular symbol of the divine presence, as in Exodus 13.21, 24.15, and 33.9. Elisha was promised the gift of Elijah's spirit, provided he witnessed his ascension (2 Kings 2.9–10); similarly, the disciples who see Jesus ascend are assured of the descent of the Holy Spirit (Acts 1.8–9). The setting of the Ascension, Mount Olivet (another name for the Mount of Olives), may echo a verse from Zechariah which speaks of the Lord on that mountain on the last day (Zech. 14.4).

The underlying message of the Ascension, as recorded in Acts, is that the incarnate Jesus ascended to heaven, but he remains on earth through the Holy Spirit until the last day and the coming of the messianic kingdom. The same basic pattern is found in John's gospel, although it is expressed in a very different way. The evangelist clearly wishes his readers to understand that Jesus rose to heaven between his appearances to Mary Magdalene in the garden of the tomb and to the disciples on the

A present-day view of the Mount of Olives, the setting of the Ascension, seen from the Temple Mount at dusk. To the left of the photograph is the Hellenistic "Tomb of Absalom" (see also p.140 and map on p.190).

The religious background to the Ascension

Apart from the Ascension of Jesus, the Bible records only one occasion when a remarkable person is explicitly said to ascend to heaven: this is the ascension of Elijah (see p.98 and main text on opposite page). Genesis says of another figure, Enoch, that "he was no more, because God took him" (Gen. 5.24), and this was also taken to mean that he ascended to heaven. However, in some non-scriptural Jewish writings, the great figures of the nation's past came to be described as having ascended to heaven: not only Enoch and Elijah, but also Adam, Abraham, Moses, Isaiah, and many others.

In ancient Greek and Roman culture, monarchs and mythical heroes were often believed to be deified or to ascend to the realm of the divine on their deaths. The most famous example in Greek mythology is the hero Herakles, who was deified after his death. The Romans deified their emperors: the last utterance of the emperor Vespasian (ruled 69–79 CE), is said to have been the wry remark: "I think I am becoming a god." Some Hellenistic religious movements possessed the concept of the redeemer or divine man, who descended from heaven and eventually returned there. Jesus is depicted in this way in the New Testament, for instance in the letter to the Colossians. New Testament writings also interpret Jesus' Ascension as his assumption of kingship over the universe, and see this foreshadowed in various Psalms that originally referred to the enthronement of the Israelite monarch.

evening of the same day. At the Last Supper, Jesus promised the disciples that they would receive the Holy Spirit after he has left the world to return to the Father (John 16.7–13). After his resurrection, Jesus forbids Mary Magdalene to touch him, "because I have not yet ascended to the Father." He sends her to the disciples with the message: "I am ascending to my Father and your Father, to my God and your God." (John 20.17; see also p.208–9). On the evening of the same day, he appears to the disciples, breathes on them, and they receive the Spirit (John 20.22). The subsequent appearances of Jesus to his disciples are therefore of a figure who has already risen to the Father: before his Ascension, he stopped Mary from touching him, but the ascended Jesus can invite Thomas to "reach out your hand and put it in my side" (John 20.26).

To support their own particular opinions, later Christian writers often claimed to know the "instructions through the Holy Spirit" (Acts 1.2) which Jesus gave to the apostles before the Ascension. What the gospels preserve of these instructions are various commands which were seen as significant for the future life and witness of the Church. These commands are regarded by most scholars as retrospective projections of early Christian belief and practice into the mouth of Jesus, legitimating them in a way analagous to the inclusion of later Israelite traditions in the Law given at Sinai. Matthew, Mark, and Acts record Jesus' final commission to his followers to make disciples of all nations, an instruction which authorized the Church's missionary activity beyond Judaism to the Gentiles (Matt. 28.19; Mark 16.15; Acts 1.8). People were to become disciples through the rite of baptism, which was to be administered "in the name of the Father and of the Son and of the Holy Spirit" (Matt. 28.19): this became the Church's regular formula. In John, Jesus bestows on the disciples – in other words, his Church – the power to forgive sins and to excommunicate (John 20.23). The leadership of the Church falls to Peter (John 21.15–17; see p.211).

IV THE EARLY CHURCH

The story of the growth of the early Church after the death of Jesus is related in the Acts of the Apostles. However, it cannot be said to give a complete picture of the early expansion of Christianity. It focuses on the career and work of the leader of the apostles, Peter, whom Jesus appointed to lead the Church, and Paul, who began his career as a fervent persecutor of Christians but became perhaps its greatest missionary after his dramatic experience on the road to Damascus (see pp.226–7).

THE ACTS OF THE APOSTLES

Above: *An early 13th-century Syriac manuscript of the Acts of the Apostles. Syriac is a dialect of Aramaic, which was probably the first language of Jesus and the twelve apostles. It became the standard language of many churches of the Near East. Although now largely supplanted by Arabic (to which, like Hebrew, it is related), it still survives in some rural areas of Syria.*

Acts is the second volume by Luke and is written as a sequel to his gospel. It begins with an account of the Ascension of Jesus (Acts 1.9; see pp.212–13). It then takes up the story of how, after the death of Judas Iscariot, the body of apostles was made up to the proper number of twelve by the choice of Matthias through sacred lot, in accordance with biblical precedent. For Luke, the title of "apostle" is reserved for the twelve, in contrast to Paul, who uses the term more widely. In Acts 1.22 Luke states that the replacement for Judas must have been a witness to Jesus' entire public ministry. The new twelve constitute the governing body of the first organized church at Jerusalem.

A more accurate translation of the title of Acts is "Acts of Apostles," because it does not describe the careers of all the twelve. Later Christian writers produced a considerable number of "Acts" of other apostles to redress this lack. Also, a significant role is played in Acts by figures who did not belong to the twelve, such as Stephen (Acts 6–7) and Philip (Acts 8). But the two leading characters in Acts are Peter, who is the hero of its first dozen chapters, and Paul, whose missionary work is recounted in the remainder of the work.

Just before the Ascension, Jesus commands his disciples to bear witness to him in Jerusalem, Judea, Samaria, and "the ends of the earth" (Acts 1.8), and this program is carried out in successive stages until the book's climax in Rome, the center of the empire of which Judea was then part. Acts is in many respects an apologia for the Christian movement, addressed to the Roman governing classes. This is seen especially in its account of the career of Paul (see pp.230–37).

Luke's two-volume writing, comprising his gospel and Acts of the Apostles, is the New Testament work that approaches most closely the standards and style of Classical (Greek and Roman) historians, although it has its own theological outlook (see sidebar, right). Luke follows common Classical convention in dedicating his books to a distinguished person, in this case a certain Theophilus, who may actually have been the author's patron. In the introduction to the gospel, which serves as an introduction to both volumes, Luke contrasts his work with that of his predecessors and hints that it is an improvement on them (Luke 1.1–4). In saying this he is writing in much the same vein as the great Greek historian Thucydides (lived ca. 475 BCE) at the beginning of his *History of the Peloponnesian War*.

ACTS AND HISTORY

In the past, the author of Acts has generally been seen as a careful and accurate recorder of events. But in more recent years his reliability has been questioned, and attention has focused on the extent to which Acts reflects Luke's own theological outlook. It is widely accepted that the author had a good knowledge of the contemporary geographical, social, and political setting of his narrative. He is strikingly accurate with respect to the names and titles of the imperial officials who appear in the course of his account, and he displays a sound knowledge of Roman legal procedure.

However, correct background information does not guarantee the reliability of all episodes and stories. It is perhaps the picture of Paul in Acts that has raised the most doubts. The author seeks to draw parallels between Peter and Paul and to show that they had a common theological outlook. But passages in Paul's letters show that their relationship was not always harmonious (see p.229). In the early Church, there were those who denigrated Paul in favor of Peter, and it may well be that Luke is consciously concerned to correct this attitude.

Christ and the Apostles; *Spanish (Catalonian), 13th century. The enthroned figure of Jesus is surrounded by the symbols of the four evangelists.*

The literary background to Acts

Acts cannot be seen simply as a work of Classical history writing, since the early Christian community, of which it is a product, had its own distinctive outlook. Nevertheless, it does employ several of the conventions of Classical historians, most notably the use of speeches, which occupy about a third of Acts.

As a famous passage from Thucydides states, Greek and Roman historians did not seek to reproduce verbatim what had been said on a particular occasion, but to give in their own words what they imagined the speaker would have said, had he been able to fully express himself. Although Thucydides claims to follow, as closely as possible, what had actually been uttered, later historians tended to treat such speeches as rhetorical exercises, paying less attention to historical probability. As they stand, the speeches in Acts are no doubt Luke's compositions, but they seem to be constructed for the particular audience to which they were addressed. Those which are written for Jewish ears are couched in the language of the Septuagint, the Greek translation of the Hebrew Scriptures, while Paul's speech to the pagan philosophers at Athens (Acts 17.22–31) is essentially a Hellenistic discourse about the knowledge of God.

The dramatic character of many episodes in Acts, especially Paul's journeys and their accompanying adventures, invites comparison with Hellenistic romances. The accounts of healing miracles, of escapes from prison, and of Paul's sea voyage to Rome are similar in form and vocabulary to many contemporary Greek writings. The voyage belongs to a great literary tradition going back to Homer's *Odyssey*.

THE DESCENT OF THE SPIRIT

The Descent of the Spirit at Pentecost, *a fresco by Giotto (ca. 1266–1337) in the Scrovegni chapel at Padua, Italy, ca. 1305.*

The gift of the Holy Spirit is of central importance to the author of Acts, because the Spirit is the motivating force of the whole story that he tells. The Spirit wins new converts, empowers the working of miracles, and gives wisdom and courage to the leaders of the young Church. It guides them to undertake significant initiatives, such as Peter's visit to the gentile Cornelius (Acts 10; see pp.228–9), and missionary journeys.

The fact that the gift of the Spirit occurred on the feast of Pentecost, the Jewish festival of Weeks (see Exod. 34.22; Deut. 16.10), may be of particular importance. Weeks was originally an agricultural celebration, marking the end of the wheat harvest and the offering of its first fruits. However, it became a commemoration of the giving of the Law on Sinai (see pp.60–61), and the depiction of the day of Pentecost in Acts has sometimes been interpreted as the moment when the old Jewish Law was replaced by a new spiritual revelation for all humanity. This theory is complicated by the fact that the association of Pentecost and the Law does not seem to be older than the third century CE, although the book of Jubilees, a text of the second century BCE, makes the Pentecost a commemoration of the "Noachide commandments," which were held to bind all humankind, gentiles as well as Jews (see p.32).

The actual account of the descent of the Spirit is couched in biblical imagery typically associated with a divine manifestation. The "sound like a rush of violent wind" (Acts 2.2) and the "divided tongues, as of fire" (Acts 2.3) recall the creation, the Flood, and many manifestations of God, especially that to Moses at Sinai. Acts also recalls the baptism of Jesus (Luke 3.21–22): the flame of the Spirit rests on each of the apostles, as the dove of the Spirit did on Jesus.

Peter's first speech

The account of the coming of the Spirit is followed by a speech by Peter, explaining what it implied (Acts 2.14–40). Following Luke's general account for the spread of the Church, this sermon is addressed to Jews. It begins by claiming that God's promises to his people have been realized through the ministry, death, and resurrection of Jesus, and various biblical passages, in particular the prophet Joel, are quoted in support. The presence of the Holy Spirit, which the hearers have witnessed, is the sign of Jesus' exaltation and glory, and the apostles' message is authenticated because they are the witnesses of his resurrection and ascension. Through the rite of baptism, they are empowered to offer forgiveness of sins and the gift of the Holy Spirit to those who will take heed and repent. This is a promise not only for Jews but for any whom God may summon.

Peter's first address sets the pattern for the general theology of the other speeches in Acts. It has often been claimed that these discourses represent the actual preaching of the first apostles. However, some specifically Lucan themes suggest that, in their present form, they are Luke's own compositions. For example, in this first sermon of Peter, proof of Jesus' divine mission is said to be provided by his miracles and signs (Acts 2.22). This is a regular theme of Luke's gospel, as is Peter's claim that everything that happened to Jesus was the result of God's "definite plan and foreknowledge" (Acts 2.23). But whatever their origin, the speeches in Acts represent important evidence for how the early Church understood the person and work of Jesus.

The Prophet Joel *(ca. 1510), by Michelangelo (1475–1564); in the Sistine Chapel, Rome. Peter's first speech includes an extensive quote from Joel (Acts 2.17–21, quoting Joel 2.28–32).*

The result of the outpouring of the Spirit onto the apostles was that they were able to "speak in other languages" (Acts 2.4), a phenomenon witnessed by "devout Jews from every nation under heaven" who are living in Jerusalem (Acts 2.5). Luke lists the nations they represent (Acts 2.9–11): they are from both inside and outside the confines of the Roman empire and are depicted primarily as Jews from the widespread Dispersion of the late centuries BCE. Luke also mentions the presence of proselytes (Acts 2.10), gentiles who had been converted to Judaism as the result of vigorous and successful Jewish propaganda. "Speaking in tongues," or "glossolalia" was a common phenomenon in the churches founded by Paul, well before the probable date of the composition of Acts. In these cases, however, it did not involve the ability to speak foreign languages, but the uttering of unintelligible, ecstatic speech which could even give the impression of madness. Luke seems to hint at this when he records that some of the onlookers thought that the apostles were drunk (Acts 2.13). Later references in Acts to speaking in tongues (Acts 10.46; 19.6) do not suggest that foreign languages were involved.

The apostles would not have needed to speak in the pilgrims' native tongues, since they would all have known a lingua franca, either Aramaic or Greek. Luke may have seen the events of Pentecost as a reversal of the Tower of Babel story (Gen. 11.1–9; see p.34). The story of the Tower of Babel tells how human presumption brought about the use of different languages which led to division in the world. The proclamation of the gospel, which all can understand, brings human beings together again, restoring their original harmony in a new community comprising, "everyone whom the Lord our God calls to him" (Acts 2.39).

PETER IN ACTS

GAMALIEL

Gamaliel the Elder, who appears in Acts 5, was a renowned rabbi of the first century CE. He was a member of the great Pharisee dynasty of Hillel and, according to tradition, served as president of the Sanhedrin. He represented the liberal interpretation of the Jewish Law that characterized the Hillel school, advocating more freedom of movement on the Sabbath and greater protection for women in divorce proceedings. According to Acts 22.3, the apostle Paul was his pupil.

Gamaliel's reputation for learning and tolerance is born out by his speech in Acts, in which he advises the Sanhedrin to leave Peter and the apostles alone, for "if this plan or this undertaking is of human origin, it will fail; but if it is of God, you will not be able to overthrow them" (Acts 5.38–39). It is entirely consistent with Luke's message that Gamaliel, a Pharisee, should show a not wholly unfavorable attitude toward the Christians. In spite of his frequent criticisms of the Jews, Luke was anxious to present Christianity as the legitimate successor of Judaism, especially as represented by the influential Pharisaic party. The Pharisees shared with the Christians a belief in the resurrection of the dead. Their opponents, the Sadducees, denied the possibility of resurrection, and twice hauled the apostles before the Sanhedrin (Acts 4 and 5). Paul later exploited the division between the two groups over the resurrection issue (Acts 23.6–10; see also pp.184–5).

In his discourse in Acts, Gamaliel refers to two messianic pretenders, Theudas and Judas, placing Theudas first. In fact, Theudas's revolt occurred long after that of Judas. This is probably a simple error by Luke and it does not affect the force of Gamaliel's argument.

The government of the early Church in Jerusalem was in the hands of the first twelve apostles, and it is Peter who appears as the undoubted leader and the spokesman of the Christian community in its confrontations with the religious authorities. He is depicted as a powerful wonder-worker: the mere falling of his shadow on the sick is said to be enough to heal them (Acts 5.15). Later in Acts, handkerchiefs and aprons which have been in contact with the apostle Paul are seen to possess similar curative properties (Acts 19.12). The first evidence of the apostles' ability to work "signs and wonders" comes immediately after the Pentecost story: this is the healing of a crippled beggar at the Beautiful Gate of the Temple, when Peter and John are attending one of the regular hours of prayer (Acts 3.1–10). The story follows the regular pattern of healing miracles in the gospels and elsewhere, mentioning the severity of the affliction, its cure, and the effect on the bystanders. The particular power inherent in Peter's person is brought out in the story: like Paul in Acts 14, he commands the cripple to stand up and walk, and his hands convey strength to the man's feet and ankles (Acts 3.6–8). It is sometimes suggested that this account betrays a knowledge of medicine, which supports the view that the author of Acts was the Luke that is referred to in Colossians 4.14 as "the beloved physician." However, the language is not really technical enough to warrant this conclusion.

The healing of the beggar is followed by Peter's second speech (Acts 3.12–26), which, like the first (see p.217), discusses the fulfillment of God's promise to send the Messiah, and the necessity to repent "so that your sins might be wiped out" (Acts 3.19). This speech stresses that the crucifixion of Jesus was the responsibility of the Jews, although Peter concedes that they acted in ignorance (Acts 3.17), as Jesus also acknowledged in Luke's gospel (Luke 23.34). As a result of this preaching, Peter and John are arrested by the Temple authorities (Acts 4.3). This marks the first of Peter's three imprisonments, which are narrated in increasingly dramatic terms. After a night in jail, the two apostles are brought before the council (Sanhedrin), where Peter makes a third speech (Acts 4.8–12). They are let off, on the condition that they are "not to speak or teach at all in the name of Jesus" (Acts 4.18). Peter and John refuse to obey this order, but they are released anyway "because of the people, for all of them praised God" for the miracle of the healed cripple.

The subsequent success of the Christian mission leads to the imprisonment of the whole apostolic college (Acts 5.18), which is then freed through the intervention of an angel (Acts 5.19). Luke's account here is modeled on a common framework for stories of prison escapes used by Classical authors such as Euripides and Ovid. Eventually the apostles appear again before the Sanhedrin, and the pattern is repeated: there is another speech from Peter (Acts 5.29–32) and, on the advice of the rabbinic teacher Gamaliel (see sidebar, left), the accused are released after an admonitory flogging and another order to stop preaching in the name of Jesus (Acts 5.40–41).

The account of Peter's third imprisonment in Acts 12 is a fine example of Luke's narrative powers. On this occasion, Peter alone is arrested, not by the Temple authorities, but as part of a persecution of leading Christians by Herod Agrippa I (ruled 37–44 CE), who, by winning the favor of the emperors Caligula and Claudius, had become client king of almost all the territory that was ruled by his grandfather Herod the Great (see map on p.143). Peter was probably detained in the Antonia fortress, the barracks next to the Temple. The vivid and detailed story of Peter's release by an angel (Acts 12.7–10) again follows the pattern of other prison escapes in both biblical and nonbiblical writings. The iron gate to the city opens "of its own accord" (Acts 12.10). There are similar occurrences in Greek literature from Homer onward, and the first-century Jewish historian Josephus recorded that during the great Jewish revolt of 66–73, the great east gate of the Temple opened by itself.

Peter's happy escape by divine help is skillfully contrasted with the divine judgment visited on his persecutor. Because he "had not given glory to God," Herod is struck down and "eaten by worms" (Acts 12.23). Being eaten by worms is a form of death that is frequently mentioned by ancient writers. In particular, Luke may have had in mind the fate of the second-century BCE Hellenistic ruler Antiochus IV Epiphanes (2 Macc. 9.9; see pp.138–9). Herod, who was popular among Jews if not Christians, did in fact die suddenly, at Caesarea, in 44 CE. The circumstances of his death clearly made a great impression on his contemporaries: the account by Josephus agrees with that in Acts, both in general outline and in some details.

Herod Agrippa I (10 BCE–44 CE), a persecutor of the early Church, portrayed on one of his own coins.

St. Peter Healing with his Shadow *(Acts 5.15), a fresco by Tommaso Masaccio (1401–1428) in the Brancacci chapel of the church of Santa Maria del Carmine, Florence; ca. 1425.*

THE JERUSALEM CHURCH

The Death of Ananias, *from the Mortlake Tapestries, based on cartoons by Raphael (1483–1520) for tapestries in the Vatican. According to Acts, Ananias and his wife Sapphira sold a property and kept some of the proceeds that they had promised to donate to the Church. The tapestry depicts the moment after Peter has publicly condemned Ananias: "Now when Ananias heard these words, he fell down and died" (Acts 5.5). Sapphira later met the same fate.*

At intervals throughout Acts, Luke gives what may be described as "editorial summaries." They are particularly frequent in the early part of the book and give an idealized picture of the general life and development of the earliest Christian community in Jerusalem. Christians are shown as devout Jews, who attend the Temple every day at the regular hours of prayer (Acts 3.1). Luke says that they often share communal meals in their homes, when they break bread together (Acts 2.46). This ritual probably took place as part of an ordinary meal, but since in his gospel Luke says that Jesus broke bread at the Last Supper (Luke 22.19), it is likely to have included the practice that was later separated out as the Eucharist. The earliest Christian community, then, was a religious fellowship group of a kind common within Judaism, and it continued the fellowship that Jesus had established with his disciples.

Luke emphasizes the unity and joy of this Christian society. The whole company of believers is said to be "of one heart and soul" (Acts 4.32). Here, Luke seems to echo similar expressions found in Greek writers, who praise the ideal of universal friendship which, they claim, existed at the beginning of human history. In the same way, Luke describes the ideal state of harmony that characterized the Church in its earliest days.

It has sometimes been asserted that Acts shows the beginnings of later Church organization, with its emphasis on tradition and succession. But, in fact, the book provides little evidence of any clear or uniform ecclesiastical structures or procedures of worship. The leadership of the Jerusalem Church at first rested with the twelve apostles chosen by Jesus, with Peter at their head and Matthias in place of the traitor Judas Iscariot (Acts 1.26). But the followers of Jesus are represented as making collective decisions on important occasions. Later, in Acts 15, authority appears to be shared by apostles and, following Jewish tradition, "elders." This is probably because persecution (Acts 11.19; 12.1) had forced several of the apostles into exile: the exiles may even have included Peter himself (see Acts 12.17 and p.229). At the Apostolic Council (Acts 15.6–29; see p.229), the leader of the Church is clearly James, Jesus' brother. There is

The seven Hellenists

Acts records a rapid increase in the number of converts in Jerusalem: in a short period, Church membership is said to have grown from one hundred twenty (Acts 1.15) to over eight thousand (Acts 2.41, 4.4). This led to problems in coping with those in need, as many believers there were poor. Subsequently, other churches had to establish a charitable fund.

Sebaste, ancient Samaria, rebuilt in Greco-Roman style by Herod the Great. Philip, one of the seven Hellenists, may have preached here (Acts 8).

Matters came to a head with a dispute between the "Hebrews" and the "Hellenists." The "Hellenists" were Greek-speaking Jews (gentiles had not been admitted to the Church at this time), as opposed to Aramaic-speaking "Hebrews." The Hellenists complained that their widows were losing out in the daily distribution of rations (Acts 6.1). The apostles decided that they themselves should be able to devote all their time to prayer, and called for "seven men of good standing" to be chosen to oversee food distribution (Acts 6.3–4). The community is then said to have elected seven men of the Hellenist party (Stephen, Philip, Prochorus, Nicanor, Timon, Parmenas, and Nicolaus), and the apostles solemnly commissioned them by the laying on of hands (Acts 6.6). The seven have been regarded as the first deacons, the clerical order which, in the early Church, was entrusted with the care of the poor and sick. But the laying on of hands conferred on the seven the authority of the apostles, and the Hellenists are not recorded as administrators but as preachers and missionaries. Two of them, Stephen and Philip, are said to have played a particularly active role (see pp.222–5), and Stephen, according to Acts, was the first Christian martyr.

evidence to suggest that, for a period of time, other members of Jesus' family exercised a kind of dynastic leadership in Jerusalem. Christian groups gathered for worship in each other's homes, and the "house church" became the typical meeting place of the early Christian communities. Worship was markedly spontaneous, and was driven by the impulse of the Holy Spirit: Acts records a prayer that was uttered in unison, when the participants were filled with the Spirit and "the place in which they were gathered together was shaken," as if by an earthquake (Acts 4.24–31).

According to Acts, one of the most striking features of the early Jerusalem Church was its common ownership of goods (Acts 2.32). This concept was a central element in the Utopian societies envisaged by many Greek thinkers, such as Plato, and it was practiced by some Jewish groups, for example the Essenes and the Qumran sect (see p.145). However, it was not as universal in the Jerusalem Church as the author of Acts appears to suggest: the mother of John Mark, for example, had her own house, with at least one servant (Acts 12.12–13). Selling one's possessions and devoting the proceeds to a common fund administered by the apostles was a voluntary, though highly laudable, act. It was done at the prompting of the Spirit with the object of providing relief to those in need. Acts recounts two contrasting stories on this theme, about the virtuous Barnabas, who sold a field and gave all the proceeds to the Church (Acts 4.37), and the wicked Ananias and Sapphira (see picture on opposite page). At Qumran, those joining the community had to hand over all their possessions to the common pool, and anyone who concealed personal belongings was temporarily excommunicated. Behind this practice lies the old biblical concept of devoting something to God: anything so dedicated could never be taken back.

THE DEATH OF STEPHEN

Stephen, one of the seven who were appointed to oversee the distribution of food among the Hebrews and Hellenists (see p.221), was the first follower of Jesus to suffer death for the faith. The story of his martyrdom in Acts 6 and 7 falls into two parts: the account of his trial and death by stoning (Acts 6.8–15, 7.54–60), and his speech to the Sanhedrin, or council, (Acts 7.1–53). The interpolated address is only loosely connected with the trial and stoning, and is probably an example of anti-Jewish polemic which Luke included in Acts as a speech of defense before the Jewish authorities. Such speeches occur regularly in Luke's work.

Stephen's death was occasioned by internal disputes among his fellow Greek-speaking Jews: whether his opponents had several synagogues or only one in Jerusalem is uncertain, but they clearly comprised Jews from North Africa and Asia Minor who had settled in the city. The narrative of the trial closely parallels that of the trial of Jesus in the gospels: the Hellenists suborn false witnesses, who accuse Stephen of blasphemy and of predicting that Jesus would destroy the Temple of Jerusalem (Acts 6.13–14). The charge against Jesus of seeking to destroy the Temple is not mentioned in Luke's gospel and Luke has probably transferred this accusation to the trial of Stephen. In the gospel, Jesus is condemned directly after his statement about the Son of Man seated at God's right hand (Luke 22.69). Similarly, the final fury of the crowd is aroused against Stephen when he has a vision of the Son of Man standing at the right hand of God (Acts 7.55). (The title "Son of Man" occurs only here in the New Testament outside the gospels.) Stephen's petition for the forgiveness of his killers, and his prayer that Jesus may receive his spirit (Acts 7.59–60), recall Jesus' words from the cross in Luke's gospel (Luke 23.34, 23.46).

How far the stoning of Stephen was legally authorized is unclear, because, under Roman rule, the Sanhedrin was probably no longer able to impose the death penalty. There is no mention of a formal sentence before the crowd rushed at Stephen and proceeded to stone him, so it may have been a simple case of lynching. But there is evidence that, the

Stephen Preaching *(left) and* Stephen Addressing the Council *(right), by Fra Angelico (ca. 1387–1455), in the chapel of Nicholas V in the Vatican, Rome.*

Sanhedrin tried to revive its power to impose capital punishment whenever it had the opportunity to do so. The historian Josephus says that the execution of James, Jesus' brother, was prompted by members of the Sanhedrin, who took advantage of the interval in 62 CE between the death of one Roman procurator and the arrival of another (see p.250). Complaints were made, and the incoming procurator deposed the high priest. In the case of Stephen, the Jewish leaders may have felt that they could gamble on the Roman power turning a blind eye to his execution.

Stephen's address to the Sanhedrin (Acts 7) is the longest speech in Acts and marks an important stage in what Luke sees as the process of separating Christianity from Judaism. Most of Stephen's speech is related to the allegation that he had foretold the destruction of the Temple. He argues that Israel's history had been a record of continual disobedience to God. The people consistently opposed Moses and turned to other gods. Finally, they constructed the Temple, which was a mistake from the beginning, since "the Most High does not dwell in houses made with human hands" (Acts 7.48). There were Jewish groups, such as the community at Qumran (see p.145), which were unfavorable toward the institution of the Temple, but Stephen seems to represent a radical strain within the Hellenist Christian community. He expresses a view that goes beyond Luke's own generally more favorable view of the Temple, and in the New Testament, Stephen's position is closest to that of the letter to the Hebrews (see pp.250–51). In his final words, Stephen turns directly on his hearers, accusing them of still resisting the Holy Spirit, just as their ancestors had done (Acts 7.51–53). Whatever the actual circumstances of Stephen's execution, for Luke it represents the rejection by the nation of the true Messiah, and therefore the nation's own rejection by God.

St. Stephen, *by Stefano Sassetta (ca. 1392–1450 or 1451). Medieval artists usually portrayed Stephen as a young man, often holding a book (the gospels, to symbolize his preaching).*

Persecution after Stephen's death

At the close of the story of Stephen's martyrdom, the author introduces "a young man named Saul," who looked after the coats of the stoning crowd (Acts 7.58) and approves of Stephen's execution (Acts 8.1). This is the first mention of the future apostle Paul (Saul was his Hebrew name). According to Acts, Stephen's death sparked off a severe persecution of the Jerusalem church, in which Paul took the lead (Acts 8.1).

This persecution was directed primarily against the particular group of Hellenists (Greek-speaking Jewish Christians) to which Stephen himself had belonged. They were forced to leave the capital, while "the apostles," presumably the twelve, were left undisturbed. Barnabas, a Greek-speaking Jew from the Diaspora, who made a large contribution to the Church (Acts 4.36–37), seems to have remained in Jerusalem, where he later introduced Paul to the apostles (Acts 9.27). Barnabas acted as a link between the Jerusalem Christians and the communities created in the gentile world as a result of missionary expansion.

The persecution was at first confined to Jerusalem, because the refugees were apparently able to preach freely elsewhere (Acts 8.4–5). According to Acts 9.31, the Church was able to build up its strength in Judea, Galilee, and Samaria (see pp.224–5). The high priest authorized Paul/Saul to go to Damascus to arrest Christians – probably some of those who had fled Jerusalem – and bring them to Jerusalem for trial (Acts 9.1–2). Damascus was outside the jurisdiction of the Jewish authorities, but there is evidence in 1 Maccabees and in the historian Josephus that the high priest had the power to request the extradition of offenders against the Jewish Law.

PHILIP AND SIMON MAGUS

For Luke, the author of Acts, the death of Stephen and the persecution that followed it (see pp.222–3) had positive consequences: the spread of the gospel from Jerusalem to Judea and Samaria. According to Acts, the Samaritan mission was undertaken by Philip, one of the seven Hellenists ordained by the apostles (see p.221). Philip preached in "a city of Samaria" (Acts 8.5) which may have been Sebaste, built in Greco-Roman style by Herod the Great on the site of the ancient city of Samaria (some versions of Acts say that Philip went to "*the* city of Samaria"; see p.221). Another likely base for his mission would have been the Samaritan religious center of Neapolis, near the ancient city of Shechem (modern Nablus).

Acts says that Philip performed many acts of healing in Samaria (Acts 8.4–7). The inhabitants of this region were particularly receptive to such miracles, since for some time they had been captivated by a practitioner of magical arts, or *magus,* called Simon, known in biblical tradition as Simon Magus (Acts 8.9). Jesus' disciples often came into competition with apparently similar wonder-workers. The Samaritans recognized Simon as someone with divine powers, but Philip's cures are said to have been more numerous and striking, and won many converts. Simon, too, was impressed by Philip's deeds: he was converted and baptized, and proceeded to join Philip in his work (Acts 8.12–13).

The second part of this story brings out the special authority of the twelve apostles. In Acts, water baptism and the gift of the Holy Spirit are closely linked, but not identical: others could baptize, but only the apostles, represented by Peter and John, could confer the Holy Spirit, which they did by laying hands on the Samaritan converts (Acts 8.16–17). This may have had some physical result, such as speaking in tongues (see p.217), because Acts relates that when Simon "saw" that Peter and John had the power to pass on the Spirit, he tried to buy the same power for himself (Acts 8.18–19 – hence the word "simony" for the buying or selling of a spiritual office). Peter severely castigated Simon for his presumption (Acts 8.20–23), and although Acts describes Simon as repenting (Acts 8.24), it was Peter's harsh denunciation of him that was always remembered in Christian tradition. Beginning with the Christian writer Justin in the second century CE, Simon Magus was seen as the root of the Gnostic heresy and as an antichrist. He was said to have been accompanied by a woman called Helena, who was originally a prostitute from Tyre, but who, in the manner of Gnostic speculation, was considered by Simon Magus to be a divine emanation from his own mind. From the confrontation with Peter in Acts, there developed a whole string of legends of wonder-working duels between Simon and the apostle in various cities of the Roman empire, until, it was said, Simon finally met an ignominious end.

A second story about Philip, his meeting with the Ethiopian eunuch (Acts 8.26–40), occurs after his return from Samaria to Jerusalem with Peter and John. The narrative appears to have been influenced by the story of the prophet Elijah (see pp.96–8). Like him, Philip is summoned to undertake a journey by the angel of the Lord (Acts 8.26; compare 2 Kings 1.3) and is mysteriously snatched away after his mission is completed (Acts 8.39; compare 2 Kings 2.11). According to Acts, Philip meets a eunuch, a high official of Candace, queen of the Ethiopians, on the road from Jerusalem to Gaza. "Candace" is actually a title used by a number of queens of the kingdom of Meroe, in what is now Sudan. The eunuch "had come to Jerusalem to worship" and "was reading the prophet

Isaiah," (Acts 8.27–28), so he was presumably either a convert to Judaism or a person on the point of conversion. At the eunuch's request, Philip explains the scriptural passage he is reading, which is from Isaiah 53.7–8. Philip interprets it in Christian terms, and proclaims to the eunuch "the good news about Jesus" (Acts 8.35). This is the first time that the New Testament explicitly identifies Jesus with the figure of the suffering servant in Isaiah 53 (see p.112), and the chapter became a key text in early Christian preaching. The eunuch was then baptized (Acts 8.38). The story of his conversion appears to reflect the baptismal practice of the early Church. The candidate was instructed in the faith (Acts 8.35), then made a formal request to be baptized (Acts 8.36), followed by a confession of belief. Some versions of Acts include the confession: "I believe that Jesus Christ is the Son of God" (Acts 8.37).

The fact that the convert is both a eunuch and an Ethiopian may have symbolic significance. The Jewish Law technically excluded eunuchs from entering the Temple precincts and taking part in worship, but Isaiah 56.4–5 says that eunuchs who keep the covenant will find their place in the house of God. Acts possibly sees this condition as now being fulfilled. Several verses in the Psalms and prophets speak of people from Ethiopia turning to Yahweh and bringing offerings: it may be significant that the Ethiopian eunuch was in charge of the royal treasury.

After baptizing the eunuch, Philip is said to have been "snatched away" by "the spirit of the Lord" (Acts 8.39). According to Acts he then reappeared in the town of Azotus (ancient Ashdod), near the Mediterranean coast, from where he conducted an evangelistic mission as far as the Roman port of Caesarea (Acts 8.40). Philip appears to have taken up residence in Caesarea, because he is later said to be living there with four unmarried daughters who possessed the gift of prophecy (Acts 21.8–9). At about this time, Peter left Jerusalem and established himself in the predominantly Greek port of Joppa (Acts 9.43). These moves by Philip and Peter may reflect the Church's desire to preach in the less strongly Jewish coastal areas of Palestine. The gospel might receive a warmer welcome here than in what had become the hostile environment of Jerusalem.

The theater at Caesarea. The city was built in Greco-Roman style by King Herod the Great and named for the emperor Augustus Caesar. An important port, after Herod's death it became the seat of the Roman procurators of Judea (see also box on p.200).

Simon Magus Offering Money to St. Peter, *by Liberale da Verona (ca. 1445–ca. 1525).*

Royal tombs in the Nabatean capital city of Petra, south of the Dead Sea. A people of Arab origin, the Nabateans dominated southern Syria and northern Arabia from ca. 90 BCE. *Their kingdom prospered as an ally of Rome until 106* CE, *when it was absorbed into the Roman empire.*

THE CALLING OF PAUL

There are three accounts in Acts of the "conversion" of the apostle Paul on the road to Damascus (see p.223). The fullest and most dramatic version occurs in Acts 9 during the story of the persecution of Christians following the martyrdom of Stephen. Paul, who is still referred to by his Hebrew name of Saul at this stage, was "breathing threats and murder against the disciples" (Acts 9.1). In his anti-Christian fervor he is said to have undertaken a mission to Damascus to bring "any who belonged to the Way" back to Jerusalem. Saul was nearing Damascus when "suddenly a light from heaven flashed around him" (Acts 9.3), causing him to fall to the ground. According to Acts, he heard a voice saying " 'Saul, Saul, why do you persecute me?' " (Acts 9.4) and, after asking who it was, received the reply: "I am Jesus, whom you are persecuting. But get up and enter the city, and you will be told what to do" (Acts 9.5). Saul is blinded by the vision and regains his sight only three days later, immediately prior to his baptism (9.17–18).

The other two accounts in Acts of the conversion are in speeches by Paul himself (Acts 22.6–11; 26.12–18). A few allusions in Paul's letters also appear to refer to this event: for example, in Galatians 1.15–16 he writes of his calling through God's grace and of seeing the risen Jesus, which set him on his mission to the gentiles (and gave him the status of an apostle). Paul's words essentially agree with the story told by Luke in Acts, but the apostle gives no connected version of what happened.

All three accounts in Acts mention the supernatural light and the voice that cause Paul and his companions to fall to the ground. This is a common feature of accounts of visionary experiences in Jewish writings of the period. The *Testament of Job*, dating from around the beginning of the Common Era, provides a close parallel with Acts. Job says: "While I was asleep, there came to me a loud voice with a very great light, saying, 'Job, Job.' And I said, 'Here am I'; and he said, 'Get up, and I will show you who this is whom you wish to know' ... and when I heard, I fell down on my bed." Rabbinic writings frequently mention the *bath qol*, a mysterious voice by which God sometimes communicates with human beings.

Scholars have attempted to provide a psychological explanation for Paul's experience, suggesting that it may have been the resolution of an interior conflict between his strict Judaism (he was a Pharisee by training; see Acts 26.5) and the impression made on him by witnessing the persecution of Christians, especially the death of Stephen. This view can be no more than speculative, but it may have some support in Paul's quotation of a popular Greek proverb in the account of the conversion he gives to Herod Agrippa: "it hurts you to kick against the goads" (Acts 26.14). This suggests that even before his conversion his conscience had been pricking him over his role in persecuting Jesus' followers. Paul's letters reveal him as a person often torn by strong inner tensions. He appears to have been subject to ecstatic and visionary experiences: Acts also records other visions (Acts 22.17–21; 27.23–24).

There are discrepancies between Acts and Paul's letters in their accounts of the immediate aftermath of his experience on the Damascus road. In Acts, Paul is blinded by the heavenly light, but is cured three days later by the disciple Ananias and baptized a Christian (Acts 9.8–18; 22.11–16). The letters do not mention any of these events. Paul writes that, after God's revelation to him, he went to Arabia without consulting a single person and subsequently returned to Damascus, only going to Jerusalem three years later to contact Peter and James, Jesus' brother (Gal. 1.16–17). This account differs from that in Acts, which records Paul as preaching in

PAUL'S BACKGROUND

In Acts, Paul is represented as saying that he was a native of Tarsus, a true-born Jew, a Pharisee, and a pupil of the famous rabbi Gamaliel (see p.218) in Jerusalem (Acts 22.3). In his letters, Paul states that he had been circumcised, belonged to the tribe of Benjamin, and had been trained as a Pharisee (Phil. 3.5). Paul's writings show considerable familiarity with the outlook and methods of first-century rabbinic Judaism.

Tarsus was rightly described by Paul as "no mean city." It was the capital of the Roman province of Cilicia and a prosperous town, known for the making of tents, which accounts for the fact that Paul earned his living as a tentmaker (Acts 18.3). Even more importantly, it was a center of intellectual life and produced many outstanding philosophers. This would account for the Hellenistic influences in Paul's writings and for his, at least superficial, acquaintance with the popular teaching of groups such as the Stoics and Cynics.

Thus, Paul was a Diaspora Jew and a native Greek-speaker, which may explain his involvement in the affair of Stephen, who came from a similar background. Paul also enjoyed the privilege of being born a Roman citizen and had a Latin name, Paul, as well as the Hebrew name Saul. It is not known how his family obtained this citizenship, but the picture in Acts of the status and privileges it conferred accords with Roman legal practice in the first century CE (see, for example, Acts 22.25–29).

Damascus and then escaping to Jerusalem, where he joined the other disciples and continued "speaking boldly in the name of the Lord" (Acts 9.19–28). It is possible that the author, Luke, has telescoped events, omitting Paul's time in Arabia, in which Luke had no particular interest.

In Acts, Paul escapes from Damascus by being lowered down the city wall in a basket, because the Jews wanted to kill him and were watching the city gates (Acts 9.24–25). This escape is also mentioned by Paul in his second letter to the Corinthians, where he attributes the danger he faced to the commissioner of King Aretas (2 Cor. 11.32–33). This was Aretas IV (ruled 9 BCE–40 CE), king of the powerful Nabatean state (see picture on opposite page and map on p.231). At this time Aretas evidently controlled Damascus, which was intermittently a Nabatean possession. The Damascus Jews may have enlisted the royal commissioner's support in their pursuit of Paul, an action paralleled elsewhere in Acts. It is also possible, assuming that Luke has left out the Arabian visit, that Paul's escape took place on his second trip to Damascus (Gal. 1.17). His preaching in Arabia, much of which formed part of Aretas's kingdom, could have provoked the Nabatean authorities, who sought to arrest him when he next set foot in their territory.

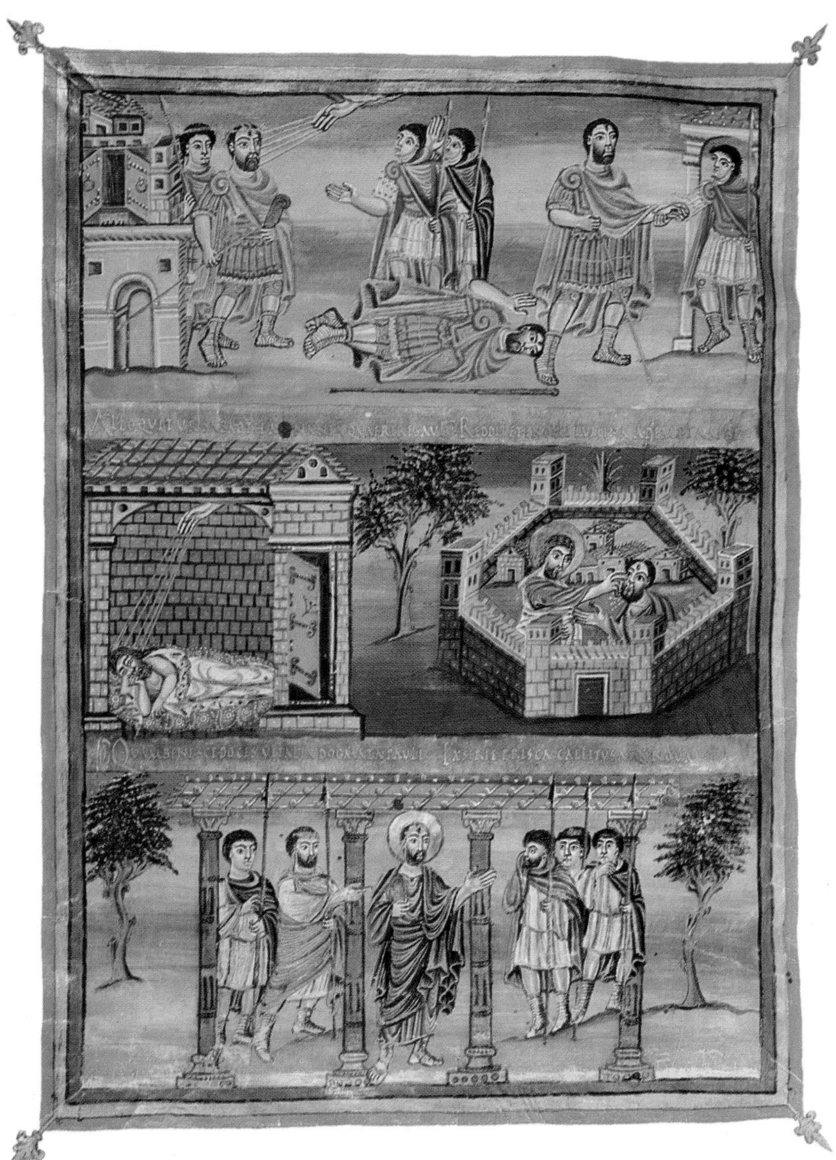

The Conversion of Paul *(top) and* Scenes of Paul Preaching, *from a bible made in 843–851* CE *for the French king, Charles the Bald. Paul is portrayed as a high official in Roman military dress.*

PETER AND CORNELIUS

Peter's conversion of a Roman centurion, Cornelius, at Caesarea is related in Acts 10. For the author of Acts, this was an important landmark in the spread of the Christian message: Luke's narrative credits Peter with the inauguration of the mission to the gentiles, and shows that he was able to defend his action successfully against the objections of the Jewish Christians at Jerusalem.

Luke's narrative deals with two related issues that faced the Church in its early days. One was the question of whether, or on what terms, non-Jews could become members of the Church. For Jewish Christians, such as the earliest Jerusalem community, the only thing that distinguished Christians from other Jews was their recognition of Jesus as Messiah. They were still subject to the Jewish Law, and gentiles who became believers had to accept all its provisions, in particular circumcision. This was always a great source of contention for non-Jews, and it was an issue with which Paul had to wrestle, particularly in his letter to the Galatians.

The other issue was the question of table fellowship. By this time, the regulations insisting on kosher meat and laying down which animals could be eaten were a central feature of Judaism, and Jews were forbid-

Roman officers of the imperial bodyguard (the Praetorian Guard); a marble relief of the 2nd century CE.

den from eating with gentiles who did not observe these rules. This created a problem when Jewish and gentile Christians were together, especially in relation to the Eucharist or when sharing a meal.

Some members of the early Church felt that Jesus had established a new order of things which superseded the strict application of Jewish dietary laws. In Mark's gospel, Jesus says "whatever goes into a person from outside cannot defile" (Mark 7.18), which the evangelist interprets: "Thus he declared all foods clean" (Mark 7.19). Paul took the same position, although he respected the consciences of those who thought otherwise. Before his meeting with the gentile Cornelius, Peter has a vision in which he sees a great sheet lowered from heaven. It contains "all kinds of four-footed creatures and reptiles and birds of the air" (Acts 10.11–12), in other words, the whole animal world. The divine voice proclaims that all these creatures have been made clean (Acts 10.15). For Acts, this nullification of the food laws means that there is now no distinction between Jew and gentile. When Peter goes to see Cornelius, he tells him: "You yourself know that it is unlawful for a Jew to associate with or visit a gentile; but God has shown me that I should not call anything profane or unclean" (Acts 10.28).

St. Peter Enthroned, *by Tommaso Masaccio (1401–1428); in the Brancacci chapel, Florence (see p.219).*

However, the real justification for the admission of gentiles into the Church is the descent of Holy Spirit on them. Acts records that while Peter is still preaching to Cornelius and his friends and family, "the Holy Spirit fell upon all who heard the word" (Acts 10.44). The circumcised – that is, Jewish – believers with Peter are amazed to hear the audience "speaking in tongues and extolling God," a sign that "the gift of the Spirit had been poured out even on the gentiles" (Acts 10.45–46; compare Acts 2.4). In Acts, the gift of the Spirit usually follows baptism, but on this exceptional occasion it precedes it.

For the author of Acts, the sequel to Peter's meeting with Cornelius serves to align Peter and Paul with respect to the mission to the gentiles, and culminates in the meeting known as the Apostolic Council in Jerusalem in Acts 15. Before this, there is an account of how gentiles were being evangelized in Antioch by refugees from persecution (Acts 11.20). Anxious, as always, to oversee all missionary operations, the Jerusalem Church sent Barnabas to Antioch as their emissary. He gave his approval to what was happening (Acts 11.22–23) and later brought Paul to Antioch to join him in preaching the gospel. Trouble arose, however, after Barnabas and Paul returned to Antioch at the end of their first missionary journey (Acts 14.26; see p.232). Some Jewish Christians came to Antioch and began teaching the necessity of circumcision, prompting Paul and Barnabas to go to Jerusalem to settle the question with the Church, then headed by the apostle James, Jesus' brother (Acts 15.1–2).

Peter reappears to address the Apostolic Council on the issue of circumcision. His appearance is rather abrupt and comes after he has been absent from the narrative since the dramatic story of his arrest and escape from prison, at the end of which it is said, a little mysteriously, that he "went to another place" (Acts 12.3–17). Before the apostles and elders of the Church, he refers to his own calling to preach to the gentiles, arguing that God "testified to them by giving them the Holy Spirit, just as he did to us" (Acts 15.8), presumably a reference to his conversion of Cornelius. He questions the need to impose special requirements on non-Jews (Acts 15.9–12). Peter is followed by Barnabas and Paul, who also report on their mission to the gentiles (Acts 15.12). In response, the council decrees that gentile converts need not be circumcised, but it takes a stricter line in the matter of table fellowship. Gentile Christians must abstain from food that has first been offered to idols, and observe the key Jewish ban on eating meat that had not been drained of blood. The Council adds another demand: abstention from fornication, probably meaning sexual immorality in general (Acts 15.19–20).

THE APOSTOLIC COUNCIL IN PAUL'S LETTER TO THE GALATIANS

Paul paints a different picture of the Apostolic Council in what appears to be his account of the same events in Galatians 2.1–10, which is generally agreed to have been written earlier than Acts. According to Paul, Peter's mission had always been aimed at Jews and Paul's at the gentiles. Paul claims that the Council, including Peter (here Paul uses Peter's Aramaic name, Cephas), had agreed that this situation should continue (Gal. 2.7–9). Paul relates that Peter had visited Antioch and had taken his meals with gentiles, until some messengers from James caused him to desist. Paul strongly rebukes Peter for this "hypocrisy" (Gal 1.13–14). The versions of Acts and Galatians can be partly reconciled if the appearance of Peter in the Acts account is spurious and he was not in fact present at the Apostolic Council (see also pp.224–5).

V THE MISSION OF PAUL

The story of the expansion of Christianity in the first century CE is largely bound up with the career of Paul, the former persecutor of Christians who underwent a dramatic calling to the faith (see pp.226–7). Little is known of the work of the other apostles, or about how the new religion reached such areas as Egypt or the East beyond the Roman empire. However, Paul's tireless propagation of the faith is vividly recorded in Acts and in the letters which he wrote to new Christian communities (see pp.238–53). He traveled widely on his mission and may have ended his career, according to one strong tradition, in the imperial capital, Rome.

A ministrant and the god Silenus, from a house at Pompeii, Italy, believed to have been used by a mystery cult.

THE ROMAN WORLD

Paul's journeys took him through the Roman provinces at a time when the empire had almost reached its greatest extent and the reigns of the emperors Augustus, Tiberius, and Claudius (see table on p.231) had created a stable and peaceful state. Compared with elsewhere in the ancient world, travel within the empire was easy and safe, as a result of a system of paved roads linking the provinces and the chief cities, as well as the extensive sea routes that were vital for the commercial life of the Mediterranean region. Paul and his companions traveled by road, apparently mainly on foot, and frequently by sea. According to Acts, Paul often came into contact with the well-established Roman system of provincial administration, which maintained law and order. A Roman citizen himself, he had dealings with the highest imperial officials, such as Gallio, the proconsul (governor) of Achaia and brother of the Roman statesman and philosopher Seneca (Acts 18; see also p.233), and Sergius Paulus, the proconsul of Cyprus (Acts 13). He also came into contact with the governors of less important provinces, such as the procurators of Judea, and with local administrators, such as the Asiarchs (officials of the province of Asia) at Ephesus, and the politarchs (city rulers) of Thessalonica.

Paul carried the gospel above all to the Greek-speaking cities of the eastern Mediterranean, and it was there that Christianity enjoyed its main success. These cities had a cosmopolitan population, and the composition of their churches reflected a wide range of social and ethnic backgrounds. It is sometimes thought that the early urban Christian communities consisted largely of the economically poor and socially deprived. However, recent studies have shown that most of the members of the early Church pursued gainful occupations and, in modern terms, they could fairly be described as "middle class."

One factor that assisted the success of the Christian mission was the wide variety of religious belief and practice in the Greco-Roman world. Official religion centered on the city temples, dedicated to the Olympian deities of the Greeks, or to the Roman deities with whom they were identified. In the provinces, the worship of the Roman emperor also grew up, as an expression of loyalty to Rome. Increasingly popular in the Roman empire in Paul's day were "mystery cults," often of Near Eastern origin, such as that of Artemis of Ephesus, which Paul encountered (see p.232). Like Christianity, they offered initiates the assurance of immortality and a more intense personal spiritual experience than the public rituals of the official religion. There were also a number of quasi-religious philosophical schools, such as the Pythagoreans and the Epicureans, whose adherents assembled in small groups in houses. Their doctrines were propagated by itinerant teachers, whom Christian preachers would have resembled in many ways. However, the new cult of Christianity enjoyed one advantage over the mystery religions and the philosophical schools: its organizational strength, which allowed each individual Christian group to see itself as part of a greater international society, the Church.

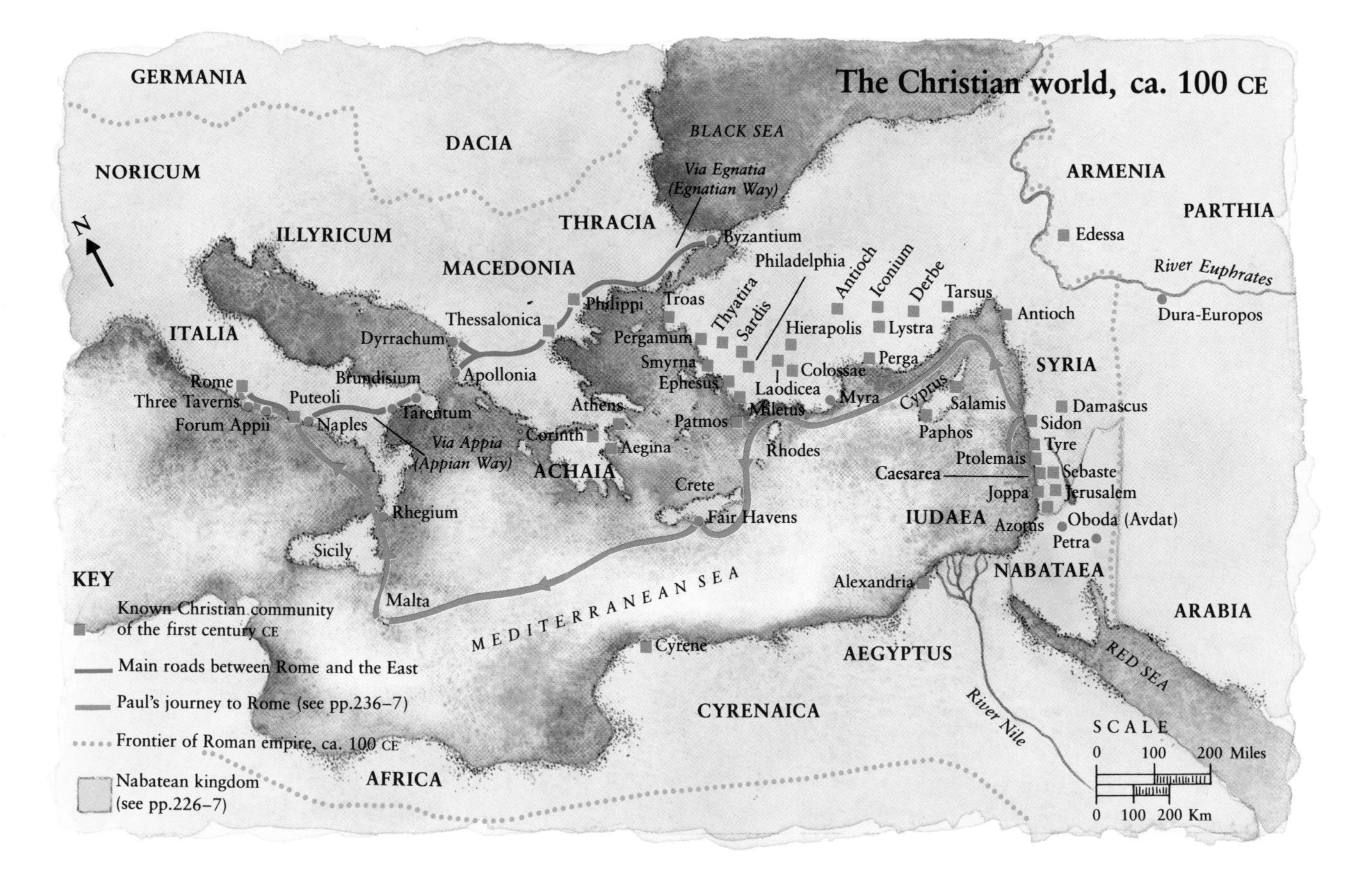

Jews in the Roman empire

Paul was born a Jew and a Roman citizen in the city of Tarsus in Asia Minor. Judaism was recognized by the Roman state as a *religio licita* ("permitted religion") and its members were allowed special privileges, such as exemption from military service and the right to observe the sabbath, although Rome disapproved of proselytism and sometimes took steps to check it. The Jews were based in the cities, where they often formed a large, wealthy, and influential segment of the population, in many cases of long standing. Generally they were strongly attached to their ancestral traditions and anxious to preserve them in a gentile environment.

The first Christians, such as Paul, were Jews. As missionary activity spread from Palestine, Christian evangelists began to preach in the synagogues, as Paul is recorded as doing in Acts (though in his letters he states that his ministry was only to gentiles). However, Acts claims that Diaspora Jews (those living outside their ancestral homeland of Palestine) were generally hostile to him and his followers. This is probably because Paul's overtures to the gentiles and readiness to dispense with distinctive Jewish practices (especially circumcision) were seen as a threat to Jewish identity. But among those attending Diaspora synagogues were those described in the New Testament as "God-fearers," gentiles attracted to Judaism by its monotheism and high ethical code, but not prepared to accept all the obligations of a full Jewish convert. Many of these converted to Christianity, which offered Jewish monotheism and morality, without what were seen as the disadvantages of circumcision and food taboos.

ROMAN EMPERORS IN NEW TESTAMENT TIMES

All dates are approximate and CE except where stated.

Augustus	27 BCE–14
Tiberius	14–37
Gaius (Caligula)	37–41
Claudius	41–54
Nero	54–68
Galba; Otho; Vitellius	68–69
Vespasian	69–79
Titus	79–81
Domitian	81–96

PAUL'S JOURNEYS IN ACTS

Journeys undertaken by heroic figures appear frequently in the mythology of ancient Greece and other cultures. In the Bible, the heroic wayfarer is Paul, who undertakes at least four journeys on his mission to spread the gospel (see maps on p.231 and p.233). In common with other epic travelers, Paul faces many threats and dangers: Acts vividly relates how Paul is stoned and left for dead at Lystra on his first journey (Acts 14.19). Like the Greek heroes Odysseus and Jason, the apostle has to overcome supernatural evil forces, represented in the Bible by the sorcerer Bar-Jesus or Elymas (Acts 14.6–11). Paul also performs miracles which boost his authority: when he makes a cripple walk, he and his companion Barnabas (see p.229) are hailed as the Greek gods Hermes and Zeus, probably local deities that were identified with those of the official Greek pantheon (Acts 14.8–18). Paul was thought to be Hermes, the divine messenger, because he took the lead in speaking of Jesus, and Barnabas was Zeus, the head of the pantheon, probably because he was physically more imposing than Paul. In one of his letters, Paul speaks of a permanent bodily "weakness" (2 Cor. 12.7–8), and the mid-second-century Acts of Paul describes him as small, balding, and bow-legged.

According to Acts, Paul's itinerary was determined by the Holy Spirit. However, the fact that Barnabas was a Cypriot may be the reason why Cyprus was the first port of call on the first journey. The remainder of the mission, related in Acts 13–14, mainly involved cities in the southern part of the Roman province of Galatia, in what is now Turkey. One member of Paul's mission, John Mark, evidently refused to go with Paul beyond Pamphylia. This subsequently led to a break between Barnabas and Paul, who refused to take John Mark on his second mission (Acts 15.37–39).

In Acts, Paul's speeches are central to his mission and each is carefully crafted by Luke to fit the occasion. His first speech (Acts 13.16–41) is delivered in the local synagogue at Pisidian Antioch and is reported at length as a model of his synagogue addresses. It is distinctly Jewish in flavor: like the speeches of Peter in Acts 2 (see p.217) and Stephen in Acts 7 (see pp.222–3), it surveys Israel's history and quotes Scripture. In contrast, the speech at Lystra, Paul's first to a wholly gentile audience (Acts 14.15–17), is brief and speaks of God in broad terms as the divine hand which "made the heaven and the earth and the sea and all that is in them" and sends "rains from heaven and fruitful seasons."

After the first journey, Paul and Barnabas travel from Antioch to attend the Apostolic Council in Jerusalem (see p.229). They return to Antioch with Judas Barsabbas and Silas, who are assigned by the council to take its decree (Acts 15.23–29) to the gentile Christians in Syria and Cilicia, which were apparently evangelized by Paul shortly after his conversion (see Gal. 1.21). From Antioch, Paul sets out on his second missionary journey, recounted in Acts 15–18. It was considerably longer than the first and its primary object was to revisit the churches established during the first journey and assess their progress. One reason for the success of the Christian movement was the creation in this way of an international organizational structure which bound the various communities together under the supervision of an apostolic founder. After falling out with Barnabas, who left for Cyprus, Paul chose Silas as his companion, perhaps implying that the terms of the council decree were now to be extended to the new congregations of southern Galatia.

On the second journey, Paul continues to preach first to Jews, with some success: Jews are converted at Thessalonica (Acts 17.4) and at Beroea (Acts 17.11–12), and the president of the synagogue was

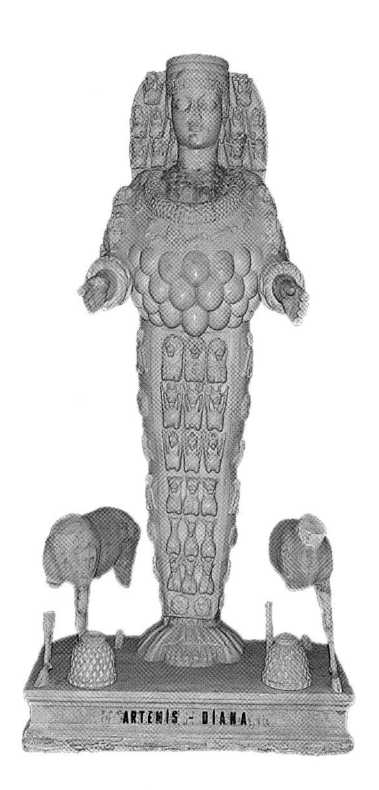

A statue of the fertility goddess Artemis of Ephesus (2nd century CE).

ARTEMIS OF EPHESUS

The account of Paul at Ephesus in Acts 19 paints a vivid picture of the popular Hellenistic cult of Artemis of Ephesus. She was originally a version of the widespread Near Eastern mother goddess, the patron of nature and fertility, and quite distinct from the Greek Artemis or the Roman Diana, with whom she came to be identified. Her original shrine was a primitive sanctuary at the mouth of the Cayster river, long predating the foundation of Ephesus on the river by Greek settlers. By New Testament times Artemis was worshiped in "all Asia," that is, Asia Minor (Acts 19.27). Ephesus was her "temple keeper" (Acts 19.35), in other words the site of her chief temple, a magnificent building that was reckoned among the Seven Wonders of the World. Ephesus was a great pilgrimage center, where, as at all such places in any age, there was a great trade in religious objects. The "silver shrines" made by Demetrius and other craftsmen (Acts 19.24) were either miniatures of the primitive sanctuary of the goddess, or of the classical temple, showing her cult image in a niche. The image, "the statue that fell from heaven" (Acts 19.35), was probably a meteorite; several other examples of meteorites becoming cult objects are known. However, Artemis was usually represented as a woman with many breasts, symbolizing her power as a fertility goddess (see picture).

converted, together with his entire household, at Corinth (Acts 18.8). Out of consideration for Jewish sensibilities, Paul has a half-Jew, Timothy, circumcised when he is recruited to his mission (Acts 16.3). However, Acts depicts the Diaspora Jews as generally hostile toward the gospel; the author often seeks to stress that the spread of the mission to the gentiles was a consequence of its rejection by Jews (as for example at Acts 18.6).

Supernatural guidance of the missionaries is even more prominent on the second journey: the Holy Spirit prevents them from traveling to Asia and Bithynia and a vision directs Paul to preach in Macedonia (Acts 16.6–9). Heathen magic is again routed when the "spirit of divination" is expelled from a slave girl at Philippi (Acts 16.16–18). The author pinpoints certain historical events, such as the emperor Claudius's expulsion of the Jews from Rome, probably around 49 CE, and Gallio's proconsulship of Achaia (southern Greece), which he held probably in 52 CE (see also p.230). Typically for Acts, the narrative of this journey portrays the Roman authorities as favorably disposed toward the Christians: when Paul and Silas are jailed at Philippi, the magistrates act against them only from fear of mob violence, and detain them briefly. When they discover that Paul and his companions are Roman citizens, they realize that they have acted illegally in having them flogged, and apologize (Acts 16.20–39). Later, Paul's appearance before Gallio at Corinth confirms that Paul's mission was not a crime in Roman eyes (Acts 18.12–17).

A calmer and more elegiac note distinguishes the narrative of Paul's third journey, related in Acts 18–21. It begins with another visit to the churches of Galatia and Phrygia, where Paul's mission had begun, but its main focus is on his stay of over two years at Ephesus, the chief commercial city of Asia Minor. Paul's fellow tentmakers, Aquila and Priscilla, who had originally joined him at Corinth on his second journey (Acts 18.2–3), were already at Ephesus. They are joined by Apollos, an eloquent and learned man from Alexandria in Egypt and follower of John the Baptist, who, however, "taught accurately the things concerning Jesus." After Paul's arrival, a group of twelve disciples of the Baptist are also integrated into the Church (Acts 19.1–7; see also p.187).

Acts portrays Paul's Ephesus period as one of considerable success. When he meets the usual Jewish hostility in the synagogue, he simply hires a lecture hall, where his message reaches a large audience (Acts 19.9–10). Rival itinerant Jewish exorcists are vanquished (Acts 19.13–16), scrolls containing magical spells – a speciality of Ephesus – are burned (Acts 19.19), and the agitation of the silversmiths who make idols of the local goddess, Artemis (see sidebar on opposite page), comes to nothing after a warning from the town clerk (Acts 19.23–41). These last three episodes are told with a touch of humor. Later, at Troas, there is a picture of the Church's distinctive commemoration of the Resurrection on the first day of the week, with a sacred meal and preaching. On this occasion, appropriately, Paul restores to life a boy who had fallen to his death from a window sill when Paul's extended preaching sent him to sleep (Acts 20.7–12).

The narrative of Paul's third journey contains three passages where the author, Luke, uses the first person "we" instead of the third person "he" and "they" (Acts 20.5–7; 20.13–15; 21.1–18). These passages seem to show that Luke accompanied Paul on some of his travels. There are other theories, such as the possibility that Luke drew on a travel diary kept by one of Paul's companions. However, there is no significant difference in style between the "we" passages and the material that precedes and follows them. The use of "we" may simply be a stylistic device, lending the story directness and realism by portraying the narrator as a participant. All three passages – together with a fourth "we" section at Acts 27.1–28.16 – give accounts of sea voyages, which were a favorite theme of Hellenistic writers.

PAUL'S FIRST JOURNEY
(Acts 13.4–14.26)

PAUL'S SECOND JOURNEY
(Acts 15.40–18.22)

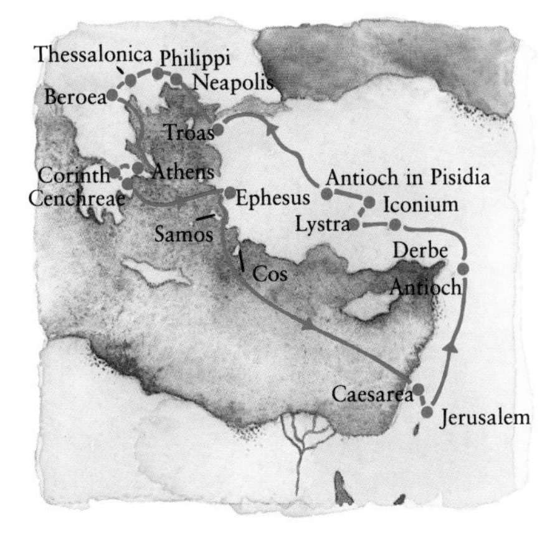

PAUL'S THIRD JOURNEY
(Acts 18.23–21.17)

St. Paul, *by Martino de Bartolommeo (1369–1434). Paul is portrayed here holding a sword, symbolizing the tradition that he was beheaded at Rome under the emperor Nero (see also p.240 and pictures on pp.237 and 239).*

PAUL ON TRIAL

A dominant theme in the account of Paul's third journey (see p.233) is the apostle's determination to revisit Jerusalem, and then go to Rome (Acts 19.31). His motives are not explained, but the story is fraught with foreboding, especially when Paul takes his leave of the Ephesian church leaders at Miletus (Acts 20.18–35). His speech contains a farewell apologia for his missionary work, a prophecy that only imprisonment and hardship await him in the future, and the sad statement that the elders will never see him again. The ominous tone grows more urgent as Paul receives warnings from the Holy Spirit at Tyre (Acts 21.4), and Caesarea, where the prophet Agabus (see Acts 11.28) dramatically warns Paul that the Jerusalem Jews will deliver him to the gentiles (Acts 21.11). But Paul remains determined to go to Jerusalem, and his companions finally have to accept that Paul's fate is in God's hands (Acts 21.14).

The account of Paul's arrival in Jerusalem parallels in some ways the earlier narrative of the Council of Jerusalem (see p.229). Like that account, but unlike Paul's own letters, it shows him as being on good terms with the leaders of the Jerusalem Church, who in turn welcome his work among the gentiles (Acts 21.19–20a). It is apparently only now (Acts 21.25) that Paul learns of the terms of the council's letter to the gentile Christians (see Acts 15.23–29). Paul is represented as a faithful Jew, although the Jerusalem leaders are aware that many doubt this (Acts 21.20b–21). To prove his loyalty to Judaism, Paul agrees to perform a special act of devotion (Acts 21.23–24). Acts records that Paul undertook this ritual at Cenchreae (Acts 18.18). In the Temple, Paul is spotted by some Jews from Asia, the scene of his third mission, who accuse him of the same crimes as those brought against Stephen (Acts 6.13). They also claim that Paul has taken gentiles ("Greeks") into the inner court of the Temple, which non-Jews were forbidden to enter on pain of death (see picture, below). Like Stephen, Paul faces being stoned by an indignant mob, but he is rescued by Roman troops stationed in the Antonia fortress, immediately north of the Temple (Acts 21.31–35; see map on p.190).

Luke's repeated emphasis on how the Romans protected Paul against Jewish hostility (see p.233) appears again in the following narrative. Claudius Lysias, the Roman tribune (commandant) allows Paul to

A Greek warning inscription from the Temple of Jerusalem, one of several fixed to the balustrade or low wall separating the Court of Gentiles from the most sacred areas of the Temple (see map on p.190). It reads: "Foreigners [gentiles] are forbidden from entering through the balustrade into the enclosure around the sanctuary. Anyone found doing so will be to blame for the death that will follow."

address the crowd. His speech goes well until he speaks of his mission to the gentiles, which infuriates his hearers so much that Lysias has to bring him into the barracks "to be examined by flogging" (Acts 22.24). But because of Paul's Roman citizenship, the tribune releases him and orders a formal hearing before the Sanhedrin, the council of Jewish elders. Paul reduces this body to disorder by skillfully playing on the theological differences between its Sadducee and Pharisee members (see pp.184–5), and Lysias once more has to rescue the apostle (Acts 23.10). The tribune then learns of a plot to kill Paul and decides to transfer the matter to a higher authority, the Roman procurator (governor) at Caesarea.

Paul is sent to Marcus Antonius Felix, who held office from 52–60 CE (see table on p.144) and was married to a Jew, Drusilla, a daughter of King Herod Agrippa I. After five days, members of the Sanhedrin arrive to lay charges against Paul, and their lawyer, Tertullus, opens the prosecution case with a speech in the Hellenistic rhetorical style (Acts 24.1–8). Paul then gives his own account of events in the Temple, denying all the accusations and claiming that the real issue is the resurrection of the dead (Acts 24.20–21). Felix adjourns the case until he can hear testimony from Lysias, but no more is said of this move and it may only have been a pretext. A few days later, Felix and Drusilla interview Paul, but when Paul turns to questions of "justice, self-control, and the coming judgment," Felix is said to be "frightened" and ends the interview. He keeps Paul under open arrest for two years, hoping to extract a bribe from him (Acts 24.26). This behavior accords with the unfavorable portrayal of Felix's character by Roman historians.

Felix is succeeded as procurator (in 60 CE) by Porcius Festus, who reopens Paul's case. He asks Paul if he is willing to be tried in Jerusalem (Acts 25.9), and in response Paul exerts his right, as a Roman citizen, to be heard before the emperor himself (Nero). Festus accepts, and the process is set in train which will bring Paul to his ultimate goal: Rome (Acts 25.10–12).

Paul and Herod Agrippa II

Shortly after Paul's interview with Festus, Acts relates that the procurator receives a visit from Herod Agrippa II, the client king of territories to the north and east of Palestine, and his sister (and lover) Bernice (Acts 25.13). Agrippa asks to hear Paul, who begins his address with a rhetorical flourish, praising the king (Acts 26.2–3). Paul summarizes his own career as presented in Acts, ending with an affirmation of Jesus' death and resurrection, and the proclamation of the gospel to all humanity. Agrippa says that Paul almost persuades him to be a Christian, and asserts the apostle's innocence. In his view, Paul could have been freed if he had not appealed to Rome (Acts 26.32). But Paul's appeal to the emperor (see main text) commits him to being taken to Rome to be heard before the imperial tribunal.

A coin of Herod Agrippa II (ruled 50–ca. 100 CE), a great-grandson of Herod the Great and the last monarch of the Herodian dynasty.

PAUL'S JOURNEY TO ROME

PAUL IN MALTA

After their escape from shipwreck, Paul and his companions are welcomed by the native inhabitants of Malta (Acts 28.2). Because of the bad weather, they light a bonfire, and as Paul puts sticks on it, a viper fastens onto his hand. The Maltese believe that this is divine punishment for a murderer and expect Paul to die, but when he shakes off the snake and is unharmed, they assume that he must be a god. Following this episode, Paul is received by Publius, the chief magistrate of the island, another example of the way in which Paul is represented throughout Acts as receiving sympathetic treatment from the Roman authorities. During a three-day stay in Publius's house, Paul lays hands on his father and eases his fever and dysentery. He also heals a number of other sick people.

Today there are no poisonous snakes in Malta, and perhaps there never were any. Paul's ability to resist snake poison and to cure illness may have been included to show him as typical missionary disciple, in accordance with Jesus' promise to the disciples at Mark 16.18: "if they handle snakes or drink any deadly poison, they will come to no harm; and the sick on whom they lay their hands will recover." The whole narrative is designed to honor Paul. Here, as elsewhere, he is recognized as a divine manifestation (Acts 14.11–12).

A clay imprint representing a Phoenician deity. In the story of Paul and the viper the local inhabitants expect "justice," in Greek diké, *to punish the apostle. The Maltese were of Phoenician origin, so* diké *probably refers to a Phoenician-Maltese deity who is given a Greek title by the author.* Diké *was the name of the Greek goddess of justice.*

The account in Acts 27–28 of Paul's voyage to Rome, especially his shipwreck off Malta (see sidebar, left), is one of the most dramatic pieces of writing in the New Testament. The narrative of the voyage, together with a few other sections of the book, is told in the first person plural, but this may simply be a stylistic device (see p.233). It reveals a good knowledge of Mediterranean geography and climate, and it has been described as one of the most informative documents for our knowledge of ancient seamanship. On the other hand, it stands in a long literary tradition, going back to Homer's *Odyssey*, in which accounts of a Mediterranean voyage almost invariably included the depiction of a storm and a shipwreck. In fact, much of the vocabulary of the *Odyssey* became part of the traditional way of recounting nautical stories, and traces of its style have been detected in Acts. The story has also been influenced by the Hellenistic genre known as the *periplus*, an account of a coasting voyage, of which a considerable number of examples remain. Luke's narrative can be compared with the work *The Ship* by the second-century CE author Lucian of Samosata, which covers some of the same geographical area (see map on p.231).

Paul and other prisoners sail for Italy, probably sometime between 60 and 62 CE (the period of office of the procurator Porcus Festus), in the custody of a military guard headed by a centurion. The ship calls first at Sidon, where Paul is allowed to visit friends, another manifestation of the friendly attitude of Roman officialdom toward him (Acts 27.3). Next, the party proceeds to Myra in Licia, where they transship into a vessel from Alexandria, which Acts 27.38 shows to have been a corn ship. It would have been sailing on a regular route from Egypt, Rome's chief source of grain, to the imperial capital. The ship reaches Fair Havens (modern Kaloí Liménes) in Crete at a time when further progress is dangerous: the narrator says that the "Fast," the Jewish feast of Yom Kippur (Day of Atonement) in October, has passed (Acts 27.9) and, according to Roman writers, the dangerous season for navigation lasted from September 14 to November 11. The owner of the ship, the state contractor for the transport of corn, is determined to press on to a safe winter haven, but the ship is overtaken by a violent storm. At this point Paul receives a divine vision to reassure him that the ship's company will survive and that he himself will reach Rome (Acts 27.23–24). After a detailed and vivid description of the storm, the vessel is finally lost, but the ship's company escape and land safely on Malta (Acts 27.44). They winter on Malta for three months (see sidebar, left), since navigation in the Mediterranean was considered impossible from mid-November to the end of January.

The following spring, Paul is loaded with marks of respect by the local inhabitants and furnished with necessary supplies to continue his voyage. He embarks on another grain ship, with the figurehead of the "Twin Brothers," a reference to Castor and Pollux, the twin sons of the god Zeus (the Roman Jupiter), who were the patrons of navigation and commonly worshiped by sailors. After calling at Syracuse in Sicily and Rhegium in the extreme south of mainland Italy, the party finally disembarks at Puteoli (modern Pozzuoli) near Naples.

Paul's itinerary from Puteoli to Rome, via Appii Forum (Forum of Appius) and Three Taverns, is well-known from the Roman authors Cicero and Horace, as well as from a world map of the third or fourth century CE. The goal and the significance of the journey is summed up in the last words of Acts 28.14: "And so we came to Rome." The program of the author of Acts, laid out in Acts 1.8, is to chart the spread of

Christianity from its original home in Palestine throughout the civilized world. This development reaches its climax when Paul is able to proclaim the gospel in Rome itself, the center of the empire.

The final chapter of Acts stresses two of the author's central themes. One is his positive attitude to the Roman power. Implicitly, Luke shows that the universal *Pax Romana*, the "Roman Peace," with the freedom and ease of travel that it ensured, facilitated the spread of the Christian faith and the consolidation of the Church. Explicitly, he shows repeatedly how the imperial representatives always treat Paul with impartial justice and protect him from his accusers, so that he is confident of a favorable hearing before the supreme tribunal of the empire. At Rome, Paul is not imprisoned but merely put under a mild form of house arrest, living for two years at his own expense and preaching the gospel "openly and without hindrance." Luke's second great theme is the rejection of the gospel by the Jews and its acceptance by the gentiles. This is the subject of Paul's final interview with the Roman Jews (Acts 28.23–29).

Acts does not relate Paul's eventual fate, but there is a strong Christian tradition that he suffered martyrdom at Rome, perhaps during the persecution of Christians that took place under the emperor Nero in 64 CE. The author's silence about Paul's death may reflect his desire to bring his work to a rhetorical climax, with Paul's unfettered proclamation of the faith in the Imperial City.

A Roman grain ship, from a wall painting at Ostia, the port of Rome, ca. 100 CE.

Christians and Jews in Rome

Paul encountered one group of Christians at Puteoli (Acts 28.14) and another at Rome. He had previously written a letter to the Roman church (see p.245), which indicates that it was already of some strength. Its existence shows how Christianity must have expanded through the activities of other apostles, and by migration: those addressed by Paul in his letter have Greek, Roman, and Jewish names.

The Roman Christians formed a respectable segment of society. They probably included slaves and a few of the aristocracy, but basically they were what today might be described as middle-class. Paul's letter mentions the "household" of two men, and greets Prisca and Aquila, who had a house large enough for a Christian assembly to meet in. It was in the houses of members of the community that Christians gathered for worship. The famous Roman catacombs were not secret meeting places, as is often suggested, but simply burial sites, which would not have provided sufficient space for corporate worship. At least three "house churches" remain in Rome beneath the present churches of San Martino ai Monti, Santa Pudenziana (named traditionally after a certain Pudens who is said to have entertained St. Peter), and the saints Giovanni e Paolo.

St. Paul, *an enamel and gilt roundel from a 10th-century Venetian book cover. Paul has been portrayed since at least the 6th century as a balding, dark-haired man (see also pictures on pp.234 and 239).*

In Paul's day the Jews of Rome were more numerous, more influential, and longer established than the Christians. The development of the community really began with Pompey's capture of Jerusalem in 63 BCE, when he brought numerous slaves back to Rome. Many subsequently gained their freedom and established a colony in Trastevere, the area on the left bank of the Tiber, and later in other parts of the city. The Jews rapidly became an important element in Roman life: in the time of Augustus (27 BCE–14 CE) they numbered in the thousands. The historian Josephus relates that eight thousand Jews greeted a Jewish delegation from Palestine in 4 BCE. Their prominence is seen in the attacks on them by Roman writers, such as the satirist Juvenal, and along with members of other oriental cults they were now and again expelled from the city by imperial decree. But the community continued to flourish – for example, many Jews arrived after the fall of Jerusalem in 70 CE – until the last years of the empire.

VI LETTERS

Of the twenty-seven books of the New Testament, no less than twenty-one are called letters, or epistles (from the Latin epistola, *a letter). Most of them are attributed to early Church leaders; thirteen of them are ascribed to the apostle Paul alone. Letters were the means by which Church leaders were able to keep in touch with the scattered Christian congregations. The letters offered guidance on theology and pastoral care, advice on local difficulties, and warnings against certain practices and abuses. The last book of the New Testament, Revelation, is also a letter, although its extraordinary visionary character places it in the genre called apocalyptic (see pp.136–7).*

NEW TESTAMENT EPISTLES

The epistles (letters) that were sent out by the early Church to the far-flung Christian community may have been influenced by Jewish practice. The New Testament epistles recall missives that were despatched by the central Jewish authorities to communities in the Diaspora, discussing the proper observance of religious festivals and fasts. One letter from the Jerusalem Jews to their kinsfolk in Egypt, enjoining them to celebrate the new festival of Hanukkah (Purification; see p.140), is preserved in 2 Maccabees 1–2; it is close in form and style to several of Paul's epistles.

Paul's correspondence would have been written on papyrus sheets, which were then rolled up and tied for transmission. The contents were dictated to a secretary, one of whom, Tertius, is mentioned by name in the letter to the Romans (Rom. 16.22). Paul sometimes added a postscript in his own hand, perhaps to make clear that the document really was from him. There was an efficient postal service in the Roman empire, but this was confined to official communications. Groups or individuals either had to employ their own couriers or entrust their letters to someone who happened to be traveling to the required destination. In ancient times, the bearer of a letter had an important role, often supplementing the written communication with further information or instructions from the sender. Paul often appointed delegates to carry his correspondence: he sent Timothy and Titus to the Corinthians (1 Cor. 4.17, 16.10–11; 2 Cor. 7.6–16, 8.16–18, 23–24). These same delegates would then bring a reply and other news back to the sender (2 Cor. 7.6–16).

The New Testament epistles make use of common Hellenistic and Jewish letter-writing forms, especially in their opening and closing greetings. They begin with the name of the sender, followed by the name of the recipient (for example, 1 Cor. 1.1–2; compare 2 Macc. 1.1). Greek letters usually employed the salutation "greetings," literally "rejoice," but Jews substituted or added the more typically Hebraic word "peace" (see, again, 2 Macc. 1.1). Paul does not use "greetings," but the distinctively Christian term "grace," which he couples with "peace" to form his most common salutation: "Grace to you and peace from God our Father and the Lord Jesus Christ" (Rom. 1.7 and so on). A prayer for the readers' health is often found, and an assurance of intercession for them in prayer (as in Rom. 1.9). There may be expressions of thanksgiving for past commendable conduct or for the faith of the communities the author is addressing (Rom. 1.8; 1 Cor. 1.4–7; Phil. 1.3–5; Col. 1.3–4; 1 Thess. 1.2–4; 2 Thess. 1.3). Significantly, Paul omits such praise in his letter to the Galatians, which is perhaps the most passionate and urgent of the epistles (see p.244). Stock polite formulas, such as "I appeal to you" and "You will do well to," introduce requests or injunctions. The letters conclude with often quite elaborate farewell greetings from the writer and his companions to those whom he is addressing. In all these respects, the New Testament writers reflect common Greco-Roman habits.

Although twenty-one of the writings in the New Testament are presented as literary epistles, they are not all of the same type and reflect the different kinds of letter that were common at the time. There was the purely personal letter, from one individual to another; Paul's letter to Philemon may belong to this category, although this is also addressed to "Apphia our sister, to Archippus our fellow soldier, and to the church meeting in your house" (see p.249). The third letter of John, from "the elder to the beloved Gaius" is another example of such personal correspondence.

There were also formal letters, intended for a wider audience, such as a particular church or community, which dealt with broader concerns and issued advice or directions to the readers. Paul's own letters combine the formal and more personal forms and were intended to be read during public worship. Some are of considerable length, and their quotations from Scripture, creeds, hymns, and other religious teachings, give them a distinctive character. At the same time, parts of them often reproduce conventional features of Hellenistic personal correspondence.

However, while Paul's letters are examples of real correspondence with individual communities, other writings that the New Testament labels "letters" are in reality more like general treatises. For example, the "letter" to the Ephesians (which may or may not be by Paul) is actually an encyclical, in other words a circular, intended for a number of churches in Asia Minor (see pp.248–9). The "letter" to the Hebrews is in fact a sermon, and contains no hint of conventional letter forms until its final verses. The "letter" of James is an example of *paranesis*, offering advice on living wisely and well (see pp.250–51). The Pastoral Letters (see pp.240–41) belong to a category of writings called "Church Order": they are largely concerned with handing down advice to particular churches on pastoral and practical matters, although they employ the literary form of a personal letter.

St. Paul, *Hungarian School, ca. 1500. The depiction of the apostle with a full head of hair and a long (rather than trimmed) beard reflects a different tradition from that represented in the image on p.237; that on p.234 appears to combine both traditions. Here, as on p.234, he carries a sword as a symbol of his supposed martyrdom (see p.240). In all three he holds the gospels, symbolizing his evangelical mission.*

Greco-Roman letters

All the letters in the Bible, and the majority of those from the Greco-Roman world, have survived only in literary contexts. But modern archaeological discoveries have produced a number of original letters, such as those from the cities of Mari, Tell El-Amarna, Lachish, Elephantine, and those dating from the first-century Jewish revolt of Simon Bar Kochba. Of particular interest for the New Testament are papyrus letters of the same period from Egypt and other arid regions, which have been preserved by the climate. They are private correspondence, often of quite ordinary individuals, and show that many of the formal elements that characterize the New Testament letters were in common use.

A more elaborate style of correspondence was produced by prominent public figures, such as the Roman statesmen Cicero and Seneca. Their letters, like those of Paul, were collected for publication. Such letters were often treatises intended for a wide circle of readers, dealing with philosophical, political, or moral issues. Letter-writing was viewed as a branch of the art of rhetoric, and handbooks have survived that provide appropriate letter forms for particular occasions. Fictitious letters could be produced for literary purposes or to sway those who might read them. Paul himself had to warn the Thessalonians not to be taken in by a missive purporting to be from him (2 Thess. 2.2). Historians felt free to make up letters and speeches that they considered to be appropriate to a particular character: examples of this kind of letter are found in the book of Revelation addressed to the seven churches of Asia (Rev. 1.4–3.22).

The Martyrdom of St. Paul, *from the workshop of Pacino di Buonaguida (active 1303–1309). According to tradition, Paul was beheaded with a sword. His head is said to have bounced three times, causing three fountains – the Tre Fontane in Rome – to spring from the earth.*

THE PASTORAL LETTERS

The New Testament does not relate the ultimate fate of either Peter or Paul. But according to a well-established tradition, especially early and strong in the case of Peter, both apostles were martyred at Rome during the emperor Nero's persecution of Christians in 64 CE. How Peter might have come to be in the city is unknown, but at the end of Acts, Paul is said to be in Rome, preaching the gospel "with all boldness and without hindrance" after his eventful journey from Palestine (see pp.236–7). He is said to have lived under a mild form of house arrest for two years (Acts 28.30–31), presumably waiting for his appeal to the emperor (see p.235) to be heard. But nothing further is said about the outcome of the appeal and Acts does not say that Paul was freed. According to one theory, at the end of the two years, he may have been freed, either because his accusers failed to press charges or through an act of imperial clemency. Another theory claims that the author may have omitted Paul's execution because he wished to end his work on a positive note. After all, the goal stated at the beginning of Acts has been achieved: the gospel is being preached in Rome.

One tradition, at least as early as the fourth-century CE Christian historian Eusebius, states that after the events recounted in the last chapter of Acts, Paul was released, resumed his missionary work, and finally returned to Rome, where he was martyred. Support for this tradition has often been found in three epistles, the first and second letters to Timothy and the letter to Titus (see box on opposite page). These are known as the "Pastoral Letters," because they offer guidance to the recipients on the pastoral care of their churches. The letters depict journeys by Paul in Asia Minor, Macedonia, Crete, and Greece. It is difficult to reconcile these travels with

The Martyrdom of St. Peter *(ca. 1550), by Michelangelo Buonarroti (1475–1564), a fresco in the Pauline chapel of the Vatican, Rome. In the traditional story, Peter is crucified under Nero in Rome. It is said that he considered himself unworthy to die in the same way as Jesus, and insisted on being crucified upside down. Like most artists from the 4th century onward, Michelangelo depicts the senior apostle as a tall, gray-haired man of great physical vigor.*

the career of Paul as related in Acts and his other letters. The journeys must therefore be ascribed to a later period of Paul's life: a passage in 2 Timothy may refer to his release from a first imprisonment in Rome (2 Tim. 4.16–17). However, this reconstruction presumes that the Pastoral Letters are in fact by Paul, a view that has been widely challenged. Most scholars now believe that they were written some considerable time after the apostle's death. The authors of the letters may have sought to reassert Pauline teaching in the face of problems in the churches that Paul had founded. Nevertheless, the Pastoral Letters may contain fragments from genuine, lost, Pauline letters, such as the names and deeds of particular individuals. There are also possibly genuine personal details, such as the reference to a cloak and notebooks left at Troas (2 Tim. 4.13). The authors may have included such details to give the letters a sense of authenticity.

The style and vocabulary of the Pastoral Letters differ considerably from those of Paul's undoubtedly genuine letters; they recall the works of Christian writers of the late first and early second centuries CE. The theological outlook of the Pastoral Letters also seems to be different in many ways from that of Paul. For example, "faith," used by the apostle in the sense of "an act of trust," becomes in the letters "*the* faith," the body of Christian doctrine. There is little mention of the work of the Spirit, which is central in the thought of the authentic Paul. Church organization appears to be considerably more developed than it does in Paul's other letters, with bishops, elders, and deacons. Again, this reflects the situation in the Church of around the end of the first century CE.

In defense of Pauline authorship, it has been said that if the Pastoral Letters date from late in Paul's life, then some of the developments in ecclesiastical structure that they indicate may already have begun. Also, the letters attack what seems to be the first real appearance of the Gnostic heresy, and Paul may have had to adapt his language and theological approach to meet this new threat to the Church.

Timothy and Titus

One argument against Paul's authorship of the Pastoral Letters is that they are addressed to individuals and not to Christian congregations, unlike all his other letters except Philemon. A later author might have employed the names of Timothy and Titus, both venerated figures in the early Church, to give authority to his own teaching.

Timothy, a native of Lystra who was chosen by Paul to accompany him on his missionary travels (Acts 16.3), appears frequently in Paul's letters (for example, Phil. 2.19–23). He was a close associate of Paul in his evangelistic work and also someone for whom the apostle felt a deep personal affection (see, for example, Phil. 2.22). Even if Paul did not write the warm words addressed to Timothy in 1 and 2 Timothy (such as 2 Tim. 1.3–4), they certainly reflect Paul's attitude. Paul employed Timothy as his personal emissary to the churches in Thessalonica and Corinth when he could not visit them himself (1 Thess. 3.2; 1 Cor. 4.17), and it is possible that Timothy could have assumed special charge of one or more of the Pauline congregations when Paul was confined in Rome. The author of the Pastoral Letters evidently knew the names of Timothy's mother and grandmother (2 Tim. 1.5).

Unlike Timothy, Titus is not mentioned in Acts, but he too was a "partner and co-worker" of Paul (2 Cor. 8.23). He accompanied Paul on an important visit to Jerusalem (Gal. 2.1), he too was an emissary of Paul's to Corinth (2 Cor. 7–8), and in particular he took the lead in the work of collecting contributions for the Jewish Christians at Jerusalem (2 Cor. 8). Whether the general picture of him in the letter to Titus can be regarded as coming from Paul is doubtful. However, the letter's reference to Titus's missionary work in Dalmatia and Crete, which is also mentioned in 2 Timothy, may represent an authentic tradition (2 Tim. 4.10; Titus 1.5). The author of the letter tells Titus that "Cretans are always liars, vicious brutes, lazy gluttons" (Titus 1.12). This is a quotation from a poem by the sixth-century BCE Cretan poet Epimenides. Paul is reported to have used the same poem in his speech at Athens (Acts 17.22–31).

LETTERS TO GREECE

A 4th-century monument on the ancient Via Egnatia (Egnatian Way), along which Paul would have traveled. It was the main road linking Thessalonica and other cities of northern Greece to the Adriatic (see map on p.231).

In the New Testament, Paul's letters are not arranged according to when they were written, but roughly in order of length. There is some dispute as to their exact chronological order, so it is difficult to use them to reconstruct Paul's life. It is also hard to see exactly how they fit into the picture of Paul that is presented in Acts.

However, it is generally accepted that the first Pauline letter is 1 Thessalonians, the first letter to the church at Thessalonica (modern Salonica) in northern Greece. It was probably written from Corinth in 50 CE by Paul and his fellow missionaries Silas (Silvanus) and Timothy. Paul visited the city during his second missionary journey, recorded in Acts 16–18. His teaching at the synagogue there attracted a good number of converts, but opposition from Jewish authorities forced him to leave (Acts 17.1–9). The letter to the Thessalonians, written subsequently, thus gives an interesting picture of a Christian community at an early stage. Addressed in the letter as a "church," the Thessalonian Christians had been suffering persecution. The letter begins with a thanksgiving, in characteristic epistolary style, in which Paul expresses his gratitude for the community's "work of faith and labor of love and steadfastness of hope in our Lord Jesus Christ" (1 Thess. 1.3), and exhorts them to persevere in the future.

In the letter, Paul offers guidance on three issues of concern to members of the Thessalonian church. He entreats them to avoid fornication, to control their bodies, and reject the sexual license of the Greco-Roman world, for "God did not call us to impurity but in holiness" (1 Thess. 4.3–7). He reminds them to be good citizens, to mind their own affairs and work to make a living, "so that you may behave properly toward outsiders and be dependent on no one" (1 Thess. 4.11–12). And he reassures them "by the word of the Lord" about the second coming of Jesus, which was seen as imminent, saying that those Christians who had already died would not be left behind but would, in fact, rise before the living, so that all would be "caught up in the clouds together ... to meet the Lord" (1 Thess. 4.13–18).

The second letter to the Thessalonians is the subject of much scholarly disagreement. Some argue that it was in fact written before 1 Thessalonians, while others hold that the differences between it and the first letter mean that it cannot have been written by Paul. Nevertheless, the second letter seems to take up two of the concerns of the first. The idea of the second coming of Christ was continuing to cause problems, but now some teachers were claiming that it had already occurred. This perhaps reflects the view, typical of later Gnosticism, that the Resurrection was a purely spiritual experience. In response, Paul reminds the Thessalonians of the apocalyptic concept of the "rebellion," which must precede the Messiah's advent (2 Thess. 2.3). The arrival of the Day of the Lord is being delayed by a "lawless one," who will enthrone himself in the Temple and claim to be God (2 Thess. 2.4). The imagery is much the same as in Mark 13, and reflects the depiction of the persecuting Hellenistic king Antiochus Epiphanes in the book of Daniel (see pp.136–9).

The second issue that echoes 1 Thessalonians is the subject of work. It would appear that some Christians thought that because the coming of Jesus was very near or had in fact come, there was no point in their continuing to earn a living. Paul rejects such an attitude, pointing to himself as an example (2 Thess. 3.9). When he and his companions were among the Thessalonians, they worked for their daily bread, even though they might have expected others to support them. Paul condemns this attitude and tells his readers in no uncertain terms that "anyone unwilling to work should not eat" (2 Thess. 3.10). He exhorts them not to be "weary in doing what is right" (2 Thess. 3.13).

The letters to Corinth

More than any of Paul's writings, the two letters to the Corinthian church provide a vivid picture of an early Christian congregation. They were written from Ephesus, probably ca. 55 CE, to the church at Corinth which Paul had planted during his second missionary journey (Acts 16–18). The letters pose various literary problems, especially 2 Corinthians, which is probably made up of several different letters which were sent at various times. The precise course of events is thus not particularly easy to determine, but the issues and debates that faced a young church in a great cosmopolitan city are vividly depicted.

The Corinthian community was rent by divisions between followers of different evangelists, and at the beginning of the first letter, Paul appeals for unity, for adherence to no one teacher except Christ (1 Cor. 1.10–17). Next, Paul addresses the issue of sexual behavior. Essentially, he asserts the authority for Christians of the Jewish moral code. He orders the excommunication of a man who, contrary to the Jewish Law, had married his father's wife, and gives his views concerning marriage and divorce (1 Cor. 5–7). He then turns to another problem posed by the pagan environment: whether Christians could eat meat that had been offered to idols. Again, Paul's concern for unity is evident. In this matter, each should act according to his own conscience and not judge others (1 Cor. 8–9).

Questions of worship and belief occupy the remainder of Paul's letter. By this time, the eucharist had clearly become the central Christian celebration, but the communal meal in which it was set was in danger of exhibiting the excesses which often accompanied Greco-Roman festivities. "Speaking in tongues" was also leading to disorder in public worship and was becoming an occasion of self-indulgence. Paul does not forbid the practice, but stresses that other gifts are necessary, especially, in a famous passage, love (1 Cor. 13). What is required is a fully communal worship, which will bind the whole congregation together, and this demands that "all things should be done decently and in order" (1 Cor. 14.26–40). To this end, "women should be silent in the churches," in accordance with the Jewish Law (1 Cor. 14.33–36). Finally, some people in Corinth were denying the resurrection: presumably, like some of the congregation at Thessalonica, they denied the resurrection of the body, and understood the concept of resurrection in a purely spiritual sense. Once again, Paul adopts a mediating position: there will, at the final end, be a real resurrection, but of a new, transformed, and spiritual body, fitted for the new age (1 Cor. 15.1–58).

The second letter to the Corinthians does not add a great deal to the general picture of the community that was provided in the first letter, except to show the continuing seriousness of its divisions. Paul's first letter had apparently had little effect and, in the last three chapters of the second letter, he abandons the polite epistolary conventions to vent his anger and frustration. He threatens the Corinthians with the force of his displeasure on another visit.

The Bema, *the ancient public tribunal at Corinth, Greece. It was probably here that Paul faced Gallio, the Roman proconsul (see Acts 18.12–17).*

GALATIANS AND ROMANS

Abraham Prepares to Sacrifice Isaac, *a mosaic floor from the 6th-century* CE *synagogue of Beth-Alpha in northern Israel. In the letters to the Galatians (Gal. 3–4) and to the Romans (see especially Rom. 4), Paul draws extensively on the story of Abraham and his faith. He says that "the promise that he would inherit the world did not come to Abraham or to his descendants through the law but through the righteousness of faith" (Rom. 4.13). This was demonstrated through Abraham's readiness to sacrifice his son at God's behest (see p.43).*

Two of Paul's letters, Galatians and Romans, are often considered together. Both are concerned with the separation of Christianity from Judaism, which was one of the most significant results of Paul's ministry. The letter to the Galatians was probably among the earliest of Paul's epistles, written ca. 48 CE, although it has also been dated as late as 55 CE. The date of the letter to the Romans (see box on opposite page) is unknown, but it may be a later elaboration of ideas first addressed in Galatians.

The separation of the two religions seems to be found for the first time in the letter to the Galatians, in which Paul argues that faith alone can lead a person to salvation. Just who the "Galatians" were is much disputed. They were possibly the inhabitants of a broad region of north and central Turkey, called Galatia after the Celts who had settled in the area ca. 250 BCE. It is more likely, however, that they lived farther south, in the more narrowly defined region which since 25 BCE had been the Roman province of Galatia. There is no evidence that Paul ever visited the northerly region, but he traveled throughout the Roman province on his three missionary journeys, evangelizing the cities of Pisidian Antioch, Iconium, Lystra, and Derbe (Acts 13.14–14.23; see pp.232–3).

The strong tone of the Galatian letter may reflect the special attachment he felt to these cities, where his work had first borne fruit, and which he now felt were in danger of being led away from the gospel. Unlike his other letters, Galatians has no introductory thanksgivings, but the opening is followed abruptly by a forceful statement of his concern: "I am astonished that you are so quickly deserting the one who called you in the grace of Christ" (Gal. 1.6). It was in Galatia that Paul's mission to the gentiles had begun, but now his gentile converts were being won over by rival Jewish-Christian teachers, telling them that they could not be saved unless they accepted the Jewish Law and, in particular, its requirement of circumcision. These teachers seem to have combined the Law with an elaborate angelology and mystical speculation, similar to that found at the community of Qumran (see p.145). Paul goes so far as to accuse the Galatians of reverting to their former belief in the "elemental spirits" that governed the universe (Gal. 4.9).

Paul's argument is that there is no reason that the Law should be imposed on gentile converts, for "a person is justified not by the works of

The letter to the Romans

Many of the themes of the letter to the Galatians recur, though in a fuller and more carefully nuanced way, in the letter to the Romans, which was probably written shortly after. Essentially a treatise or essay, rather than a letter proper, it sums up Paul's theology and, as such, stands apart from his other undisputed writings.

Romans is also unique because it was not sent to a community that Paul himself had founded. He writes that it is his ambition to take the gospel to places where the name of Christ has not been heard, not wanting to "build on someone else's foundation" (Rom. 15.20). This raises the question of the exact occasion and purpose of his writing. In the first chapter, Paul appears to be assuming responsibility for evangelizing the whole Greco-Roman world, and writes that he has long planned to visit Rome and would now hope to do so (Rom. 1.11, 13). Paul may have felt that the Christians in Rome lacked a proper organizational structure: Paul does not use the word "church" outside the last chapter (the authenticity of which is disputed). It may be that the Roman community had been disrupted by the expulsion of the Jews from Rome that was ordered by the emperor Claudius in 41 or 49 CE. Those who were expelled could well have included Jewish Christians still linked with the synagogues there.

The letter may be seen as an attempt to lay a "foundation" for the Roman church and, not least, to avoid the tensions between Jewish and gentile Christians which had arisen elsewhere. In it, Paul develops further the doctrine of justification by faith alone, one of the key themes of Galatians. His famous message that "a person is justified by faith apart from works prescribed by the law" (Rom. 3.28) is to be understood in the context of the relationship of Jew and gentile in the church: the Jewish Law divided them, but a common faith in Jesus would unite them. Jewish Christians and gentile Christians are sinners (Rom. 2.12– 3.20), yet all can achieve salvation (Rom. 3.21–31). Toward the end of the letter is a discussion of Israel's future. Paul likens the gentiles to wild olive shoots grafted onto an olive tree to replace branches that had broken off, that is, Jews who have rejected the gospel. In the end, however, "all Israel will be saved" (Rom. 11.17–26).

After setting out the theological basis for the Church, Paul turns to the attitudes of the Roman community in its everyday dealings: mutual harmony and tolerance among its own members; loyal obedience to the state; and the fulfillment of social obligations.

Top: *A 1st-century bust of the emperor Claudius (ruled 41–54 CE).*

Bottom: *It is possible that Paul wrote Romans from Corinth. This 1st-century inscription from the city bears the name of an official called Erastus, who may be the treasurer mentioned in Romans 16.23.*

the law but through faith in Jesus Christ" (Gal. 2.15). He does not disown his own Jewishness (Gal. 2.15), and he recognizes that the Law had its value "before faith came," in the sense that it served to convict human beings of their sinfulness (Gal. 3.19–24). On the other hand, he calls it a curse (Gal. 3.10–12), because, as he sees it, it is impossible to observe its commands fully. With the coming of Christ (Gal. 5.2–4), Paul writes, this curse has been lifted (Gal. 3.14), and the authority of the Law has come to an end (Gal. 3.25). Salvation is attained only through faith in Christ, a teaching developed more fully in Romans. In Galatians, Paul sets out a radical concept of Christian, and indeed human, freedom. The believer is delivered from old allegiances, whether Jewish or gentile, because the gospel transcends all human differences: "there is no longer Jew or Greek, there is no longer slave or free, there is no longer male or female; for all of you are one in Christ Jesus" (Gal. 3.28).

THE CAPTIVITY LETTERS

A cell in the Roman-period prison of Philippi, believed to be on the same site as the cell in which Paul was detained, according to Acts 16.23–24.

Four of Paul's letters, Philippians, Philemon, Colossians, and Ephesians, are known as the "Captivity Letters," because in each case the author portrays himself as being imprisoned (Phil. 1.7, 12.14; Philem. vv. 1, 10; Col. 4.3; Eph. 3.1). The letters are related in other ways: Colossians and Ephesians, in particular, are connected both by their ideas and wording (see pp.148–9). They mention names that also occur in the letter to Philemon (see box on opposite page); these include Epaphras (Philem. v. 23; Col. 4.12), who may have founded the church at Colossae. The letter to Philippi lacks such close ties with the other three, except for the mention of Timothy in its opening greeting (compare Col. 1.1).

Much debate has centered on the place of composition, especially in the case of the letter to the Philippians. It is generally agreed that Paul was the author of the letter. He visited Philippi, a leading city of the Roman colony of Macedonia, on his second missionary journey (Acts 16), and the church he established there was his first foundation on the mainland of Europe. Some have suggested that the letter may have been sent from Rome or Caesarea, both places in which Paul was imprisoned. If the references in the letter to the "imperial guard," or Roman praetorium (Phil. 1.13), and to "those in the emperor's household" are taken to mean the imperial palace and Caesar's entourage, they would point to Rome as the point of origin. However, these references could equally apply to the official residence of a provincial governor and thus be applicable to Caesarea, the seat of the procurator of Judea.

Another school of thought claims that the letter was written from Ephesus, where Paul could have been imprisoned during his two-year stay (Acts 19.10). In the letter, Paul writes that he hopes to send his companion Timothy to the Philippians, "so that I may be cheered by news of you" (Phil. 2.19). Such a return journey by Timothy would have been difficult if he had had to travel from as far away as Rome or even Caesarea. If Paul were imprisoned at Ephesus, however, a return journey to Philippi would have been much more straightforward: Paul had traveled the same route himself on his third missionary journey (see p.233). Timothy was evidently closely associated with Ephesus, overseeing the Church's mission there according to 1 Timothy 1.3. Also, the Philippian church heard of Paul's imprisonment and felt that it was worthwhile to send one of its own members, Epaphroditus, as an aide to the apostle. Epaphroditus became ill and news of his illness reached Philippi, because Epaphroditus subsequently heard that his colleagues were worried for him (Phil. 2.25–26). The passage seems to imply that traffic between Philippi and the place of Paul's imprisonment was fairly regular and straightforward: there is no mention of the kind of travelers' perils faced by Paul in the account of his journey to Rome in Acts.

The warm and intimate tone of the letter to the Philippians contrasts with that of the correspondence to Galatia and Corinth (see pp.243–5). Paul speaks of his prayers for the Philippians (Phil. 1.3–11), and assures them that he will return to Philippi after his release from prison (Phil. 1.19–36). He thanks the Philippians for the financial support that they, alone, had given him in the early days of his ministry (Phil. 4.22), and expresses his appreciation of Epaphroditus (Phil. 2.25–30). Although he has to warn the church at Philippi against "evil workers" who insist on circumcision (Phil. 3.2–3), the general picture in the letter is of a harmonious and active community. Uniquely in Paul's indisputably genuine letters, "bishops" and "deacons" are addressed as the church leaders (Phil. 1.1), pointing to an already clearly organized ecclesiastical structure, at least in

Philemon and slavery

The letter to Philemon is addressed not only to Philemon but also to the church that meets in his house and to Apphia and Archippus. However, the entire content of the letter concerns Philemon and his slave, Onesimus.

Onesimus had run away from, and perhaps also robbed, his master. He had been put in prison, where he had met Paul and been converted to Christ. Paul had become deeply attached to him: he writes of how Onesimus had become "useful" (Philem. v. 11), playing on the slave's name, which means "useful" or "beneficial" in Greek. However, Paul believed strongly that all Christians should be good citizens and respect the legal and social order. This is probably why he decided to return Onesimus, bearing this letter to his master Philemon, who apparently lived at Colossae in Asia Minor: Paul sends greetings (v. 23) from five fellow prisoners who are also mentioned at the end of the letter to the Colossians.

Roman law prescribed the harshest penalties for runaway slaves, sometimes even death. But Paul does not simply ask for clemency or seek to tell Philemon his duty. Rather, he appeals to the common faith shared by both master and slave, and asks that Onesimus be received back as a brother "both in the flesh and in the Lord" (v. 16). Paul promises to repay Philemon for whatever Onesimus may have stolen, and seems to hint that Philemon might even consider returning the slave to him to continue to help him in prison (vv. 13–14).

Slavery was a widespread and accepted feature of the Roman world, but in the first century CE attitudes were changing. The emancipation of slaves, which gave them full civil rights and the possibility of Roman citizenship, occurred on a considerable scale, and the increasing egalitarianism preached by philosophers, such as the Stoics, affected the way in which the legal institution of slavery came to be regarded. As the letter to Philemon reflects, the early Church accepted the institution of slavery and taught that slaves should be obedient to their masters and perform their duties continuously (see also Eph. 6.5–6; Col. 3.22). But masters were to treat their slaves fairly and to give them their freedom if the opportunity arose (for example, Col. 4.1; 1 Tim. 6.1–2; Titus 2.9–10). Christian authors denounced the more brutal aspects of slavery, such as the slave trade (1 Tim. 1.10). Christianity also preached a much more radical doctrine which was to lead to the repudiation of slavery. For Paul, "in Christ" there was no distinction between slave and free (Gal. 3.28). Jesus had himself adopted the "form of a slave" (Phil. 2.7) and this was to be the pattern for Christians. Paul describes himself as a servant and prisoner of Christ (Rom. 1.1; Phil. 1.1) and proclaims that any slave who received his call to be a Christian would be free in the Lord, and, equally, a free person is a slave in the service of Christ.

Top: *A 1st-century CE Greek inscription recording the foundation of a synagogue, possibly the one for freedmen (liberated slaves) mentioned in Acts 6.9.*

Bottom: *A 2nd-century Roman bronze and lead vase portraying a Syrian slave.*

Philippi. Philippians 2.6–11, which is the most famous passage in the letter, contains an early Christian hymn, which Paul adopts and modifies in order to demonstrate how the imitation of Christ is the model for the life of the believer. The hymn is important as showing how rapidly the doctrine of the person of Christ developed in the early church. Jesus is seen as pre-existent, "in the form of God" (Phil. 2.6), but in spite of his divinity, he chose to take "the form of a slave" and to be born in human form (Phil. 2.7). The hymn appears to combine the Jewish concept of divine Wisdom (see pp.126–7) with the figure of the Suffering Servant in the book of Isaiah (see p.113), whom Christians understood as prefiguring the suffering Christ.

ETHICAL LISTS

In Martin Luther's German translation of the Bible, two sections have the heading *Haustafeln*, household tables or codes (Col. 3.18–4.1; Eph. 5.21–6.9). Very similar passages occur elsewhere in the New Testament, particularly in 1 Peter, and all seem to follow a broadly common pattern. Known as "ethical lists," they address believers in pairs according to their position in the household – wives and husbands, children and fathers, slaves and masters – and set out the mutual duties called for within each relationship. They are derived from catalogs of such duties in both Classical and Jewish sources, but the motivating clauses attached to them are distinctively Christian. In writings that may have been produced considerably later than Paul's lifetime, such as the Pastoral Letters, there are lists dealing specifically with positions in the Church, whether officers (bishops, elders, and deacons) or Church members (the old, the young, and widows).

Alongside the household and community catalogs, there are also lists of vices to be shunned and virtues to the embraced. These, too, reflect a widespread Greco-Roman ethical background, but again they are infused by the particular Christian concepts of love and faith, as in Colossians 3.14, which bids: "Above all, clothe yourself with love, which binds everything together in perfect harmony."

COLOSSIANS AND EPHESIANS

The letters to the churches of Colossae and Ephesus (see box, below) are closely related: the latter appears to incorporate about a quarter of the contents of the former, and both are perhaps the most disputed documents in the New Testament. Scholars differ sharply as to whether they were written by Paul or by a later author or authors who used and developed genuine Pauline concepts.

Colossians, at least, can claim strong grounds for Pauline authorship. The city of Colossae in Asia Minor had not been evangelized by the apostle himself but, according to the letter, by his fellow worker Epaphras (Col. 1.7–8). This probably occurred during the time of Paul's ministry in nearby Ephesus (Acts 19; see pp.232–3), probably ca. 52–55 CE. Several other individuals mentioned in Paul's letters are associated with the place, including Philemon (see p.247). Colossians has the regular form of one of Paul's indisputably genuine letters and contains authentic details such as a mention of the neighboring towns of Laodicea and Hierapolis.

The purpose of the letter was to alert the Colossians to the danger of false teachers, who sought to captivate them "through philosophy and empty deceit, according to human tradition, according to the elemental spirits of the universe, and not according to Christ" (Col. 2.8). This teaching was probably a kind of Jewish mysticism such as affected other Pauline churches. One verse appears to accuse the author's opponents of angel-worship (Col. 2.18), but this may mean that they claimed to have a visionary experience of angels in their worship. There is a precedent for this in the so-called "Angelic Liturgy" of the community at Qumran (see p.145). The earthly worshipers there viewed themselves as sharing in the worship conducted in heaven by the angels.

The letter to the Ephesians

The Pauline authorship of Ephesians is highly debatable. Regular epistolary features are lacking and the letter does not appear to address the problems in a particular congregation, in the way that Paul's other correspondence does. The words "in Ephesus" in the opening verse are not found in some of the most important early manuscripts of the New Testament, and this has led to the view that the document was really an encyclical, intended for circulation to a group of churches in Asia Minor, of which Ephesus was perhaps the center. It is the first of the seven churches of Asia Minor addressed in Revelation (see p.254). Also, the vocabulary is uncharacteristic: the letter contains some ninety words that are not found in Paul's other writings.

The 1st-century CE odeon *(small theater) at Ephesus in western Asia Minor (modern Efes, Turkey).*

In some respects, Ephesians seems to reflect a situation later than that of Paul's time. The conflict of Jew and gentile, which so much exercised Paul, is long past: the very "secret" of the gospel is now the union of Jew and gentile in the one body of the Church.

Ephesians carries even further the teaching found in Colossians 1 about the cosmic Jesus (Eph. 1.20–23; see main text on opposite page). The Ephesian letter contains reminiscences of almost every one of Paul's letters, but it seems to be looking back on them, and reads like a digest of the great themes of his preaching. Probably, then, the work should be viewed as a preservation and interpretation of Paul's ideas, created to meet the needs of the church when the living witness of the first apostles had passed away.

To counteract this teaching, and in order to reject the divinity of any figure but Jesus, the writer of the letter presents a picture of the cosmic Christ, especially in Colossians 1.15–20. These verses represent a hymn, originally in praise of Wisdom and probably Jewish, which is reinterpreted to refer to Jesus. Jesus is described as "the firstborn of all creation" (Col. 1.15) and depicted as the head of the universal Church (Col. 1.18). It has been argued that the language and theology of this passage differ significantly from that of Paul's authentic letters – for instance, elsewhere Paul only employs the term "church" to refer to a local congregation.

Another aspect of Colossians that has raised doubts about its Pauline authorship is its teaching about the end of the world and the universal resurrection of the dead. Elsewhere in Paul's writings, this final consummation of Jesus' victory is regarded as still to come, but Colossians appears to take the view that resurrection is already realized for the believer (Col. 3.1–4). It is possible that a later writer is attempting to cope with the problem of the delay in the second coming of Christ, whose imminent return had been expected in the Church's early days.

A host of angels (see main text on opposite page) hovers above the head of Christ in The Crucifixion *by Giotto (ca. 1266–1337), one of his frescoes for the Scrovegni chapel, Padua, ca. 1305.*

JEWISH CHRISTIANITY

The letter to the Hebrews and the letter of James are both, in different ways, closely linked to Jewish traditions. The distinctive characteristic of Hebrews is that it seeks throughout to show that Jesus is the true fulfillment of Judaism's history and faith. The title "to the Hebrews" was not part of the original work and the letter's author is not known; it does not claim to be by Paul. The writer was clearly very learned, trained in Hellenistic rhetorical methods and in using the Septuagint, the Greek translation of the Hebrew Scriptures.

The work is both a profound theological treatise and a genuine letter, addressed to a particular community and its problems. However, just what this community was and where it was situated remains the subject of much discussion. Because the letter sends greetings from "those from Italy" (Heb. 13.24), and because its earliest attestation is at Rome around 90 CE, it has often been suggested that it was sent to a Jewish Christian group in Rome. Attention has also been drawn to the similarities between Hebrews and the literature that came from the Greek-speaking Jewish community at Alexandria in Egypt, such as the Wisdom of Solomon and the work of the philosopher Philo (see p.132). More recently, a study of the texts from Qumran (see p.145) has led some experts to suggest that Hebrews is addressed to members of the Church in Palestine who had converted to Christianity from a strand of Judaism that stood outside the mainstream, such as the Qumran community and the Essenes (see p.145). If this theory is correct, these converts had probably not adequately understood the real nature of Christianity and were in danger of returning to their old beliefs.

Angels were central to the thinking of the Qumran community and it was widely believed that angels would usher in the end of the world (see, for example, Revelation). The recipients of Hebrews appear to have shared a similar belief. Hence, the author of Hebrews begins by demonstrating that Jesus is superior to all angels, and that God "did not subject the coming world" to them (Heb. 2.5). He then proceeds to argue that Jesus is also greater than Moses. Above all, Jesus transcends the whole

The Martyrdom of James, *by Lorenzo Monaco (ca. 1370–ca. 1422). According to the historian Josephus, the brother of Jesus was martyred in ca. 62 CE on the initiative of the high priest Ananus, who took advantage of the gap between the death of one Roman procurator, Festus, and the arrival of another, Albinus. It is probable that otherwise only the Romans had the authority to impose capital punishment (see p.198).*

Melchizedek

An important feature of the letter to the Hebrews is the role of Melchizedek, whose significance, says the author, is to be understood only by those well instructed in the faith. The story of Melchizedek, the priest king of Salem who blessed Abraham (Gen. 14.18–20; see p.38), is said to demonstrate the superiority of Melchizedek's office, because "it is beyond dispute that the inferior is blessed by the superior" (Heb. 7.7). The letter also claims that Jesus is "a priest forever, according to the order of Melchizedek" (Heb. 7.17, quoting Ps. 110). This means that Jesus is superior to Abraham and also to the Levitical priesthood, for "Levi himself was still in his ancestor's loins when Melchizedek met him" (Heb. 7.9–10).

The writer of Hebrews describes Melchizedek in terms that go well beyond Genesis: he has "neither beginning of days nor end of life, but resembling the son of God, he remains a priest forever" (Heb. 7.3). This language appears to reflect a strand of Jewish apocalyptic writing. For example, a fragmentary text discovered at Qumran portrays Melchizedek as a heavenly savior and an angelic being, immortal and nonhuman. He is referred to as *elohim*, the usual Hebrew word for "God," but here meaning one of the members of the divine heavenly assembly (see sidebar on p.27). In Hebrews, however, the word "resembling" shows that Melchizedek is similar to but not the equal of Jesus, whose superiority to all angels has been established earlier in the letter (see main text).

Abraham Presents Prisoners and Booty to Melchizedek *(Gen. 14.20), from the Psalter of St. Louis (French, ca. 1270).*

Jewish priestly and sacrificial system. He alone is the "great high priest" (Heb. 4.14), and only his sacrifice, not the sacrifices of Judaism, can truly make an atonement for human sins. Faith in Jesus and in his saving work is the way to salvation. The significance of the heroic figures of Israel is that they lived by faith, looking to the promises now fully realized in Christ. The letter's recipients are exhorted always to look to Jesus; this will both sustain them under persecution in this life and give them the perfect reward of their faith in the hereafter.

The letter of James is traditionally attributed to James, one of the brothers of Jesus, and the leader of the early Jewish Christian community in Jerusalem (Acts 15; see pp.220–21). It may date from around the early forties CE. The attribution is not impossible, but many scholars think that the work was written by someone else and dates from the end of the first century CE. It is addressed to "the twelve tribes," a reference to the traditional tribes making up the nation of Israel, and so, like Hebrews, it is probably directed at Jewish converts to Christianity. Although the style contains many characteristics reminiscent of the Hebrew or Aramaic languages, the letter is written in good Greek, and has many points of contact with literary and philosophical forms current in the Hellenistic world. For example, it contains what is called *paranesis*, exhortation of an ethical, practical nature, which was common in philosophical treatises and in Jewish writings of the Diaspora. The letter is also a diatribe, a short ethical treatise of the sort that was typical of the Cynic school of philosophers.

The author's prime concern is the unity and harmony of a community that is evidently under a degree of stress. The letter emphasizes the positive value of the testing of faith and endurance, and the danger to the wellbeing of the community that arises from gossip, backbiting, and envy. Rich people who seek to behave as if they were superior to their poorer fellows are roundly condemned.

FAITH AND WORKS

In James 2.14–26, the author apparently puts good works before faith, claiming: "faith by itself, if it has no works, is dead" (James 2.17). This passage, which led Martin Luther to tear the letter to James from every Bible he came across, has often been seen as a direct attack on Paul's central doctrine of "justification by faith," whereby people are saved by their faith, not by their deeds. But it is possible that the author of James was using the terms "works" and "faith" in a different way from Paul. In James, "works" could mean acts of charity, rather than Paul's sense of strict (but not necessarily sincere) adherence to the letter of the law. Similarly, "faith" in James may simply mean belief in the existence of God, rather than the total commitment to and trust in God that is Paul's understanding. Viewed in this light, the issue in James becomes less theological than practical: essentially, the author is discussing the merits of hospitality in a Christian, as at James 2.15–16.

St. Peter Baptizing New Converts, *a fresco of ca. 1427 by Tommaso Masaccio (1401–1428).*

PETER, JUDE, AND JOHN

The last five of the New Testament epistles are of uncertain authorship, but are attributed to well-known figures of the early Church: the apostle Peter; the apostle John, son of Zebedee (see box on opposite page); and Jude, the brother of Jesus. The first of the two letters of Peter is written to scattered communities over a wide area of Asia Minor. They are addressed as "the exiles of the Dispersion" (1 Pet. 1.1), which suggests that they were mainly Jewish Christians living outside Palestine. The letter is clearly an encyclical (a circular for wide distribution), written from the city of Rome, which goes by the name "Babylon," as in Revelation (see pp.254–6). If it really is by Peter, it provides evidence for his missionary activity, of which little is heard elsewhere in the New Testament (see pp.228–9). It would also lend support to the very early tradition that he was martyred during an otherwise unrecorded visit to Rome. However, some have objected that the literary Greek of the letter is too sophisticated for a man who had begun his career as a Galilean fisherman. But this fact does not necessarily disprove Peter's authorship, because his style could have been corrected by the scribe Silvanus, or Silas, who actually wrote the letter down (1 Pet. 5.12). Besides, Peter had long since left his first vocation behind by the time of the letter's composition.

The letter is written to encourage Christians who evidently feel isolated in a hostile environment, where they are "aliens and exiles" (1 Pet. 2.11). They have suffered persecution from their neighbors and the civic authorities, who "are surprised that you no longer join them in the same excesses of dissipation" (1 Pet. 4.4). The letter mentions "the fiery ordeal that is taking place among you" (1 Pet. 4.12), the choice of words perhaps indicating that the persecution was localized. However, if this refers to a more severe and general attack by the Roman state, the author could mean the persecution instigated by the emperor Nero in 64 CE, in which Peter himself is traditionally said to have died. It has also been suggested that the author lived much later than Peter at the time of the persecution of 112 CE in Asia Minor under the emperor Trajan (98–117 CE).

In view of their trials, Peter exhorts his readers to lead lives that give no cause for offense, to bind together and love one another, even in the face of further attacks (1 Pet. 2.11–12; 4.7–11). Their calling is to follow Jesus, who is presented as the figure of the Suffering Servant in the book of Isaiah (see p.113). Peter assures them that their afflictions will not endure for long, for Christ will soon return (1 Pet. 1.5–6; 4.7; 5.10).

It is widely accepted that 2 Peter is not by the apostle but is one of many writings produced within the early Church and deliberately attributed to one of its famous founders. It reflects a situation that existed only after Peter's death, when Christianity faced the problems of moving from a Jewish to a predominantly gentile environment. There is no general consensus as to its date, although 2 Peter 3.4 says that the false teachers who deny Christ were first condemned "long ago": this implies that the letter was written after the first generation of Christians, ca. 80–90 CE or later. The reference to Paul's letters as "scriptures" (2 Pet. 3.16) also suggests that 2 Peter was written at least after Paul's death (which took place perhaps ca. 60–65 CE), and therefore probably after Peter's death as well.

The work opens with a standard form of address used in letters of the period (see p.238). It also takes the form of a farewell testament, a genre that was common in Jewish literature. Typically, these "testaments" claimed to be the final thoughts of biblical figures. In the style of such testaments, 2 Peter purports to have been written at a time when Peter saw his death as imminent (2 Pet. 1.14). It includes a summary of his teaching

(2 Pet. 1.3–11), as well as a final message, here bitterly denouncing the work of false teachers (2 Pet. 2.1–3.7). These teachers seem to have been people who wished to rid Christianity of certain features, particularly the apocalyptic hope of Jesus' return (2 Pet. 3.3–4). The author explains that the delay in the second coming appears lengthy only from a human perspective, whereas "with the Lord one day is like a thousand years, and a thousand years are like one day" (2 Pet. 3.8). God, he says, is providing time for sinners to repent (2 Pet. 3.9).

One reason for viewing 2 Peter as a comparatively late work is the use it makes of the letter of Jude (for example, compare 2 Pet. 2.1–18 and Jude vv. 4–13). The writer of Jude describes himself (v. 1) as the "brother of James," generally taken to mean James the brother of Jesus. Whether or not the author really was another of Jesus' brothers, Jude almost certainly has a Jewish Christian and Palestinian origin. It is not a real letter but a general tract, devoted entirely to combating the false teaching which it believes has infected the Christian community. This teaching is not very clearly described, but it involves licentious behavior (Jude vv. 4, 7) and disrespect toward angels (v. 8). The polemic against false teaching is based on the old Scriptures, and on the Hebrew rather than the Greek Septuagint. Jude sees biblical episodes in which God punished his opponents (vv. 5, 7, and 11) as prophesies of the fate awaiting the heretics of his own day. He is familiar with apocalyptic Jewish literature from Palestine, and quotes from one such work, the first book of Enoch (vv. 14–15). He also refers to a legend about the death of Moses that may once have formed the ending of a work known as *The Testament of Moses* (v. 9).

The letters of John

The three letters traditionally attributed to the author of the Fourth Gospel contain no real clues as to their authorship and background. Apart from its traditional title, 1 John is anonymous, and the writer of the other letters describes himself only as "the elder." All three have shared themes and vocabulary, and they have been credited to John because of their many resemblances to John's gospel. The opening of 1 John recalls the prologue to the gospel, and significant terms in the gospel, such as "life," "light," and "love," are prominent in the letters, especially 1 John. However, there are also important differences, and authorship should perhaps be ascribed to a "circle of John" rather than to a single author.

The first letter of John is much longer than the other two and deals with those who deny that Jesus "has come in the flesh" (1 John 4.2). This appears to anticipate the second-century CE Docetic heresy, which held that Christ's body was only apparent, not real: 1 John strongly asserts Jesus' full and real humanity and pleads for the church to be a harmonious body, bound by mutual love. Both themes are also emphasized in 2 John, which is addressed to a church personified as "the elect lady." It is linked to 3 John by their common author, "the elder," although the third letter is written to an individual called Gaius. He is asked to provide the elder's envoys with hospitality, refused by Diotrephes, the leader of the congregation to which they had been sent.

St. John the Evangelist, *by Pedro Berruguete (Spanish, ca. 1450–1504). The evangelist, to whom the letters of John are traditionally ascribed, is accompanied here by his traditional symbol, an eagle.*

THE SEVEN CHURCHES
Revelation begins with individual letters to seven congregations in the Roman province of Asia, opposite the island of Patmos (see map, above). Apart from the great church of Ephesus, founded by Paul, only here in the New Testament are we given any details of Christian life in these centers. They were all important administrative centers and they are listed in a circle, from south to north and back again. However, the number seven is a symbol of perfection and completeness (see main text, right) that is meant to represent the entire Church and, although each congregation has its own letter, this is also intended for all "the churches."

The author shows a good knowledge of the setting and traditions of each particular city, and he has a specific message for every Christian congregation. Three of the churches are warned against false teachers, described as Nicolaitans. Little is said of their vices, but they are linked with Balaam and Jezebel (see p.67), and accused of advocating sexual license and eating food sacrificed to idols, problems which also faced Paul. The reference to Pergamum as "where Satan is enthroned" may be a warning against emperor worship: Pergamum was the first center of such worship in Asia Minor. But the general picture is of reasonably peaceful and well-established groups which are, however, in danger of complacency and succumbing to the temptations of their religious and social environment.

REVELATION

The Revelation to John, also called the book of Revelation and the Apocalypse, has always proved to be a difficult book to interpret, and it has been interpreted in many ways over the centuries. A modern reader may be puzzled by the apparently bizarre symbolism, and even repelled by the intense hatred and bloodthirstiness which it displays. Many people have seen the work as an authoritative prediction of the future, and have endeavored to deduce from it a date for the second coming of Christ and the end of the world, and have sought to identify the evil beings that it depicts with various historical figures, such as Napoleon or Hitler.

Biblical scholars now generally try to clarify Revelation with reference to the particular historical background against which it was written, and see it in the context of a long literary tradition. The book is, in fact, a letter, intended to be read in the churches to which it is addressed (see sidebar, left). It gives an account of what it claims is a revelation or vision of the imminent future, which God has given the author. In Greek, "revelation" is *apokalypsis* and the work is an example of the genre called "Apocalyptic," which was common in Judaism at around the beginning of the Common Era. It is found elsewhere in the Bible, for example in the book of Daniel (see pp.134–7). Revelation is, however, distinctively Christian, and the only instance of an entire apocalypse in the New Testament, although apocalyptic passages occur in several other writings there (see, for example, box on p.191).

The author names himself as "John," and says that he received his revelation on the island of Patmos, where he claims to have been imprisoned "on account of the word of God and the testimony of Jesus." Patmos is known to have been used by the Romans as a penal colony. John (Hebrew *Y[eh]ohanan*, Greek *Iohannes*) was a common name, and the author of Revelation is possibly best thought of as an early Christian prophet. However, nearly every apocalypse claimed some great figure from the past as its author, and Revelation may deliberately want the reader or listener to think that it was the work of the apostle John, son of Zebedee. In fact, since at least the second century CE, the "John" of Revelation has traditionally been identified with the apostle, who is also supposed to have written the Fourth Gospel and the letters of John (see box on p.253). Quite apart from the problem of authorship in the case of the letters, it is difficult to see the same hand in all these works, although there are certain links between them.

The language of Revelation is unique. The author can write good Greek, but his style is often ungrammatical and includes many turns of phrase derived from Hebrew or Aramaic, probably with the object of aligning the style of his book with the older Scriptures. Revelation is carefully constructed and pervaded by a symbolic numerology, which was common both in Jewish writings and in the Greco-Roman world. The basic numbers are three, four, and seven, all of which signify completeness and perfection, as does twelve (three times four) which represents the twelve tribes of Israel and the twelve apostles of Jesus. In Revelation, groups of seven feature many times: seven letters to scholars (Rev. 1–3), seven seals (Rev. 5.1), seven visions, seven bowls of wrath (Rev. 16.1), as well as seven angels and seven trumpets (Rev. 8.2). Three disasters ("woes": Rev. 9–16) are followed by the final deliverance (Rev. 19–22) to make up the number four. There are four living creatures around God's throne (Rev. 4.6) and four destructive horsemen (Rev. 6.2–8), figures derived from the four winds. The "New Jerusalem,"the perfect messianic city, has twelve gates for the twelve tribes (Rev. 21.12) and twelve foundations for the twelve apostles (Rev. 12.14), and the number twelve is also the basis of

Rome in the New Testament

The attitude of the book of Revelation to the Roman state contrasts in certain respects with that found elsewhere in the New Testament. In spite of the arguments of some scholars, Jesus does not seem to have aligned himself with groups such as the Zealots who advocated armed resistance to Rome: he advocated the paying of taxes to Caesar and the co-existence with the Empire of God's kingdom, because it was not of this world. The enigmatic instruction of Jesus in Luke to his followers to buy a sword may suggest that some Christians participated in the great Jewish revolt of 66–74 CE, but it is clear that most of them did not. Acts consistently brings out the comparatively friendly attitude of Roman officials toward Christians in the face of Jewish hostility. For Paul, the authorities are God's agents for the maintenance of justice and order, and the general teaching of the New Testament letters is that Christians should be loyal, law-abiding citizens, and pray for the welfare of their rulers.

A coin of the emperor Vespasian issued after the fall of Jerusalem to his son Titus in 70 CE. It shows a female slave and bears the Latin words IUDAEA CAPTA – "Captive Judea."

At the same time, Christians believed that the second coming of Christ would mean the end of all earthly authority and this view finds extreme expression in Revelation. The author views Rome exclusively as the embodiment of emperor worship, to accept which is to deny the sovereignty of God: the "beast" which symbolized the empire is the agent of Satan. Christians owe no loyalty to Rome and soon only respond with endurance and suffering: they are not to fight against it, for God's action alone will soon excuse its fall. Here the author may also be expressing the outlook of marginalized groups and people in the empire for whom the Roman state appeared not as a civilizing authority but as an instrument of oppression.

its dimensions: twelve thousand furlongs and one hundred forty-four (twelve times twelve) cubits (Rev. 21.16–17). There are twenty-four elders (Rev. 1–3) and one hundred forty-four thousand of the elect, twelve thousand from each of the tribes of Israel (Rev. 4–8).

The famous "number of the beast," six hundred sixty-six (Rev. 13.18), is an example of Jewish numerology. Hebrew letters also function as numbers, and the number 666 is reached by adding up the Hebrew letters which transliterate *Kaisar Neron* – the Greek for "Emperor Nero" (see sidebar on p.256). The number six falls short of the perfect number, seven: the Roman emperor might claim divine status, but he is in fact no more than an imperfect human being (see also box, above).

The rich and all-pervading symbolism of Revelation is an amalgam of three sources: the Bible, Jewish apocalyptic literature, and the mythology of the ancient Near East. Particularly important is the author's use of Scripture. There are allusions to and reminiscences of almost every book of the Hebrew Scriptures on every page, but the imagery from Daniel (see pp.134–7) and Ezekiel (see p.114) is particularly powerful. For example, the description of Jesus in the opening vision (Rev. 1.7; 1.13–16) is a combination of features of Daniel's "Son of Man" and the divine figure in Ezekiel's first vision. The author reinterprets, for his own purposes, many great biblical themes, such as the Creation and the Exodus from Egypt, and he also draws on Isaiah's understanding of the return of the exiles from Babylon as a new Exodus and new Creation. Rome becomes Babylon (Rev. 14.8 and later), a symbol of evil political power in the Bible (see p.35), and it is predicted that, like Babylon, Rome will fall. In contrast, Jerusalem, the holy city, becomes a heavenly city and the center of the new age (Rev. 21.1–2).

Although all the symbols of Revelation are found in the old Scriptures, their presentation echoes the way in which they were elaborated in Jewish

THE DATE OF REVELATION

The early Church believed that Revelation was written during the reign of the Roman emperor Domitian (81–96 CE). Domitian certainly insisted on the divinity of the imperial line, and emperor worship, the supreme object of attack in Revelation, was universal throughout Asia Minor in his time. However, it is very doubtful if Domitian, as is often thought, persecuted Christians who refused to worship him, at least on any large scale. The attacks on believers mentioned in the letters to the Seven Churches (see sidebar on opposite page) are attributed to Jewish hostility, not the Romans (see, for example, 2.9).

The description of Nero, "the beast," reflects a widespread legend that he had not really died (in 68 CE), but was about to return with a great army from the East (see Rev. 13.3). Often the author seems to present future events in terms of the past: the depiction of the gentile attack on Jerusalem seems to refer to the Jewish War of 66–74 CE. The list of seven kings, of whom five have fallen (Rev. 17.9–10), may refer to the chaotic conditions in the Roman empire directly after Nero's death (there were five emperors in the space of a year). Hence the reign of Domitian and conditions in Asia Minor in his time remain a likely background for Revelation.

literature around the beginning of the Common Era. The central role played by angels reflects the developed angelology of such texts and the author knows the myth of the fallen angels, which is prominent in the traditions of Enoch. His picture of worship in heaven (Rev. 8.3–5) recalls similar concepts found in the Dead Sea Scrolls from Qumran (see p.145). It is possible that some of the details are modelled on the Christian liturgy of his day, which he may have viewed as patterned on the angels' worship. Images such as the Lamb (Rev. 5.6), the beast from the sea (Rev. 13), and the dragon (Rev. 12) also derive from Jewish traditions. Behind them lie the even older traditions of ancient mythology. The dragon is the chaos monster of Mesopotamian myths (see pp.22–3; pp.26–27), identified in Revelation with Satan (Rev. 12.9). The episode of the woman clothed with astral symbols whose child is threatened by the dragon (Rev. 17) is based on a myth about a mother goddess, and this concept may underlie the contrasting depictions of Babylon the harlot (Rev. 17), and Jerusalem the bride (Rev. 21). The author is also influenced by the Greco-Roman world, especially by its astrology and cosmogony. It has been suggested that the work is structured on the form of the Greek drama, which would have been performed in the theaters of the cities to which it was addressed.

A bronze bust of the emperor Nero, the "beast" of Revelation (see below).

THE NUMBER OF THE BEAST

In Revelation 13.18, the author says: "Let anyone with understanding calculate the number of the beast, for it is the number of a person. Its number is six hundred sixty-six." In Hebrew, letters are also used as numbers; the figure 666 is reached by adding up the numerical values of the Hebrew letters *Qsr Nrwn*, which transliterate the Greek *Kaisar Neron* ("Emperor Nero"):

Q	=	*100*
S	=	*60*
R	=	*200*
N	=	*50*
R	=	*200*
W	=	*6*
N	=	*50*
		666

The Apocalypse (Judgment Day), *a 15th-century Russian icon (Novgorod School).*

A BOOK-BY-BOOK SUMMARY OF THE BIBLE

THE HEBREW SCRIPTURES or OLD TESTAMENT

The books of the Hebrew Scriptures are summarized below in the order generally followed in Christian Bibles.

Genesis

The title of the first book, Genesis, comes from Greek and means "origin," a good indication of its character: it describes the origins of the world and of the Israelite nation. God creates the universe in six days (see pp.26–7). In the Garden of Eden, the first humans commit the Original Sin, and are cast out into the world (see pp.26–7). Their firstborn son, Cain, murders his shepherd brother, Abel (see pp.30–31). Generations later, Noah and his family are deemed by God to be the only humans worthy of being spared in a universal deluge. Noah's family and the animals he has been instructed to save repopulate the earth.

After attempting to build the Tower of Babel, the people of the world are scattered (see pp.34–5), and Genesis moves on to the story of the ancestors of Israel (see pp.36–51). The first of them, Abraham, leaves Mesopotamia for Canaan. God promises Abraham and his descendants, the Israelites, the land of Canaan as a homeland in perpetuity (see p.37). Abraham's faith is tested in the so-called "Sacrifice of Isaac" (see p.43). Isaac marries Rebekah, and the family prospers in hard times. The younger of their two sons, Jacob, tricks his brother Esau out of his inheritance and flees to the land of his relatives (see pp.44–5). Jacob dreams of a ladder leading to heaven (see p.45). He first marries Leah and then Rachel, and, by them and their two female servants, fathers twelve sons, who become the founders of the twelve tribes of Israel (see p.47). Jacob himself is renamed "Israel" (see p.46).

Moses at the Burning Bush (Exod. 3.2).

Jacob-Israel's eleventh and favorite son, Joseph, is sold by his jealous brothers and ends up in Egypt, where he rises to prominence by interpreting Pharaoh's dreams (see pp.48–9). During a famine, Joseph's brothers arrive to buy grain and do not recognize Joseph at first. Eventually they are reconciled, and the brothers settle down to a prosperous new life in Egypt.

Exodus

The title is from the Greek *exodos*, meaning "departure," and the book relates the events surrounding the departure of the Israelites from Egypt and the years of wandering that followed, during which God handed down most of their laws.

Many years after the death of Joseph and his brothers, their descendants have become no better than slaves in Egypt. Pharaoh is alarmed at their numbers and orders the death of all newborn males. One Israelite couple has a son, Moses, who is abandoned on the Nile in a reed basket. However, Pharaoh's daughter rescues Moses and adopts him (see p.52).

Moses is forced to flee after killing an Egyptian, and following a dramatic encounter with God (Yahweh) at the Burning Bush (see pp.54–5), he becomes the leader of his people. God gives him and his brother, Aaron, miraculous powers with which to convince Pharaoh to let the Israelites leave Egypt for the Promised Land of Canaan. Pharaoh is obdurate in the face of nine plagues imposed on his land, and only releases the Israelites after the tenth plague, the death of the firstborn of each Egyptian family. (see pp.56–7). This last plague sees the institution of the Israelite festival of Passover (see also p.58).

The liberated Israelites head for the wilderness that lies between Egypt and the Promised Land, but they are pursued by Pharaoh's troops. The waters miraculously part, enabling them to cross the sea, but Pharaoh's pursuing troops are swept away (see pp.58–9).

In the wilderness, God gives the Israelites manna and quails to eat, and water springs from a rock for them to drink. Joshua becomes the Israelite military commander.

Moses goes up to meet God on Mount Sinai, where he receives the Ten Commandments or Decalogue (see p.60) and a series of other laws. He comes down to find the people feasting around an idol, the Golden Calf. Enraged, he breaks the tablets on which the divine Laws are inscribed. Some Israelite leaders are purged. Moses once more goes up Mount Sinai, and two more tablets of Law are divinely inscribed, renewing the Covenant. Subsequently, the construction, furnishings, and consecration of the Tabernacle – God's sanctuary – are described in detail, as are the vestments of its priests (see pp.60–65).

Leviticus

The book of Leviticus is primarily concerned with the laws given by God to Moses on a range of ritual and other religious matters (the title of the book means "of the Levites," the Israelite priestly tribe). These laws cover burnt offerings and sacrifices (see pp.122–3); the investiture and duties of priests; ritual cleanness; the commemoration of the Day of Atonement (Yom Kippur; see p.119); and the preservation of holiness in everyday life and in the feasts of the liturgical calendar.

Numbers

God orders a census of the twelve tribes of Israel in Sinai: the census is taken and the results are given at length (hence the name of this book). Further divine laws are promulgated.

The tribal leaders under Moses make ritual offerings, and the Levites are formally consecrated as a priestly clan. The decision is taken to leave Sinai and cross the desert wildernesses in the direction of the Promised Land (see pp.66–7). The Israelites set out, following the "Cloud of Yahweh." Soon, however, there are rumblings of discontent among the people, who are dissatisfied with their lot. Despite reports from scouts sent out by Moses that the land of Canaan is "flowing with milk and honey," there is outright rebellion against Moses and God. Yahweh is angered, and Moses only manages to appease him after God makes it clear that some of the senior Israelites, including Moses himself, will not reach the Promised Land.

A second serious rebellion among the Israelites is divinely punished by earthquake and plague, which kill thousands. Trouble occurs yet again when the Israelites stop for a time at Kadesh-barnea, where they have to withstand military attacks from the local Canaanites. Some of the Israelites begin to worship local idols, and God sends a plague that kills twenty-four thousand Israelites.

The military commander, Joshua, is appointed Moses' deputy. The Israelites invade the land of Midian (see p.53), and most of its inhabitants are slaughtered. More divine laws are promulgated.

Deuteronomy

Previous Israelite history and ritual observance is recapitulated and elaborated, hence the title of the book, which means "Second Law" (see p.107).

At the end of the book, Moses issues orders to the Israelites to prepare to enter the Promised Land, and makes a battle speech enjoining them to have faith in Yahweh. On Mount Nebo, Moses blesses the twelve Israelite tribes, and then dies, within sight of the Promised Land (see p.68).

Joshua

Following Moses' death, God confirms Joshua's command over his chosen people, and preparations are made to enter the Promised Land. Spies are sent into the city of Jericho, and although their presence is suspected, they are kept hidden at the house of Rahab the prostitute. After the spies have returned to Joshua, the Israelites cross the Jordan into Canaan (see pp.68–9). Jericho is the first city to fall to the Israelites, and all its inhabitants, apart from Rahab and her family, are put to the sword (see pp.70–71). Next to be taken is the city of Ai, followed by other cities and towns in Canaan. The whole territory is divided among the Israelite tribes, and the book concludes with Joshua leading the people in a reaffirmation at Shechem of the Covenant with God (see p.68). At the end of a long life, Joshua dies.

Judges

The opening chapters of Judges (see pp.72–3) present a picture of the early period of settlement in Canaan. Sometimes the Israelites fight and conquer the indigenous peoples, sometimes they live peacefully alongside them. Sometimes they abandon Yahweh and adopt the gods of their new neighbors, but each time this happens, God is angered and inflicts punishment on the Israelites.

Jephthah and his daughter (Judg. 11.30–40).

Eventually, however, God hears his people's pleas and sends a heroic deliverer, or "judge," to reverse Israel's misfortunes. The first two judges, Othniel and Ehud, are dealt with briefly. The third judge, Deborah, and her military commander, Barak, rout the forces of their enemy Sisera, who is killed by Jael, the wife of Heber the Kenite. Sisera's death evokes the celebrated Song of Deborah and Barak (see p.74).

A generation later, the judge Gideon (see p.74) similarly routs the Midianites, but afterward Israel relapses into idolatry. Another crisis is ended through Gideon's successor, Jephthah, who defeats the Ammonites. However, in his triumph Jephthah vows to sacrifice the first creature that he sees on his return home: unhappily for him it is his daughter, his only child.

The judge Samson (see p.75), is a more solitary hero. He takes on the Philistines, although he also marries one, Delilah. She discloses the secret of Samson's great strength – his long hair – to her own people, who seize Samson and blind him. He gains his revenge in one last act of strength, which causes the Philistine temple to collapse, killing himself and all his enemies.

A time now comes when there is neither judge nor king, and the solidarity of the Israelites weakens. Following a mass rape at Gibeah by members of the tribe of Benjamin, the Benjaminites are almost wiped out by the other tribes.

Ruth

The folktale-like story of Ruth is placed among the historical books between Judges and 1 Samuel in Christian Bibles, because of its thematic links to both: it is set in the time of the judges and tells the story of a great-great grandmother of King David, who features prominently in Samuel. In Christian belief, she is therefore an ancestor of Jesus. Ruth is also a moral work and in the Jewish Bible it usually appears after Proverbs.

Ruth is the Moabite widow of an Israelite husband. She accompanies her mother-in-law, Naomi, back to Naomi's home town, Bethlehem in Judah. There, Ruth comes under the protection of their kinsman, Boaz, who, after addressing himself to the legal and social formalities involved, marries her. Their son, Obed, is King David's grandfather (see p.124).

1 Samuel

The history of the Israelites resumes with the birth of Samuel, the last and greatest of the judges, to the previously barren Hannah (see pp.76–7). At an early age, Samuel is placed with the priest Eli for religious training, and it is soon evident that he has been called to serve God.

David and Goliath (1 Sam. 17.49).

When Samuel reaches adulthood, the Israelites are defeated by their coastal neighbors, the Philistines (see p.77), who capture the ark of the Covenant, the dwelling place of God. They take the ark to the temple of their god, Dagon, in the city of Ashdod. But a mysterious plague of cancers immediately strikes the citizens, and the ark is hastily moved to another city. But the plague breaks out there too, and also in a third city. Finally the Philistines send the ark back to the Israelites, complete with gold offerings of appeasement.

It is another twenty years before Samuel becomes the judge or leader of Israel. Many years after that the people decide that they would prefer to be ruled by a king. God, through Samuel, warns them of the disadvantages of monarchy, but they are adamant. God reluctantly consents to their request, and sends Samuel to select Saul, of the tribe of Benjamin, to be the first king of the Israelites (see pp.78–9). Even before he is officially made king, Saul proves his worth by leading a successful attack on the Ammonites.

Saul triumphs over a great number of Israelite enemies, including the Philistines. Nevertheless, there is friction between Samuel and Saul, which ultimately leads to the latter's downfall (see pp.80–81). Saul is mentally tormented, and he employs the shepherd boy David, son of Jesse, to be both his armor bearer and the harpist that soothes him in times of stress. Unknown to him, God has already instructed Samuel to anoint David as Saul's successor.

The Philistine armies return, and their champion, the giant Goliath, challenges the Israelites. Armed only with a slingshot, David defeats the giant and kills him. David becomes so popular that Saul is both jealous and afraid of him. David inflicts further defeats on the Philistines, and his success enrages Saul even more. Finally, Saul tries to kill David, who flees the court with the assistance of his great friend Jonathan, Saul's son. About two years later, Saul and Jonathan are both in battle against the Philistines on Mount Gilboa. Jonathan is killed and Saul, badly wounded, kills himself.

2 Samuel

David laments the deaths of Saul and Jonathan. He is anointed king of the southern tribe of Judah. Seven years later, after a war between Judah and the followers of Saul's dynasty, David becomes king of the united Israelite nation. He captures Jerusalem from the Jebusites, makes the city his capital, and installs in it the ark of the Covenant (see pp.82–4).

Various military victories follow and David acquires a small empire (see p.83). He commits adultery with Bathsheba, the wife of Uriah, one of his officers, and ensures that Uriah is killed in battle. He incurs divine anger for this behavior, and God's displeasure is conveyed by the prophet Nathan. God's punishment is that David's first child by Bathsheba dies; but she bears another, Solomon.

David's last years are marked by court intrigue and rebellions fomented by his many offspring (see p.84). At one stage, his son Absalom is in virtual control of the kingdom, but he is killed by David's generals, despite David's continued love for him. There are also times of famine, and times when David addresses psalms of praise to God. The book ends with David, following divine command, erecting an altar on the threshing floor of Araunah the Jebusite – the site of the future Temple of Jerusalem.

1 Kings

Before his death, David appoints Solomon his successor. After a succession struggle (see p.85), Solomon is anointed king. He effects multiple political marriages, notably with the daughter of the pharaoh of Egypt, and permits foreign faiths to be practiced, even in Jerusalem. But he also builds a magnificent Temple to God in the city (see pp.90–91). Its construction takes thirteen years before it is finally dedicated.

Solomon's wealth and wisdom (see pp.86–7) become renowned, and the queen of the prosperous Arab state of Sheba comes to visit him (see p.90). But Solomon also institutes forced labor, and there are rumblings of popular discontent. A revolt under Jeroboam is quelled only with difficulty before Solomon dies (see pp.92–3).

After Solomon's death, political and religious factionalism divide the kingdom, with the northern tribes breaking away from the southern tribal areas of Judah and Benjamin, the heartlands of David and Solomon's dynasty. Two kingdoms are established: Judah in the south, and Israel in the north (see p.95). Compiled and edited primarily by people with connections to Judah and its capital, Jerusalem, the biblical history is generally more favorable toward the southern kingdom.

During the reign of King Ahab of Israel, the prophet Elijah becomes a prominent figure (see pp.96–7). He foretells a drought and relates it to the ungodliness of the king of Israel. Ahab is a fine warrior who is married to a Phoenician, Jezebel, a devout worshiper of the god Baal. Elijah performs miracles both to reinforce his authority and to confound the rival prophets of Baal.

Ahab displays great military prowess against Israel's enemies. But God is angered by his deceitful appropriation of a vineyard through the murder, at Jezebel's instigation, of its owner, Naboth. Elijah predicts terrible deaths for both king and queen, which come to pass as foretold. Jezebel outlives Ahab, but when her death comes it is particularly grisly (see p.97).

2 Kings

After a miraculous crossing of the Jordan, the prophet Elijah is taken up to heaven, and his mantle passes to his servant Elisha, who recrosses the Jordan and proceeds to perform further miracles. Elisha accurately forewarns the king of Israel against military raids by enemies. Elisha can do nothing, however, to avert the famine caused by the Aramean siege of the capital of Israel, Samaria.

Elisha then becomes resident prophet at the courts of a succession of kings. One is Jehu, who is responsible for the assassination of surviving members of Ahab's dynasty, including his son, Jehoram, and his widow, Jezebel, together with all the priests and prophets that serve Jezebel's god Baal. After Elisha dies, according to the Bible, there is no prophet in the land for another sixty years or so.

The kingdom of Israel falls to the Assyrian king, Shalmaneser. The residents of the capital, Samaria, are deported and the city is repopulated with eastern foreigners (see pp.100–101). The Bible relates that, about a decade later, King Hezekiah of Judah (see pp.104–5) is also attacked by the Assyrians. The prophet Isaiah, who is active during Hezekiah's reign (see p.105) reassures the king that, for a period at least, Judah is safe, but also warns that a time is coming when the residents of Judah will be carried off into exile in Babylon.

After Hezekiah, the lengthy reign of King Manasseh is dominated by the shadow of the Assyrians, and the Bible accuses Manasseh of idolatry (see p.106). His successor, King Josiah (see pp.106–8) undertakes a program of religious reform, during which the "book of the law of Moses," possibly the core of what is now the book of Deuteronomy, is said to be "rediscovered" in the Temple (see p.107).

Josiah falls in battle against the Egyptians, who set up a puppet regime, which is overthrown by the Babylonians (see p.108). The deportation of the citizens of Jerusalem and the confiscation of most of the city's treasures follow. Within eleven years, however, the remaining people of Judah rebel against Babylon. Jerusalem is at once besieged, captured, pillaged, and destroyed. This time, the Bible reports, all of its people are taken away into captivity in Babylon (see pp.112–13).

1 Chronicles

The first and second books of Chronicles are largely based on the earlier historical books, but they are considered of value because of the way in which they re-present and reinterpret the events of Israel's history in the light of the author's contemporary world (Judah under Persian rule after the return of the Babylonian Exiles; see pp.114–17). The first third of 1 Chronicles gives genealogies from Adam to the time of the return from Exile. The remainder concentrates on the reign of David.

2 Chronicles

The last book in the Jewish Bible. The first nine chapters of 2 Chronicles recount the reign of Solomon. The remainder relates the history of the two kingdoms until the Exile and the beginning of the restoration of Judah under the Persian king Cyrus, who allows the Exiles to return to rebuild the Temple (see pp.114–17).

Ezra

The books of Ezra and Nehemiah deal with the history of Judah after the return from Exile. They were considered as one book until around 300 CE, when they were divided as they are now. Both men were probably active in the reign of Artaxerxes I of Persia (465–425 BCE; see pp.116–17). Much of Ezra and all of Nehemiah are recounted in the first person.

Artaxerxes commissions Ezra (Ezra 7).

Ezra opens with the decree of Cyrus, king of Persia, that those who wish to may leave Babylon and return to Jerusalem to rebuild the Temple. The returning exiles are listed.

Rebuilding work on the Temple begins. There is some opposition to the returned exiles, and work is halted temporarily. But, encouraged by the prophets Haggai and Zechariah, the Jewish authorities appeal to Cyrus's successor, Darius, whose archives confirm the original royal decree. The Temple is completed and rededicated.

Darius's successor, Artaxerxes, sends Ezra, a Jewish official of the Persian government, to Judah to ensure that the Jewish law is being strictly observed. Ezra finds that many of the formerly exiled families have been intermarrying with non-Jews, and puts a stop to the practice.

Nehemiah

Some eighty-five years after the return from Exile, Nehemiah, the Jewish cupbearer to the Persian king Artaxerxes, is moved to hear of the reputedly dilapidated state of Jerusalem. The king commissions him to go and supervise repairs. He surveys the damage and organizes those who are to be responsible for specific areas of work. However, Nehemiah encounters opposition from other Persian officials and there are delays in the rebuilding work in the city.

Other problems include social privations and various intrigues. But, finally, the city wall is once more in place, with gates, watchtowers, and a citadel.

Ezra is invited to rededicate the city by reading from the book of Moses. The occasion turns into a mass celebration of the Feast of Succoth (Tabernacles or Booths), followed, some days later, by a ceremony of atonement.

Nehemiah reports back to Artaxerxes. The next time Nehemiah returns to Jerusalem, he moves to counteract numerous irregularities in the operation and upkeep of the Temple.

Esther

Like Ezra and Nehemiah, the book of Esther is a story set in the time of the Persian empire (see pp.130–31). However, there is little to link it to authentic history apart from the name of the king in the story, "Ahasuerus," who could be any of several Persian rulers called Xerxes or Artaxerxes (see p.115).

Ahasuerus gives a grand banquet at his capital, Susa, but his queen, Vashti, refuses to attend. The king summarily puts her aside. The beautiful Esther, a young Jewish woman brought up by her uncle Mordecai, is chosen to replace her. But Mordecai makes an enemy of the king's chancellor, Haman, who determines to eliminate all Jews in the empire. He persuades the king to decree death to "a certain people" who keep their own laws.

Mordecai prompts Esther to go to the king and invite Ahasuerus and Haman to her own banquet. There, she reveals that she, the queen, is under threat of death from the king's own decree, and that it is Haman's doing. Horrified, the king rescinds his decree and orders Haman to be hanged on the very gallows he has set up for the Jews. Mordecai is promoted to chancellor, and the Jews are given license to take revenge on their enemies in the empire. In plotting to destroy the Jews, Haman had cast lots (*purim*); Esther's triumph is therefore celebrated in a new Jewish festival, called Purim.

Job

God allows Satan to test Job to see if he will remain unquestioningly faithful in the face of adversity. Job subsequently suffers various misfortunes, deprivations, and bereavements. Three of his friends arrive to offer sympathy, or at least intellectual discussion on the subject of divine justice. Job protests his innocence and rails against his fate. A fourth character, Elihu, appears, and

Job defends himself (Job 3–31).

attempts to vindicate his own view of God's mysterious ways. God himself intervenes to speak to Job. He tells him that for humans to discuss how God functions is presumptuous, since God is utterly beyond all mortal understanding. Job is humbled. God also reproaches the three friends and orders them to make a special sacrifice. Finally, God restores Job, bringing him even greater happiness and prosperity than he had enjoyed before (see pp.128–9).

Psalms

The Psalter or book of Psalms is a collection of varied types of liturgical works, such as hymns, supplications, praises, and songs of thanksgiving. They are addressed to God and expressed in poetic language, often of great majesty and beauty. Some are accompanied by instructions for musical accompaniment or indications as to their purpose. Many are attributed to King David (see p.121).

Proverbs

A collection of traditional sayings and popular maxims on such themes as correct behavior, godly thoughts, purity in mind and worship, the avoidance of sinners, and the quest for wisdom. This book includes a lengthy list of aphorisms attributed to King Solomon and his counsellors (see p.126).

Ecclesiastes

The author, who claims to be Solomon himself, discourses on the nature of existence in God's universe, and on eternal values. Everything about one's life on earth is "vanity and a chasing after wind." A world order exists, but we cannot know it; our only certainty is death. The conclusion to this searching, skeptical, and intensely personal work is that, while people can never know the secret of existence, human integrity makes our stay on earth tolerable. This allows us to enjoy to the full any benefits God may send.

King David singing Psalms (see the preambles to Pss. 3, 18, and others).

Song of Solomon

A poetic dialogue between a woman and her male lover – sometimes called the bride and the groom – expressed in heightened erotic language, and with interpolations by a chorus of "daughters of Jerusalem." It may or may not have been intended as a spiritual allegory (see p.129).

Isaiah

The book of Isaiah is in three distinct parts, although it purports to be the prophecies of one man, Isaiah, who was active in the reign of Hezekiah (see p.105, p.113, and p.116). The first part probably represents the work of Isaiah himself. In it, he warns that Judah will inevitably suffer because of its religious inadequacy. Jerusalem, now a den of vice and idolatry, will be destroyed, but later renewed through the visitation of Yahweh. He foretells the coming of a savior, and describes the awesome power of God. Although Judah is to be overwhelmed and its population deported, a remnant will in time return. In due course, Judah's enemies will suffer for their actions, as will the inhabitants of many of the regions around Judah, even as far away as Egypt.

The prophet attacks the wickedness of humanity and declares that the people of Zion will triumph. Salvation is in the hands of God.

The Assyrian king Sennacherib attacks Judah, but is driven off. King Hezekiah recovers from illness.

Isaiah 40–55 is a work of the Babylonian Exile, by an anonymous prophet who is referred to as "Second Isaiah." King Cyrus of Persia, he claims, will be God's instrument. There is a description of a figure called "the servant of the Lord" who, it is said, will rise up from insignificance among the people to bring justice to the nations. Israel will be freed, the land restored, and monotheism reinstituted: Yahweh will be God of all, judge of the world. There is to be a new covenant for a new Jerusalem, a new heaven and a new earth in which all worship Yahweh.

Chapters 56–66 are another distinct section, "Third Isaiah," dating in origin from the period just after the Exile. It depicts widespread religious and moral abuses, and promises divine retribution, so that the glory of God and his chosen people be shown in its true greatness.

Jeremiah

God calls Jeremiah, who lived in the period just before the Babylonian Exile (see p.109), to preach of the doom and destruction of Israel on account of the religious and moral impurity of its people. Even the priesthood is lax and corrupt. The Israelites are urged to repent and turn once more to God. Foreign invasion is regrettably inevitable: the Jews will suffer, their weak faith providing no security.

Jeremiah is threatened for his words but continues his diatribe. He lambasts the people and graphically depicts the horrors of war and deportation. Babylon is to be the instrument of Yahweh's wrath.

But all is not lost. There will come a "branch" of David – a member of the royal line of Judah – who will reign as true king. Israel and Judah will be restored and will once more worship God and revere their king. There will be a new covenant for a renewed Jerusalem, rebuilt in splendor.

Jeremiah is arrested for his outspoken views and is imprisoned until Jerusalem is besieged and falls to the Babylonians. Many of the city's inhabitants are deported, but Jeremiah is allowed to remain. However, following the assassination of the Babylonian governor of Judah, the prophet is taken off to Egypt by remnants of the former army of Judah "for his own protection." Jeremiah protests that all this means is that Egypt will in turn be invaded by Babylon. This, according to the Bible, is exactly what takes place.

Lamentations

Also called the Lamentations of Jeremiah, reflecting its traditional attribution to that prophet (see p.112). The author utters five carefully constructed dirges for those who have died in the fall of Jerusalem. The city mourns, the people wail in anguish, but there remains hope and trust in God.

Ezekiel

Before the fall of Jerusalem, Ezekiel (see p.112) is commissioned by God to speak out against the wickedness and idolatry of Israel. He lambasts false prophets and lists the sins committed by the nation throughout its history. Yahweh's sword is sharpened and polished, and the Babylonians will wield it. Many nations will suffer, even Egypt. Jerusalem falls, but in a vision Ezekiel sees the bones of Israel rise from the ground and acquire new flesh. The nation, he predicts, will be restored by Yahweh to its own land, and David's line will reign again. Ezekiel then relates his vision of the reconstructed Temple in Jerusalem. He describes rituals for festivals that are to be reinstituted, and appends some notions of the allocation of territory among the remnants of the twelve tribes of Israel.

Daniel

Daniel (see pp.134–7) is said to be one of four young Jewish boys taken in the sack of Jerusalem by King Nebuchadnezzar and brought up

Daniel interprets Nebuchadnezzar's dream of the great statue (Dan. 2.31–45).

up during the Exile at the royal court in Babylon. Through prayer, Daniel is able to reveal to the king the meaning of a dream about a great statue. Daniel and his friends gain prestige in Babylon. The king has a golden idol made and demands that all should worship it. But three of Daniel's friends (Shadrach, Meshach, and Abednego) refuse, and are thrown into a blazing furnace. To the king's amazement, they survive unharmed through divine intervention.

Nebuchadnezzar has another dream. Daniel interprets it as a warning that the king will lose his reason until he recognizes God. This prediction soon comes to pass.

Some time later, during a feast given by King Belshazzar, fiery writing mysteriously appears on the wall of the banqueting hall. Daniel interprets it to mean that the king will die and his empire will fall to the Persians and Medes. That night, Belshazzar is assassinated.

Belshazzar's successor, the Persian king Darius, is inveigled by his courtiers into issuing a decree that all prayer should be addressed to him. When Daniel refuses to pray to him, he is thrown into a pit of lions – but emerges safely, preserved by God's angel. The penitent king throws the courtiers to the lions instead.

Then it is Daniel's turn to dream: he has apocalyptic visions, most of which are messianic in content (the "Son of Man" is a recurrent expression). The visions probably reflect the turbulent events in Judah under the Seleucid empire in the second century BCE, the time when the book of Daniel is thought to have been put together.

Hosea

The prophet Hosea (see p.103) was probably active ca.740–730 BCE. He is told by God to take a prostitute as a wife. He is then asked to love a woman as God loves Israel. Hosea's experience with his wife or wives (whether one or two women are involved is open to discussion) is a symbol of God's relationship with his people. God's bride, Israel, has become a faithless whore, corrupt, occasionally professing insincere repentance, though more usually ruled by sinful desires (especially true of her kings, nobles, and priests). But after a display of his anger and jealousy, God will yet restore the one nation he loves above all others.

Joel

According to the book of the prophet Joel, God promises to rid Judah of a plague of locusts, and to restore the prosperity of the devastated land. The plague is symbolic of God's even more devastating judgment of the world on the last day. However, for those who repent, this will herald a great outpouring of divine blessing (see also p.217).

Amos

Amos (see p.102) was active under Jeroboam II of Israel (786–784 BCE), a prosperous time. The prophet denounces the evil committed by the peoples on every side of Judah and Israel. But he is also aware of the corruption, ostentatious luxury, faithlessness, fornication, and obstinacy of his own people. There cannot be salvation without repentance. The "day of the Lord" is coming, and dire punishment is on its way: but the glory of the house of David will one day be restored.

Obadiah

The book of Obadiah is the shortest work in the Hebrew Scriptures/Old Testament. The prophet is scathing about the Edomites for invading Judah after the destruction of Jerusalem by the Babylonians in 587/6 BCE. But the coming "day of the Lord" will turn the tables on Edom, when Israel is restored to its full might and glory.

Jonah

God instructs Jonah (see p.124) to go to Nineveh in Assyria and warn the inhabitants that their wickedness will lead to their destruction. But Jonah is reluctant to obey his call, flees to Joppa, and takes a ship in the opposite direction. A storm blows up, for which Jonah accepts blame, and, at his own request, is thrown overboard. Swallowed by a giant fish, Jonah is vomited ashore, and heads for Nineveh. There he preaches the imminent destruction of the city.

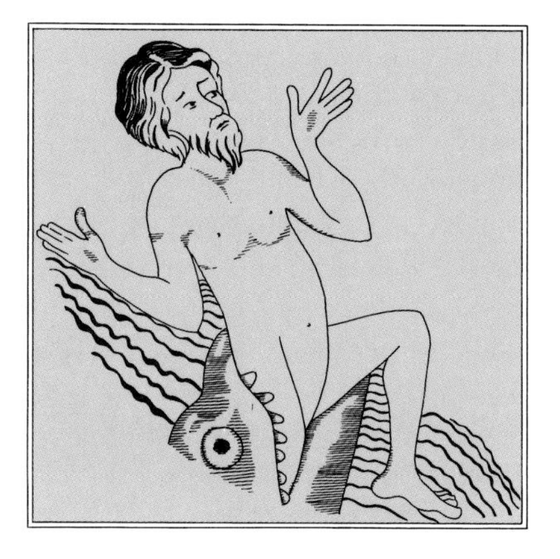

Jonah emerges from the fish (Jon.1.10).

The citizens repent and renounce their ways, and God does not destroy the city. Jonah indignantly demands to know of God what the point of all his efforts has been. Through a sign, Yahweh demonstrates that he is the God of all nations, who will show concern even for those who are not Israelites if they repent of their sins.

Micah

Micah (see p.103) prophesied in the last decades of the northern Israelite kingdom of Israel (ca. 735–725 BCE).

The book of Micah, which includes writing by, among others, an anonymous contemporary, forecasts destruction for both Israel and the southern kingdom, Judah. But afterward there will be liberation and restoration, led by a national savior who will be born, like King David, in Bethlehem in Judah. Present problems are the result of corruption and faithlessness, for which the people should seek God's forgiveness.

Nahum

The book of Nahum is the only prophetic writing that actually calls itself a book in the text; it may be a liturgy of the Jerusalem Temple (see p.101). Nahum is a brilliant poetic expression of joy at the prospect of the destruction of Nineveh, the capital of the hated Assyrian empire, which fell to the Babylonians and Medes in 612 BCE. Nineveh, a city steeped in blood and debauchery, will be utterly wiped out; Judah and Israel will be restored.

Habakkuk

Habakkuk complains to God for letting lawlessness go unpunished. He foresees invasion by the Babylonians (Chaldeans) as the instrument of divine retribution (he was probably active during the period of Babylonian expansion, ca. 610–600 BCE). In reply, God reassures the prophet of the coming judgment. The prophet then utters a public prayer expressing his trust and joy in God in the face of anguish.

Zephaniah

The Bible says that Zephaniah was a descendant of King Hezekiah of Judah who was active in the reign of King Josiah (640–609 BCE; see pp.106–8). The prophet foresees the approach of the "day of the Lord," on which Judah will be destroyed and Jerusalem sacked and pillaged. Even humble repentance for sin may not be enough to mitigate God's anger against every citizen. But in due course, Israel will be raised up again to triumph over her present enemies and oppressors. There will be shouts of joy and psalms of exultation in Zion. The exiles will return, and all those remaining alive on earth will be Yahweh's people. Zephaniah 1.14–15 inspired a famous medieval Latin poem called *Dies Irae*, which became part of the Latin Requiem Mass.

Haggai

Haggai and Zechariah prophesied in the period after the return from Exile in Babylon, in the early years of Darius (522–486 BCE). Haggai (see p.114) is appalled that people continue to live in fine houses while God's Temple still lies in ruins. Zerubbabel, the governor of the Persian province of Judah and a person of Judean royal descent, organizes a work party to start rebuilding. Haggai declares that God has chosen Zerubbabel for a position of greatness.

Zechariah

Zechariah (see p.114) receives several divine visions that together indicate God's intention to restore the Promised Land to his people, and to see the Temple rebuilt under the direction of Zerubbabel. Zion must rejoice in the renewed interest of God and listen to his words. Jerusalem will be great again, and God's people will be universally respected and admired. A savior-king is coming who will rule a kingdom of peace stretching to the ends of the earth.

Zechariah's final vision is apocalyptic, and features the final battle of Jerusalem. God is victorious: from that time, all nations worship him and do his will.

Malachi

Malachi, who was probably active ca. 475 BCE (after Haggai and Zechariah but before Ezra and Nehemiah), rages against the corrupt worldliness of the priesthood, whose teachings and customs are causing many to stumble from the path of righteousness. Moreover, idolatry and divorce are permitted. The "great and terrible day of the Lord" is coming when every sinner will be purified like molten gold and silver, or burned up like stubble. God declares that he will send a messenger before him to pave the way for his arrival: the great prophet Elijah will appear to turn people to God so that they may be saved. Christians have traditionally understood the book of Malachi as a work containing important prefigurations of the coming of Jesus. For this reason, the Old Testament places the prophetic books last, so that Malachi immediately precedes the New Testament.

THE APOCRYPHA/ DEUTEROCANONICAL BOOKS

Tobit

Tobit, a godly man, becomes blind. His son, Tobias, sets out for Media, to his kinsman Gabael's house, to collect some family savings. His guide and companion is the mysteriously omniscient Azariah. At the Tigris River, Azariah helps Tobias to catch a great fish, the innards of which, Azariah says, have powerful properties. Azariah reveals that they are to spend the first night of the journey at the house of another kinsman, Raguel, whose daughter Sarah is by right available for marriage to Tobias. But Tobias knows that Sarah is plagued by an evil spirit who has already killed seven suitors, and demurs. Azariah tells Tobias how to drive off the demon using the fish's heart and liver, and on arrival at Raguel's house, Tobias duly asks for Sarah's hand in marriage. He is accepted, and goes to

Tobias and Azariah (the angel Raphael) catch the fish in the Tigris (Tob. 6.4).

spend the night with her. Meanwhile, Raguel digs a grave outside for Tobias, whom he supposes must be the demon's next victim. The next morning, Azariah's charm has worked and the couple are formally betrothed. Azariah is sent on to collect the savings and to bring Gabael and his family back for the wedding. After a full fortnight of celebrations, Tobias, Sarah, and Azariah set out for Tobias's home.

Tobit and his wife Anna are overjoyed and relieved at their son's return. They are even happier when Tobias, on Azariah's instruction, uses the fish's gall to cure Tobit's blindness. At the ensuing feast, Tobit and Tobias wonder how they can possibly reward Azariah. But Azariah reveals that he is in fact the archangel Raphael, who has been both testing Tobit's faith and rewarding his virtue. He ascends to heaven as his hosts prostrate themselves in awe.

Tobit, Anna, Tobias, and Sarah live happily until the end of their days.

Judith

Probably composed during the oppressive rule of Antiochus Epiphanes (see pp.138–9), the book of Judith relates how King Nebuchadnezzar (here called the ruler of Assyria) defeats the Medes. He then sends his general, Holofernes, to punish the local Israelites for refusing to support him. Holofernes lays siege to the town of Bethulia. One of its citizens, the godly widow Judith, is determined not to give in. After fervent prayer she puts on her brightest clothes, leaves the town, and makes her way to Holofernes' camp. She offers to help him defeat her people, and proposes staying in the camp until this is achieved. Four nights later, alone in a tent with the drunken Holofernes, Judith decapitates him with his own sword. She runs back to Bethulia with his head and rouses the citizens, who attack and vanquish the Assyrians. All give thanks to God (see also p.131).

Additions to Esther

The Septuagint, the Greek version of the Jewish Bible, includes six passages in the book of Esther for which no Hebrew originals survive:

1. Mordecai dreams of coming events and warns the king of two treacherous eunuchs (inserted before Esther 1.1).
2. The text of the king's decree

Judith with Holofernes's head (Jud. 13.9).

eliminating the Jews in the empire is cited (inserted after 3.13).
3. Mordecai and Esther pray to God for deliverance (inserted after 4.17).
4. Esther approaches the king to seek his understanding (5.1–2).
5. The text of the king's edict annulling his former decree is cited (inserted after 8.12).
6. Mordecai realizes the meaning of his initial dream (inserted after 10.3).

Wisdom of Solomon

The author, who claims to be King Solomon, contemplates the nature of wisdom in relation to religious faith and to everyday life. He contrasts the earthly lives of the godless and the faithful, and their different destinies. The author understands God as the ultimate Wisdom, and reviews Israelite history in the light of that understanding (see p.126 and p.131).

Sirach (Ecclesiasticus)

A collection of sayings on a wide range of subjects. Wisdom, personified as a woman, is extolled. The author also expounds on the glory of God and ends with a hymn of thanksgiving (see p.127).

Baruch

Baruch, a follower of Jeremiah, is in Babylon with the Exiles. He prays to God, acknowledging the faithlessness and corruption that he assumes to be the reason for the Jews' present sufferings, and in repentance entreats God's mercy. God has chastised Israel, but will surely also restore the nation and rebuild Jerusalem.

Letter of Jeremiah

This text is usually included as the last chapter of Baruch and purports to be from the prophet Jeremiah. The Babylonian Exile, Jeremiah says, will be long but not permanent. The exiles should not pick up bad habits, such as idolatry. Foreign idols cannot tell right from wrong; they cannot even be relied upon to sanctify a vow or curse. They are simply inanimate objects and should be shunned.

Additions to Daniel

1. The Prayer of Azariah and the Song of the Three Jews (or **Three Young Men**)

Shadrach, Meshach, and Abednego, Daniel's three friends, address a hymn of praise and repentance to God as they sit in the fierce heat of the furnace to which they have been condemned (inserted after Dan. 3.32 in the Septuagint).

2. Susanna and the Elders

Two lecherous elders, thwarted in their desire to have intercourse with the virtuous Susanna, have her brought to trial for her life on false charges of adultery. She is sentenced to die, but Daniel intervenes and demonstrates that the elders' evidence is false. The elders are put to death (Dan. 13 in the Septuagint).

3. Bel and the Dragon (or **Serpent**)

King Cyrus is aghast at Daniel's scorn for the statue Bel, to which great quantities of food and drink are offered daily. Daniel shows the king that it is not the idol that consumes the offerings but the priests and their families, who sneak into the sanctuary by night through a secret door. The angry king has the priests executed. Daniel destroys the idol. He later annihilates a dragon worshiped by the Babylonians, for which the king is forced by the mob to throw him into a den of lions. But the lions do not harm Daniel, and he is freed by Nebuchadnezzar, who throws Daniel's enemies to the lions instead. They are devoured at once. (Dan. 14 in the Septuagint).

---

1 Maccabees

An account of the history of the Jews under Alexander the Great and his successors, especially the Seleucid king Antiochus IV Epiphanes (reigned 175–164 BCE). The book recounts his persecutions and the rebellions led by the Maccabees (see pp.138–41). Antiochus defiles the Temple of Jerusalem and persecutes the Jews. Two years later he issues a decree abolishing most of Judaism's distinct features. The old priest Mattathias takes up arms to defend the faith. Following his death, his son Judas Maccabeus takes over command and wins a succession of victories, after which the Temple is purified and reconsecrated, a ceremony celebrated annually in the Jewish festival of Hanukkah. Judas is killed in battle, after having concluded a treaty with the Romans. His brother Jonathan succeeds him, and is appointed high priest, but the Seleucids retake Jerusalem and the Temple, and Jonathan is killed. His brother Simon succeeds, regains Jerusalem, and inaugurates a brief period of prosperity before being treacherously assassinated. The book ends with the accession of John (Hyrcanus) to the leadership of the Maccabee dynasty (see pp.140–41).

---

2 Maccabees

Several episodes recounted in 1 Maccabees are embellished for moral and religious effect. The book's central theme is the eternal sanctity of the Temple and of the true faith. Steadfastness in faith is exemplified by gruesome martyrdoms. A reference to the resurrection of the dead (2 Macc. 12.43–45) influenced Christian theology.

---

3 Maccabees

The events of 3 Maccabees precede those of 1 Maccabees by about half a century. Ptolemy IV Philopator, the Greek king of Egypt, defeats Antiochus III of Syria in battle and then tours the temples of his domains. By a miracle, he is prevented from entering the Temple of Jerusalem, and, enraged, he orders a persecution of the Jews in Egypt. In Alexandria, the Jews are rounded up and five hundred elephants are maddened with wine and let loose on them. But another miracle takes place and the elephants trample the Egyptians instead. Ptolemy finally repents of his persecution and frees the Jews in his kingdom.

---

4 Maccabees

A more philosophical tone pervades this work. Its author sets out to show not only how total obedience to the Jewish Law corresponds to the cardinal virtues, but also how such obedience may represent the triumph of spirit over bodily sensation. Martyrdoms recounted in 2 Maccabees are held up as examples of how unwavering faith leads to salvation.

---

1 Esdras

An alternative rendering of eleven chapters of 2 Chronicles, the whole of the book of Ezra, and a chapter or so of Nehemiah. It also contains some additional material, in particular an ancient Zoroastrian parable about "the strongest thing in the world," which, after lengthy discussion, turns out to be Truth.

---

2 Esdras

A partly Jewish, partly Christian book that properly belongs to the genre called apocalyptic (see pp.136–7). The work discusses the resurrection of the dead and the Messiah who will appear in glory and reign for four hundred years, at the end of which he and all humanity will die. But within another seven days, the righteous dead of all times will be resurrected and welcomed into the new and glorious Jerusalem.

Prayer of Manasseh

A penitential psalm purporting to be a prayer addressed to God by King Manasseh of Judah (see p.106), in which he regrets his former apostasy and hopes for divine forgiveness.

Psalm 151

A psalm supposedly composed by David after his defeat of the giant, Goliath (see p.80). David extols himself as God's champion, the leader of a resurgent Israel.

THE NEW TESTAMENT

Matthew

The gospels of Matthew, Mark, and Luke are closely related and are referred to collectively as the Synoptic Gospels (see pp.148–51). Each gospel recounts the life, works, teachings, death, and resurrection of Jesus. Matthew is the work of a Jewish Christian who sees in Jesus the Messiah. Jesus' coming, Matthew claims, has been foretold in the Scriptures. Jesus represents the fulfillment of God's promises, the inauguration of a new Covenant, and the coming of the kingdom of heaven. Matthew opens with a genealogy of Jesus. Unique to his gospel are the stories of the visit of the Magi after Jesus' birth, the Massacre of the Innocents, the Flight

Jesus rises from the tomb (Matt. 28.6), as the guards (Matt. 27.66) sleep.

to Egypt (see pp.158–9), and the Sermon on the Mount (although Luke has a similar but much shorter passage; see pp.168–9). The evangelist includes frequent quotations from the Hebrew Bible.

Mark

The shortest and possibly the earliest of the gospels. Its author restricts himself to the bald facts as much as possible. Nonetheless, Jesus' major burdens – hostility from the authorities and incomprehension on the part of his disciples – are well depicted. Mark's account begins with Jesus' baptism and ends abruptly after the appearance of Jesus to the two Marys at the tomb (16.8). From 16.8 the gospel has been completed by a different author.

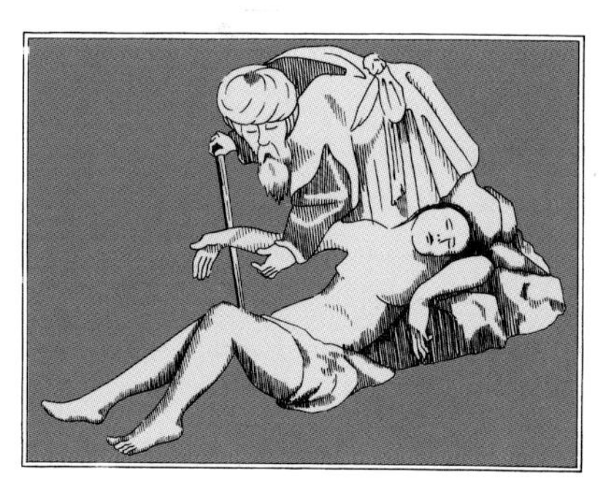

The Samaritan tends the injured man on the Jericho road (Luke 10.30–37).

Luke

Luke is a literary Hellenistic author, whose gospel is the first of two volumes, the second being Acts (see below). Together they recount the history of the faith from the birth of John the Baptist to the preaching of the apostle Paul in the capital of the Roman empire. Luke strives throughout to lend his account historical authenticity. His message is primarily addressed to gentiles, and he therefore tones down the specifically Jewish aspects of Jesus' life and teachings. A major theme is the compassion and forgiveness of God through Jesus, by means of which salvation is open to all.

Numerous episodes are unique to Luke, such as the annunciation of the birth of the Baptist (see pp.152–3); the presentation of Jesus in the Temple (see p.157); the story of Jesus among the Temple teachers (see p.161); and many of the best known parables, such as the stories of the Good Samaritan and the Prodigal Son (see p.174).

John

The most mystical of the evangelists, John – traditionally identified as the apostle John, son of Zebedee – seems to have drawn on a different tradition of Jesus' life and teachings than the Synoptics (see pp.148–51). The first part of John's gospel (chs. 2–12) revolves largely around six extended discourses that explain six miracles. The rest of the gospel is notable for a further series of speeches by Jesus (see pp.176–7). Unlike the other gospels, John portrays Jesus as someone who regularly visited Jerusalem for the important Jewish festivals; in the Synoptics, Jesus makes only one main visit, at the end of his life.

Acts of the Apostles

Luke the evangelist (see above) continues his story in his second volume, Acts. He describes how the apostles receive the Holy Spirit at Pentecost and at once begin preaching to the nations (see pp.216–17). Several of the apostles (although not all of the twelve are mentioned in Acts) perform miracles (see pp.218–19), and soon the authorities are trying to stop them. The first follower of Jesus to be martyred is Stephen, a prominent

Paul escapes from Damascus by being lowered down the city wall (Acts 11.25).

activist, who is arrested, condemned, and stoned to death. The young Saul/Paul, a Pharisee from Tarsus in Asia Minor, is present at the stoning and approves of it (see pp.222–3).

On his way to arrest Christians in Damascus, Saul/Paul experiences a vision that brings about a dramatic change of heart: he is baptized and begins preaching the gospel, becoming himself the target of persecution (see pp.226–7).

Congregations of followers of Jesus spring up outside Palestine, notably at Antioch in Syria, from where Paul and his colleague Barnabas embark on a missionary journey to Asia Minor. Paul makes further missionary journeys to Asia Minor and Greece (see pp.232–3). Before his second journey, he attends an important council of Church leaders at Jerusalem. The council agrees that gentile converts to Christianity in Syria and Asia Minor need not comply with strict Jewish requirements such as circumcision and the food laws (see p.229).

After his third journey, Paul returns to Jerusalem, where he is persecuted by Jewish authorities. As a Roman citizen, Paul appeals to the Roman authorities and eventually wins the right to put his case to the emperor (see pp.234–5). He sails for Rome (see pp.236–7), and Acts ends with Paul under mild house arrest in the imperial capital, preaching the gospel. Luke says nothing further about his appeal or ultimate fate (see pp.240–41).

Romans

In his letter to the Roman church, Paul says that faith in God's truth and justice leads to the salvation of all Christians, whether their background be Jewish or gentile. The new Israel is the successor to the old Israel, but is not by any means restricted to Israelites. Faith guarantees salvation – deliverance from sin and death. People must love one another, and love those who act against them. The secular law should be obeyed (see p.245).

1 Corinthians

Paul writes to the church at Corinth calling for unity in the face of factionalism. He addresses sexual morality, marriage, divorce, the Eucharist, and the importance of love (see p.243).

2 Corinthians

Probably a composite of several Pauline letters, 2 Corinthians expresses, among other things, Paul's frustration that his first letter has apparently had little effect (see p.243).

Galatians

Paul launches into this letter with an expression of astonishment that the Galatian church should be turning from the true gospel so soon. The Galatians should ignore those who say that Christians must follow the Jewish Law and be circumcised. Paul insists that Jewish traditions are not compulsory for Christians: faith in Jesus is more important than the works of the Law (see pp.244–5).

Ephesians

Paul is traditionally said to be the author of this letter, but this is doubtful. Its author states that believers are neither Jews nor gentiles: all are Christians, part of God's household. The author calls for mutual respect between owners and slaves, employers and laborers.

Philippians

The letter to the Philippian church is generally agreed to be by Paul. It is personal and affectionate in tone. He thanks God for the work of the Philippians in spreading the gospel, and prays that their love for each other continues to increase so that, in unity and humility, they may all work together for salvation. But he also warns them against "evil workers" who are insisting on the necessity of circumcision (see pp.246–7).

Colossians

The letter to Colossae is probably, but not certainly, by Paul. If he is the author, it was possibly sent to Colossae at the same time as the letter to Philemon (see below). Colossians portrays Jesus as a cosmic figure, "the firstborn of all creation," the supreme head of the universal Church. The letter enjoins its recipients to hold fast to Jesus, because, it says, Jesus has redeemed all people and forgiven them their sins. Paul calls upon the members of the Colossian church to abandon the desires and temptations of the flesh, which represent a person's past existence (see p.248).

1 Thessalonians

The first letter to the church at Thessalonica is often believed to be the first of Paul's letters. He thanks God for the faith shown by those he addresses, who he says have been an example to everyone in the face of adversity. He also calls upon them to refrain from fornication and sexual license, and to be upright, law-abiding citizens who are prepared to work for a living. He gives reassurance about the imminence of Jesus' second coming (see p.242).

2 Thessalonians

The authorship of the second letter to the Thessalonians has aroused much debate, with many rejecting Paul as the author. However, it picks up on themes covered in 1 Thessalonians, such as the second coming of Jesus. The second coming is being delayed, the author claims, by a "lawless one," who will occupy the Temple and claim to be God (see p.242).

1 Timothy

The first of the "Pastoral Letters," which deal with the running of the Church and care of the Christian faithful (see p.240). The authorship of this and the following two letters is disputed, unlike most of Paul's undoubtedly genuine letters. The author tells Timothy that Jesus came into the world to save sinners, and if Jesus could show mercy to the greatest sinner (Paul, who began as a persecutor of the faith), then he can do so to anybody who has faith in him.

Congregational leaders must manage people and be able to teach by example and through mutual respect. The author gives instructions about widows, elders, and slaves. In fighting for their faith, the faithful will win eternal life (see p.241).

2 Timothy

The second of the Pastoral Letters. The author, who is possibly not Paul, calls upon his correspondent, whom he once more addresses in very warm terms, to use the strength that comes with the grace of Jesus to bear witness to the gospel. As a teacher of the gospel, he should be prepared to endure suffering, as the author does, for if the faithful hold firm they will reign with Jesus. The author entreats Timothy to avoid the sort of specious debates and discussions that lead people astray. He should abide by what he has been taught and knows to be true (see p.241).

Titus

The third and last of the Pastoral Letters. Again, Paul's authorship is open to question. The author tells his correspondent to appoint church elders in Crete as he had formerly instructed. The elders must teach sound doctrine in order to counter the teachers of Jewish myths. He goes on to give examples of such sound doctrine and the behavior that goes with it.

The author gives instructions to live a life of decency; lack of decorum, the author writes, scarcely befits those who are awaiting the second coming of Jesus. His correspondent should insist on respectful behavior (see p.241).

Philemon

Paul wrote the letter to Philemon from prison, where he had met and converted Onesimus ("Useful"), Philemon's runaway slave. Paul sent Onesimus back home with the letter, which asks Philemon, a leader of the church at Colossae, to take Onesimus back (the letter may have been sent to Colossae at the same time as the letter to the Colossians). He urges Philemon to treat Onesimus not as a

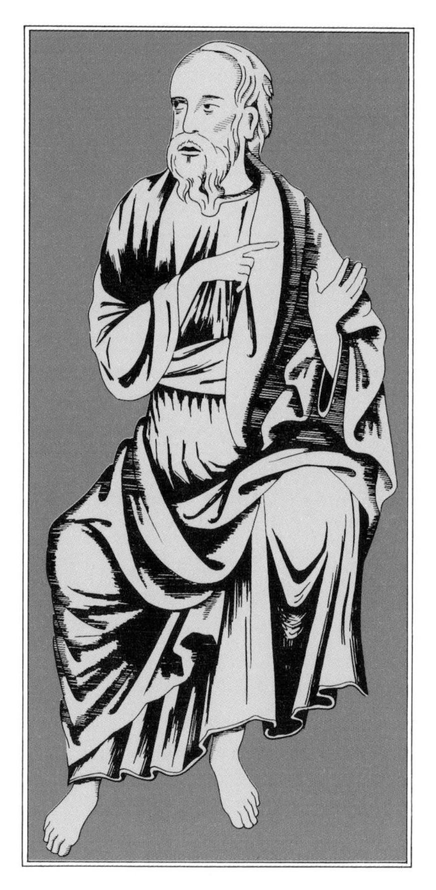

The apostle Peter, to whom the letters of Peter are attributed.

slave but as a new brother in Jesus.

Paul offers restitution for anything Onesimus may have stolen when he escaped, and perhaps hints that Philemon might free the slave to continue to work for the author in prison (see p.247).

The two-horned beast, and its worshipers on earth (Rev. 13.11–12).

Hebrews
The author of the letter to the Hebrews is unnamed. The letter shows how Jesus represents the fulfillment of God's promises to Israel. His sacrifice alone atones for the sins of humankind. Salvation comes through faith in him and his works (see pp.250–51).

James
This letter is traditionally said to be from James, the brother of Jesus. The author addresses a group of Jewish Christians, who are told to consider their trials as a privilege, and temptation as an opportunity to do right. They should assist the poor, especially if they themselves are comfortably off (see p.251).

1 Peter
The first of two letters traditionally attributed to the apostle Peter. The author offers encouragement to the faithful in the face of local hostility. They should love one another in total honesty and follow Jesus (see p.252).

2 Peter
This letter is generally accepted not to be by Peter. The author claims that he does not have long for this world. He denounces false teachers and says that the expectation of Jesus' second coming must never be abandoned.

1 John
The first of three letters traditionally attributed to the apostle John, son of Zebedee. The author denounces those who deny that Jesus appeared in the flesh, and calls upon those addressed to be bound together in mutual love (see p.253).

2 John
This brief letter is from an author calling himself "the elder." Believers must love one another and not be deceived by those who say that Jesus never came in the flesh (see p.253).

3 John
The author of 3 John also calls himself "the elder." It is addressed to "Gaius" and accuses the leader of Gaius's church of lack of hospitality and of spreading false accusations.

Jude
The author is said to be Jude, a brother of Jesus. He exhorts those addressed to keep the faith and beware of false teachers (see p.253).

Revelation (The Apocalypse)
Traditionally attributed to the apostle John, son of Zebedee, Revelation is in the form of a letter to seven churches in Asia Minor. The author sees an apocalyptic vision of a great sacrificial Lamb. The Lamb receives a scroll from God and breaks the seven seals on it one by one: the first four seals produce four apocalyptic horsemen, the next two raise the dead, and the last brings silence. Seven angels then blow their trumpets in sequence: the first six blasts cause destruction on earth; the seventh joins heaven together with what remains. A woman in childbirth appears: the child is taken to heaven. The angels under the archangel Michael defeat a seven-headed dragon. A terrible beast emerges. Its lesser companion (see picture, above) deceives the world into worshiping it.

The day of judgment breaks. Seven angels empty bowls of plague representing God's anger on to the world and over the beast. The cities of the world collapse and vanish. Babylon, the "Great Whore," burns and the world laments her passing. Songs of victory resound in heaven. The new age begins – after which Satan is released for a short time before his ultimate end. Then there is a new heaven and a new earth, and on the earth a new, messianic city of Jerusalem (see pp.254–6).

GLOSSARY

Apocrypha Writings that are not universally regarded as part of the canon of Scripture. These include Jewish Apocrypha, such as Judith, Tobit (see pp.130–31), and the books of the Maccabees (see p.141), which stand outside the Jewish canon and the canons of some Christian churches. In most cases they are works found in Greek in the Septuagint (see below) but for which no Hebrew original existed at the time of the formation of the canon, even if they were probably originally written in Hebrew, such as Sirach (or Ecclesiasticus; see p.127) and 1 Maccabees. In the Roman Catholic canon many of these writings are called Deuterocanonical, that is, they are acknowledged as distinct from the main body of Scripture but not as inferior in value. Christian Apocrypha include early Christian writings not incorporated into the New Testament, such as apocryphal gospels, acts, and letters (see p.149).

Aramaic The Semitic language of Aram (Syria). It is closely related to Hebrew (see below), which it had largely displaced as the everyday language at least of northern Palestine (see below) by the first century CE. It was almost certainly the language spoken by Jesus (see p.170) and his disciples. Aramaic and its dialect, Syriac, are still spoken in some rural areas of Syria. Syriac is the language of the Syrian Christian church (see also p.98 and p.214).

Canaanites In the Bible, the term can mean any of the pre-Israelite inhabitants of Palestine. More specifically, it often refers to the Phoenicians, the seafaring people of the coastal area that is now part of Syria, Lebanon, and Israel. Before the discovery of many Canaanite writings at the ancient site of Ugarit (modern Ras Shamra, Syria), the Bible was the chief source of knowledge about Canaanite beliefs and deities, such as Baal (see p.72).

Christ see **Messiah**

Confession A profession of faith, such as that made by the apostle Peter to Jesus (see p.188).

Deuterocanonical see **Apocrypha**

Deutero-Isaiah An alternative name for the biblical author known as Second Isaiah (Greek *deuteros*, "second"; see also p.113).

Diaspora, Dispersion The settlement of Jews outside their ancestral homeland of Palestine, a process that began with the Babylonian conquest of Jerusalem in the early sixth century BCE (see pp.110–13).

Eschatology A doctrine that looks for and anticipates the imminent end of the world. Adjective: eschatological.

Evangelist One who tells the "good news" (Greek *evangelion*) of the coming of Christ. The term is applied in particular to the authors of the four canonical gospels (see pp.148–51).

Hebrew The language of the Israelites. In the later centuries BCE, it was predomimantly displaced as the vernacular language of northern and coastal Palestine (see below) by its Semitic relative, Aramaic (see above), and by Greek. However, as the original language of the Jewish Scriptures, it remained in liturgical, priestly, and scholarly use. In the late nineteenth century, Hebrew was revived as a spoken language as part of the Zionist movement. Today, as the official language of the modern state of Israel, it serves to unify a Jewish population of many linguistic backgrounds.

Hellenism The widespread adoption of Greek language and culture in the Near East in the period following the conquests of Alexander the Great (356–323 BCE), when the region was largely ruled by the Greek dynasties that succeeded him. The so-called Hellenistic period was the era from ca. 300 BCE until the coming of the Roman empire (see pp.132–3; pp.138–9).

Messiah (Hebrew, "Anointed One.") In Jewish and Christian theology, the Messiah is the savior figure who was sent by God to herald the end of the present world and the beginning of the divine kingdom (see p.171). In Christian theology the Messiah is Jesus, who is therefore known as Christ, from *Christos* (or *Khristos*), a Greek title that translates the word Messiah. Adjective: messianic.

Mosaic Of or pertaining to Moses, as in "Mosaic Law."

New Testament or **New Covenant** The second part of the Christian Bible. It consists of writings which, in Christian belief, bear witness to the covenant (testament) made between God and humanity in the person, ministry, death, and resurrection of Jesus. The New Testament is said to represent God's final and supreme covenant with the world, supplementing and at the same time superseding the covenant of the Old Testament (see below). The books of the New Testament fall into three main categories: gospels (see pp.148–51); Acts of the Apostles (pp.214–37); and letters (the epistles and Revelation; pp.238–56).

Old Testament The first part of the Christian Bible, which, in Christian belief, bears witness to the covenant (testament) made between God and humanity before the coming of Jesus. The Old Testament consists of the books of the Jewish Bible, with or without the Apocrypha (see above), depending on tradition. There is some variation in the order of books between the Jewish Bible and the Old Testament (see also **New Testament).**

Palestine The region bounded roughly by the Jordan River to the west, the Negeb Desert to the south, and the Golan Heights of Syria to the north. The name means "land of the Philistines" and was originally given by the Greeks to the coastal area occupied by the Philistines (see p.77). "Palestine" and "Palestinian" are used in this book as purely geographical terms without political connotations.

Passion The suffering (Latin *passio*), both mental and physical, of Jesus in the last twenty-four hours of his life, from the agony in Gethsemane (see pp.196–7) to the Crucifixion (see pp.206–7).

Pauline Of or pertaining to the apostle Paul.

Pentateuch (Greek *Pentateuchos*, "Five-Part Work") The first five books of the Bible, comprising the Law (see pp.60–63; pp.118–19). It is also known as: the Torah (see below); the book of Moses; the book of the Law; and the book of the Law of Moses.

Postexilic Of the period after the return to Judah, in the late sixth century BCE, of the Jewish exiles in Babylon (see pp. 114–15).

Rabbinic, rabbinical Of or pertaining to the rabbis (Jewish teachers and interpreters of the Law) and their teachings.

Septuagint The Greek translation of the Hebrew Bible, so called because it was supposedly undertaken by seventy scholars (Latin *septuaginta*, "seventy") at Alexandria in Egypt (see p.132). The Septuagint, which contains writings regarded as non-canonical by Jews and as apocryphal or deuterocanonical by some Christian traditions, is the Old Testament of the Greek Orthodox church.

Synoptic Gospels The collective name given to the gospels of Matthew, Mark, and Luke (see pp.148–9).

Talmud Commentary on the second-century CE codification of the Mishnah, the "oral Torah" (see p.118). There are two Talmuds, one produced in Palestine, ca. 400 CE, and one in Babylonia, ca. 550 CE.

Torah (Hebrew, "Law.") The first five books of the Bible, also called **Pentateuch** (see above), the books containing the Jewish Law.

Zion, Mount The hill of Jerusalem upon which stand the City of David and the Temple Mount. By extension, Zion can also mean Jerusalem.

ABBREVIATIONS

The following abbreviations are used in this book:

General abbreviations

CE	Common Era (the equivalent of AD)		
BCE	Before the Common Era (the equivalent of BC)		
ca.	*circa*, about		
p.	page;	pp.	pages
ch.	chapter;	chs.	chapters
v.	verse;	vv.	verses
NRSV	New Revised Standard Version		

Quotations from the Bible

The books of the Bible are referred to according to the list of short forms and abbreviations set out below; the abbreviations are used only in parentheses within the text. Chapter and verse are separated by a period (.), and a sequence is indicated by a dash (–). Thus, 1 Samuel 9.24 (or, in brackets, 1 Sam. 9.24) = chapter 9, verse 24 of the first book of Samuel; Acts 3.7–10 = chapter 3, verses 7 through 10 of the Acts of the Apostles; Matthew 5–10 (or Matt. 5–10) = chapters 5 through 10 of the gospel according to Matthew. Very occasionally, subdivisions of verses are indicated by letters. Thus, Genesis 2.4a = the first part of verse 4 of Genesis chapter 2.

Short form	Abbreviation
Acts of the Apostles	Acts
Additions to Esther	Add. Esth.
Amos	Amos
Baruch	Bar.
Bel and the Dragon	Bel
1 Chronicles	1 Chron.
2 Chronicles	2 Chron.
Colossians	Col.
1 Corinthians	1 Cor.
2 Corinthians	2 Cor.
Daniel	Dan.
Deuteronomy	Deut.
Ecclesiastes	Eccles.
Ephesians	Eph.
1 Esdras	1 Esd.
2 Esdras	2 Esd.
Esther	Esther
Exodus	Exod.
Ezekiel	Ezek.
Ezra	Ezra
Galatians	Gal.
Habakkuk	Hab.
Haggai	Hag.
Hebrews	Heb.
Hosea	Hos.
Isaiah	Isa.
James	James
Jeremiah	Jer.
Job	Job
Joel	Joel
John *or* Fourth Gospel	John
1 John	1 John
2 John	2 John
3 John	3 John
Jonah	Jon.
Joshua	Josh.
Jude	Jude
Judges	Judg.
Judith	Jth.
1 Kings	1 Kings
2 Kings	2 Kings
Lamentations	Lam.
Letter of Jeremiah	Let. Jer.
Leviticus	Lev.
Luke	Luke
1 Maccabees	1 Macc.
2 Maccabees	2 Macc.
3 Maccabees	3 Macc.
4 Maccabees	4 Macc.
Malachi	Mal.
Mark	Mark
Matthew	Matt.
Micah	Mic.
Nahum	Nah.
Nehemiah	Neh.
Numbers	Num.
Obadiah	Obad.
1 Peter	1 Pet.
2 Peter	2 Pet.
Philippians	Phil.
Philemon	Philem.
Prayer of Azariah and Song of the Three Young Men (Jews)	Song of Thr.
Prayer of Manasseh	Pr. of Man.
Proverbs	Prov.
Psalm(s)	Ps(s).
Revelation	Rev.
Romans	Rom.
Ruth	Ruth
1 Samuel	1 Sam.
2 Samuel	2 Sam.
Sirach (Ecclesiasticus)	Sir.
Song of Solomon	Song of Sol.
Susanna	Sus.
1 Thessalonians	1 Thess.
2 Thessalonians	2 Thess.
1 Timothy	1 Tim.
2 Timothy	2 Tim.
Titus	Titus
Tobit	Tob.
Wisdom of Solomon	Wisd. of Solomon
Zechariah	Zech.
Zephaniah	Zeph.

FURTHER READING

Introductions

Alter, Robert, and Frank Kermode. *The Literary Guide to the Bible*. Cambridge, Mass.: Bellknap, 1987.

Anderson, Bernard W. *Understanding the Old Testament*. Englewood Cliffs, New Jersey: Prentice-Hall, 1985.

Bartlett, John R. *The Bible: Faith and Evidence*. London, England: British Museum, 1990.

Childs, Brevard S. *Introduction to the Old Testament as Scripture*. Philadelphia: Fortress, 1979.

——. *The New Testament as Canon: An Introduction*. Philadelphia: Fortress, 1985.

Clements, R. E., ed. *The World of Ancient Israel*. New York: Cambridge University, 1989.

Court, John M., and Kathleen M. Court. *The New Testament World*. Englewood Cliffs, New Jersey: Prentice-Hall, 1990.

Eissfeldt, Otto. *The Old Testament: An Introduction*. New York and Evanston: Harper and Row, 1965.

Koester, Helmut. *Introduction to the New Testament*. Philadelphia: Fortress, 1982.

Kümmel, Werner G. *Introduction to the New Testament*. Nashville, Tenn.: Abingdon, 1975.

Metzger, Bruce M. *The New Testament: Its Background, Growth, and Content*. Nashville, Tenn.: Abingdon, 1983.

Rendtorff, Rolf. *The Old Testament: An Introduction*. Philadelphia: Fortress, 1986.

van der Woude, ed. *The World of the Old Testament*. Grand Rapids, Mich.: Eerdmans, 1989.

Biblical Interpretation, Textual Criticism, and Translation

Ackroyd, P. R., C. F. Evans, G. W. H. Lampe, and S. Greenslade. *The Cambridge History of the Bible*. 3 vols. Cambridge, England: Cambridge University, 1988.

Anderson, G. W., ed. *Tradition and Interpretation*. Oxford, England: Clarendon, 1979.

Coggins, R. J., and J. L. Houlden. *A Dictionary of Biblical Interpretation*. Philadelphia: Trinity, 1990.

Epp, Eldon Jay, and George W. MacHae, eds. *The New Testament and its Modern Interpreters*. Atlanta: Scholars, 1989.

Kenyon, Frederic. *Our Bible and the Ancient Manuscripts*. New York: Harper and Brothers, 1958.

Knight, Douglas A., and Gene M. Tucker, eds. *The Old Testament and its Modern Interpreters*. Atlanta: Scholars, 1985.

Kraft, Robert A., and George W. E. Nickelsburg, eds. *Early Judaism and its Interpreters*. Philadelphia: Fortress, 1986.

Kugel, James, and Rowan A. Greer. *Early Biblical Interpretation*. Philadelphia: Westminster and John Knox, 1986.

Levenson, Jon D. *The Hebrew Bible, the Old Testament, and Historical Criticism*. Louisville, Ky.: Westminster and John Knox, 1993.

Metzger, Bruce M. *The Text of the New Testament: its Transmission, Corruption, and Restoration*. New York: Oxford University, 1992.

Morgan, Robert, and John Barton. *Biblical Interpretation*. New York: Oxford University, 1988.

Neill, Stephen, and N. T. Wright. *The Interpretation of the New Testament, 1861–1986*. New York: Oxford University, 1988.

Nida, E. A. *The Book of a Thousand Tongues*. Rev. ed. New York: American Bible Society, 1972.

Orlinsky, Harry M., and Robert G. Bratcher. *A History of Bible Translation and the North American Contribution*. Atlanta: Scholars, 1991.

Wurthwein, E. *The Text of the Old Testament*. Grand Rapids, Mich.: Eerdmans, 1979.

Literary Background

Ackroyd, P. R., A. R. C. Leaney, and J. W. Packer, gen. eds. *Cambridge Commentaries on Writings of the Jewish and Christian World 200BC to AD200*. New York: Cambridge University, 1985–.

Barrett, C. K., ed. *The New Testament Background: Selected Documents*. Rev. ed. San Francisco: Harper and Row, 1989.

Charlesworth, James H., ed. *The Old Testament Pseudepigrapha*. 2 vols. Garden City, New York: Doubleday, 1983, 1985.

Dalley, S. *Myths from Mesopotamia: Creation, the Flood, Gilgamesh, and Others*. New York: Oxford University, 1989.

Gibson, J. C. L., ed. *Canaanite Myths and Legends*. Edinburgh, Scotland: T. and T. Clark, 1978.

Gray, John. *Near Eastern Mythology*. London, England: Hamlyn, 1969.

Hennecke, E. *New Testament Apocrypha*. 2 vols. Nashville: Westminster and John Knox, 1991.

Leick, G. *A Dictionary of Ancient Near Eastern Mythology*. New York: Routledge, 1991.

Ogilvie, R. M. *The Romans and their Gods*. London, England: Hogarth, 1986.

Pritchard, James B., ed. *Ancient Near Eastern Texts Relating to the Old Testament; The Ancient Near East in Pictures Relating to the Old Testament*. Rev. ed. Princeton: Princeton University, 1970.

Ringgren, Helmer. *Religions of the Ancient Near East*. Philadelphia: Westminster, 1973.

Saggs, H. W. F. *The Greatness that was Babylon*. London, England: Sidgwick and Jackson, 1988.

Shafer, B., ed. *Religion in Ancient Egypt: Gods, Myths, and Personal Practice*. New York: Routledge, 1991.

Stone, Michael E., ed. *Jewish Writings of the Second Temple Period: Apocrypha, Pseudepigrapha, Qumran, Sectarian Writings, Philo, Josephus*. Philadelphia: Fortress, 1984.

Vermes, Geza. *The Dead Sea Scrolls in English*. Sheffield, England: JSOT, 1987.

Vernant, J. P. *Myth and Society in Ancient Greece*. New York: Zone Books, 1989.

Wiseman, D. J., ed. *Peoples of Old Testament Times*. Oxford, England: Clarendon, 1975.

History and Archaeology

Ben-Tor, Amnon, ed. *The Archaeology of Ancient Israel*. New Haven, Conn.: Yale University, 1991.

Garbini, Giovanni. *History and Ideology in Ancient Israel*. London, England: SCM Press, 1988.

Herrmann, Siegfried. *A History of Israel in Old Testament Times*. Philadelphia: Fortress, 1981.

Jagersma, Henk. *A History of Israel from Alexander the Great to Bar Kochba*. Philadelphia: Fortress, 1986.

Kenyon, Kathleen M. *The Bible and Recent Archaeology*. Rev. ed. London, England: British Museum, 1987.

Miller, J. Maxwell, and John H. Hayes. *A History of Ancient Israel and Judah*. Philadelphia: Westminster, 1986.

Schürer, Emil. *The History of the Jewish People in the Time of Jesus Christ*. Rev. ed. by Geza Vermes and Fergus Miller. Edinburgh, Scotland: T. and T. Clark, 1973–87.

Soggin, J. Alberto. *An Introduction to the History of Israel and Judah*, London, England: SCM Press, 1993.

de Vaux, Roland. *The Early History of Israel*. Philadelphia: Westminster, 1978.

Wilkinson, John. *The Jerusalem Jesus Knew: An Archaeological Guide to the*

Gospels. New York: Thomas Nelson, 1983.
Wright, G. Ernest. *Biblical Archaeology*. Philadelphia: Westminster, 1957.

Anthropology, Sociology, and Folklore

Gaster, Theodor H. *Myth, Legend, and Custom in the Old Testament*. New York and Evanston: Harper and Row, 1969.
Gottwald, N. K. *The Tribes of Yahweh: A Sociology of the Religion of Liberated Israel, 1250–1000BC*. Maryknoll, New York: Orbis, 1979.
Lang, B., ed. *Anthropological Approaches to the Old Testament*. Philadelphia: Fortress, 1985.
Lemche, N. P. *Early Israel: Anthropological and Historical Studies on the Israelite Society before the Monarchy*. Leiden: Brill, 1985.
Meeks, W. A. *The First Urban Christians: The Social World of the Apostle Paul*. New Haven, Conn.: Yale University, 1983.
Niditch, Susan. *Folklore and the Hebrew Bible*. Minneapolis, Minn.: Fortress, 1993.
Rogerson, J. W. *Anthropology and the Old Testament*. Sheffield, England: JSOT, 1984.
Stambaugh, John E., and David L. Balch. *The New Testament in its Social Environment*. Philadelphia: Westminster, 1986.
de Vaux, Roland. *Ancient Israel: Its Life and Institutions*. New York: McGraw-Hill, 1965.
Wilson, R. R. *Sociological Approaches to the Old Testament*. Philadelphia: Fortress, 1984.

Geography

Aharoni, Yohanan. *The Land of the Bible: A Historical Geography*. Philadelphia: Westminster, 1979.
Bahat, Dan. *The Illustrated Atlas of Jerusalem*. New York: Simon and Schuster, 1990.
Baly, Denis. *The Geography of the Bible*. Philadelphia: Fortress, 1987.
Brueggemann, W. *The Land: Place as Gift, Promise, and Challenge in Biblical Faith*. Philadelphia: Fortress, 1986.
May, Herbert G. *Oxford Bible Atlas*, 3rd ed. rev. by John Day. New York: Oxford University, 1984.

Religion, Theology, and Literature

Blenkinsopp, Joseph. *The Pentateuch: An Introduction to the First Five Books of the Bible*. Garden City, New York: Doubleday, 1992.
Coggins, R., A. Phillips, and M. Knibb. *Israel's Prophetic Tradition*. New York: Cambridge University, 1982.
Conzelmann, Hans. *An Outline of the Theology of the New Testament*. New York: Harper and Row, 1969.
Cross, Frank Moore. *Canaanite Myth and Hebrew Epic: Essays in the History of the Religion of Israel*. Cambridge, Mass.: Harvard University, 1973.
Dunn, J. G. D. *Unity and Diversity in the New Testament*. Philadelphia: Westminster, 1977.
Eichrodt, Walther. *Theology of the Old Testament*. 2 vols. Philadelphia: Westminster, 1961–7.
Engnell, Ivan. *A Rigid Scrutiny*. Nashville: Abingdon, 1969.
Golka, Friedemann W. *The Leopard's Spots: Biblical and African Wisdom in Proverbs*. Edinburgh, Scotland: T. and T. Clark, 1993.
Hanson, P. D., ed. *Visionaries and their Apocalypses*. Philadelphia: Fortress, 1983.
Hasel, G. F. *New Testament Theology: Basic Issues in the Current Debate*. Grand Rapids, Mich.: Eerdmans, 1978.
Hengel, Martin. *Judaism and Hellenism*. 2 vols. Philadelphia: Fortress, 1974.
Hooke, S. H., ed. *Myth, Ritual and Kingship*. Oxford, England: Clarendon, 1958.
Kraus, Hans-Joachim. *Worship in Israel: A Cultic History of the Old Testament*. Richmond, Va.: John Knox, 1966.
Mowinckel, S. *The Psalms in Israel's Worship*. 2 vols. Oxford, England: Blackwell, 1962.
Nickelsburg, George W. E. *Jewish Literature between the Bible and the Mishnah*. Philadelphia: Fortress, 1981.
Otzen, B., H. Gottlieb, and K. Jeppesen. *Myths in the Old Testament*. London, England: SCM Press, 1980.
Patrick, Dale. *Old Testament Law*. Atlanta: John Knox, 1985.
von Rad, Gerhard. *Old Testament Theology*. 2 vols. New York: Harper and Row, 1962–5.
Smith, Mark S. *The Early History of God: Yahweh and the Other Deities in Ancient Israel*. San Francisco: Harper and Row, 1990.
de Vaux, Roland. *Studies in Old Testament Sacrifice*. Cardiff, Wales: University of Wales, 1964.
Whybray, R. N. *The Intellectual Tradition in the Old Testament*. Berlin and New York: de Gruyter, 1974.

Reference

Achtemeier, Paul J., ed. *Harper's Bible Dictionary*. San Francisco: Harper and Row, 1985.
Botterweck, G. Johannes, and Helmer Ringgren, eds. *Theological Dictionary of the Old Testament*. Grand Rapids, Mich.: Eerdmans, 1977.
Buttrick, George A., and K. Crim, eds. *The Interpreter's Dictionary of the Bible*. 5 vols. Nashville, Tenn.: Abingdon, 1976.
Freedman, David Noel, *et al. The Anchor Bible Dictionary*. 6 vols. New York: Doubleday, 1992.
Kittel, Gerhard, and Gerhard Friedrich, eds. *Theological Dictionary of the New Testament*. Grand Rapids, Mich.: Eerdmans, 1985.
Metzger, Bruce M., and Michael D. Coogan. *The Oxford Companion to the Bible*. New York: Oxford University, 1993.
Peake, A. S. *Commentary on the Bible*. Edinburgh, Scotland: Thomas Nelson, 1962.
Roth, Cecil, ed. *Encyclopaedia Judaica*. New York: Macmillan, 1972.

INDEX

Page references to main text, boxed text, and sidebars are in roman, to captions in *italics*. References to the glossary (p.272) have the suffix g. Where reference material occurs in text and captions on the same page, the text reference is given preference.

A

D

E

I

J

K

L

M

Q

R

S

T

U

V

W

X

Y

Z

PICTURE CREDITS

The publishers wish to thank the following for their kind permission to reproduce the illustrations in this book:

KEY:

b below; c center; t top; l left; r right

BAL: Bridgeman Art Library
DBP: Duncan Baird Publishers
RHPL: Robert Harding Picture Library
HPL: Hutchison Picture Library
PEF: Palestine Exploration Fund
ZF: Zefa Picture Library
ZR: Zev Radovan, Jerusalem

Page 1 BAL/Scrovegni (Arena) Chapel, Padua, Italy; **2** BAL/Pierpont Morgan Library, New York; **3** BAL/Museo Basilica San Marco, Venice; **6** ZR; **7** BAL/British Library, London, Cott Nero CIV f.209v; **8** BAL/British Library, Add 14761; **9** BAL/British Library, Add 14761; **11** BAL/Lambeth Palace Library, London, Ms 209 f.3v; **12** BAL/National Gallery, London; **14** BAL/Bible Society, London; **15** BAL/Bible Society, London; **16** ZR; **17t** ZR; **17b** PEF; **18** ZR; **19** Sonia Halliday Photographs, England; **21** BAL/Museo Basilica San Marco, Venice; **22** ZR; **23** ZR; **24** BAL/British Library, Or 1404; **26** BAL/British Library, Roy 19 D III f.3; **27** BAL/British Library; **29** BAL/Index/Museo Diocesano de Solsona, Lerida, Spain; **30** British Museum; **31t** DBP; **31b** BAL/Giraudon/Valley of the Nobles, Luxor (Thebes), Egypt; **33l** BAL/Victoria & Albert Museum, London; **33b** Staatliche Museen, Berlin; **34** BAL/Private Collection; **35** British Museum; **37** RHPL; **39t** DBP; **39b** PEF; **40t** ZF; **40b** ZF; **41** BAL/British Museum, Add 54180 f.136b; **42** BAL/San Vitale, Ravenna, Italy; **43** e.t. Archive/Museum of Turkish & Islamic Arts, Istanbul; **44t** BAL/Österreichische Nationalbibliothek, Vienna; **44b** PEF; **45** BAL/British Library; **46** ZR; **47** PEF; **48** BAL/Fitzwilliam Museum, Cambridge, England; **49** BAL/Fitzwilliam Museum, Cambridge; **51** Sonia Halliday Photographs; **53l** BAL/Giraudon/ Valley of the Kings, Luxor (Thebes), Egypt; **53r** BAL/Louvre, Paris; **55** BAL/Giraudon/Musée Condé, Chantilly, France, Ms 9/1695 f.12; **57t** BAL/Fitzwilliam Museum, Cambridge; **57b** BAL/Victoria & Albert Museum; **58** BAL/Giraudon/Egyptian National Museum, Cairo; **60** BAL/Dahlem Staatliche Gemäldegalerie, Berlin; **61** ZF; **62** Sonia Halliday Photographs; **63t** BAL/Index/ Monastery of St. Catherine, Mount Sinai, Egypt; **63b** e.t. Archive/Louvre, Paris; **64t** e.t. Archive/Louvre, Paris; **64b** BAL/Private Collection; **66** ZF; **67t** BAL/Giraudon/Musée de la Chartreuse, Douai, France; **67b** ZR; **68** ZR; **70t** BAL/Giraudon/Bibliothèque Nationale, Paris; **70b** RHPL; **71** ZF; **72t** British Museum; **72b** RHPL; **73** ZR; **74t** e.t. Archive/ Archaeological Museum, Aleppo; **74b** PEF; **75** BAL/British Library, Kings 5 f. 20r; **76** ZR; **77t** ZR; **77b** BAL/Giraudon/Medinet Habu, Egypt; **79** BAL/Victoria & Albert Museum; **80l** BAL/British Museum, Arundel 155 f.93; **80r** HPL/Francis; **81** BAL/Kunsthistorisches Museum, Vienna; **82t** ZF, **82b** ZR; **84t** ZR; **84b** BAL/Fitzwilliam Museum, Cambridge; **85t** BAL/Bibliothèque Nationale; Grec 139 f.136v **85b** ZR; **86** BAL/British Museum, Roy 15 D III f.285; **87t** BAL/Agnews & Sons, London; **87b** ZR; **88t** BAL/British Museum; **88b** ZF; **89** Rex Features/Sipa/Pignières; **90** BAL/British Library, Add 11639 f.522r; **91t** PEF; **91b** From an original drawing by Dr. Leen Ritmeyer; **92** HPL/Regent; **94** ZL; **96** BAL/St. Peter's, Leuven, Belgium; **97t** British Museum; **97b** British Museum; **98** BAL/Museum of the History of Religion, St. Petersburg, Russia; **99t** BAL/British Museum, Louvre, Paris; **99b** PEF; **100t** & **b** e.t. Archive; **101** e.t Archive/Louvre, Paris; **102** BAL/Museo Catedralico, Cuenca, Spain; **103l** BAL/Bibliothèque Nationale, Paris, Fr 13091 f.29v; **103r** BAL/British Library, Harl 2803 f.264; **104** PEF; **105** ZF; **106** BAL/Victoria & Albert Museum; **107** British Museum; **108** ZF/Goebell; **109** BAL/Musées Royaux des Beaux-Arts de Belgique, Brussels; **110** BAL/Index/Museo Real Academia de Bellas Artes, Madrid; **111** ZR; **113t** BAL/Lambeth Palace Library, Ms 3f. 258; **113b** BAL/Giraudon/ Louvre, Paris; **114** e.t. Archive; **115** ZF/Schörke; **117t** BAL/Biblioteca Medicea-Laurenziana, Florence; **117b** RHPL/Jackson; **120t** BAL/British Library, Ms 368.f.66; **120b** ZF; **121** ZR; **122** ZR; **123t** BAL/British Library, Add 27210 f.11; **123b** BAL/Chartres Cathedral, France; **124t** e.t. Archive/Museum of Turkish & Islamic Arts, Istanbul; **124b** BAL/Louvre, Paris; **125** ZR; **126** BAL/Louvre, Paris; **127** BAL/Louvre, Paris; **128t** BAL/Museo e Gallerie Nazionali di Capodimonte, Naples; **128b** British Museum; **129** DBP; **130** BAL/Victoria & Albert Museum; **131** BAL/Giraudon, Musée Condé, Chantilly; **133t** HPL/Pate; **133b** Scala, Museo Nazionale, Napoli; **134** BAL/Kunsthistorisches Museum, Vienna; **135** BAL/Christie's, London; **136** BAL/British Library, Add 11695 f.240; **137** British Museum; **139t** BAL/Giraudon; **139b** BAL/Bibliothèque Nationale, Paris; **140t** BAL/Giraudon/Musée des Beaux-Arts, Nantes, France; **140b** ZR; **142** ZR; **144l** ZF; **144r** ZF; **145** Sonia Halliday Photographs; **146** BAL/Giraudon, Musée Condé, Chantilly; **148** BAL/St. Mark f.14v, Bibliothèque Municipale, Epernay, France; **149** BAL/Museo dell'Opera del Duomo, Siena, Italy; **150t** BAL/Österreichische Nationalbibliothek, Vienna; **150b** BAL/Lichfield Cathedral, Staffordshire, England; **151** BAL/Lambeth Palace Library; **152** BAL/Musée Jacquemart-André, Paris; **153** ZF; **154t** BAL/University of Liverpool Art Gallery, England; **154b** Oriental Institute, Chicago; **155** BAL/Fitzwilliam Museum, Cambridge; **156** BAL/Kunsthalle, Hamburg; **157** BAL/Santa Maria in Trastevere, Rome; **158** Impact/le Garsemur; **159l** BAL/Giraudon/Musée Condé, Chantilly, Ms 65/1284 f.52r; **159r** ZR; **160** BAL/York City Art Gallery, England; **161** RHPL; **163l** BAL/Christie's, London; **163r** PEF; **164** RHPL; **165** ZF; **166** ZF; **167l** BAL/ Christie's, London; **167r** ZR; **168** ZR; **169** BAL/Johnny van Haeften Gallery, London; **170** BAL/Fitzwilliam Museum, Cambridge; **172l** ZF; **172r** BAL/Museo d'Art de Catalunya, Barcelona; **173** BAL/Giraudon/Musée Condé, Chantilly, Ms 81/1057 f.72; **174** BAL/Christie's, London; **175** ZR; **176b** BAL/National Gallery, London; **177** ZR; **178t** DBP; **178b** RHPL; **179** BAL/British Library, Harl 1527 f.27; **180** British Museum; **181l** BAL/Scrovegni (Arena) Chapel, Padua; **181r** ZF; **182** Scala/S. Apollinare Nuovo, Ravenna, Italy; **183** BAL/Museo dell'Opera del Duomo, Siena; **184** BAL/York City Art Gallery; **185** BAL/Private Collection; **186** BAL/Universiy of Liverpool Art Gallery; **187** RHPL; **188** Sonia Halliday Photographs; **189** BAL/British Library, Add 17738 f.4; **191** BAL/Museo dell'Opera del Duomo, Siena; **192** ZR; **193t** BAL/Göreme, Turkey; **193b** Sonia Halliday Photographs; **195t** ZR; **195b** BAL/Christie's, London; **196l** ZF; **196r** BAL/Museo dell'Opera del Duomo, Siena; **197** ZF; **198** ZF; **199** British Library, Ms OR481 f104v; **200** ZR; **201l** BAL/Prado, Madrid; **201r** ZR; **202** From an original drawing by Dr. Shimon Gibson of the PEF; **203t** BAL/Victoria & Albert Museum; **203b** BAL/Christie's, London; **204** ZR; **205** BAL/Giraudon/Louvre, Paris; **206** BAL/Musées Royaux des Beaux-Arts de Belgique, Brussels; **207** ZF; **208** BAL/Victoria & Albert Museum; **209** BAL/Museo dell'Opera del Duomo, Florence; **210** BAL/Magyar Nemzeti Galeria, Budapest; **211t** ZF; **211b** BAL/Louvre, Paris; **212t** BAL/Index/Osma-Soria Chapter House, Soria, Spain; **212b** ZF; **214** BAL/Bible Society, London; **215** BAL/Museo d'Art de Catalunya, Barcelona; **216** BAL/Scrovegni (Arena) Chapel, Padua; **217** BAL/Sistine Chapel, Vatican; **219l** BAL/Brancacci Chapel, Sta. Maria del Carmine, Florence; **219r** ZF; **220** BAL/Belvoir Castle, Leicestershire, England; **221** RHPL; **222** BAL/ Nicholas V Chapel, Vatican; **223** BAL/Pushkin Museum, Moscow; **225t** ZR; **225b** BAL/ Fitzwilliam Museum, Cambridge, Ms 36-1950 f.49.v; **226** ZF/Goebbell; **227** BAL/Bibliothèque Nationale, Lat 1 f.386v, Paris; **228** BAL/Louvre, Paris; **229** BAL/Brancacci Chapel, Sta. Maria del Carmine, Florence; **230** BAL/Villa dei Misteri, Pompeii; **232** Sonia Halliday Photographs; **233** Sonia Halliday Photographs; **234t** BAL/York City Art Gallery; **234b** PEF; **235** ZR; **236** ZR; **237t** C. M. Dixon; **237b** BAL/Biblioteca Marciana, Venice; **239** BAL/Magyar Nemzeti Galeria, Budapest; **240t** BAL/Fitzwilliam Museum, Cambridge; **240b** BAL/Pauline Chapel, Vatican; **242** ZR; **243** ZR; **244** ZR; **245t** BAL/Louvre, Paris; **245b** ZR; **246** ZR; **247t** DBP; **247b** BAL/Louvre, Paris; **248** ZR; **249** BAL/Scrovegni (Arena) Chapel, Padua; **250** BAL/Louvre; **251** BAL; **252** BAL/ Brancacci Chapel, Sta. Maria del Carmine, Florence; **253** BAL/Convent of S. Tomas, Avila, Spain; **255** ZR; **256B** BAL/Tretyakov Gallery, Moscow. Artwork on pp.256–71 ©DBP. Sources: **257–62, 266tl, br** Nuremberg Bible, 1483/BAL/Victoria & Albert Museum; **263**; **264**; **268–71** BAL/DBP.